# Rapid Guide to Hazardous Chemicals in the Workplace

**Second Edition**

Edited by

**Richard J. Lewis, Sr.**

VNR VAN NOSTRAND REINHOLD
New York

DISCLAIMER
Extreme care has been taken in preparation of this work. However, neither the publisher nor the author shall be held responsible or liable for any damages resulting in connection with or arising from the use of any of the information in this book.



Library of Congress Catalog Card Number 89-25087
ISBN 0-442-23804-5



Printed in the United States of America

Van Nostrand Reinhold
115 Fifth Avenue
New York, New York 10003

Van Nostrand Reinhold International Company Limited
11 New Fetter Lane
London EC4P 4EE, England

Van Nostrand Reinhold
480 La Trobe Street
Melbourne, Victoria 3000, Australia

Nelson Canada
1120 Birchmount Road
Scarborough, Ontario M1K 5G4, Canada

16 15 14 13 12 11 10 9 8 7 6 5 4 3 2 1

**Library of Congress Cataloging-in-Publication Data**

Rapid guide to hazardous chemicals in the workplace/edited by Richard J. Lewis, Sr.—2nd ed.
p. cm.
ISBN 0-442-23804-5
1. Hazardous substances—Handbooks, manuals, etc. I. Lewis, Richard J., Sr.
T55.3.H3R37 1990 89-25087
604.7—dc20 CIP

*With love to Velma W. Ross*

*Thanks to Gracie for help with everything. My gratitude to Susan H. Munger and Alberta W. Gordon of Van Nostrand Reinhold for their professional assistance with this book.*

# Contents

# Introduction

The massive revision of the OSHA Permissible Exposure limits mandated this revision. Most of the standards contained in 29CFR 1910.1000 were added or changed. Some substances have Final Rule limits which became effective on September 1, 1989. Others have Transitional Limits with Final Rule limits becoming effective on December 31, 1993. All OSHA Transitional and Final Rule Limits are listed in this guide. The ACGIH, DFG MAK, and DOT entries were updated as well. All Safety Profiles were rewritten to remove cryptic codes and supply more information in clear, easy to comprehend language.

This book continues to fulfill the need for a rapid reference to the most frequently encountered hazardous materials. This edition contains almost 800 entries, an increase of about 100 over the previous edition. Each entry was selected because its dangerous properties prompted regulation by government agencies or consideration by consensus groups.

Each entry has a recommended safe workplace air concentration or other workplace control recommendation from either the U.S. Occupational Safety and Health Administration (OSHA), the American Conference of Governmental Industrial Hygienists (ACGIH), or the German Research Society (DFG). These standards and recommendations constitute the most comprehensive set of workplace air level guidelines and are presented in one convient format in this book.

For assessment of transport hazards, the U.S. Department of Transportation (DOT) hazard class number and description are included. This information serves as an index to the transportation regulations of the United States and most international shipments. The hazard class number is in most cases an internationally agreed upon United Nations number. These values are useful guides to the control of workplace atmospheres, but are not the complete solution.

The Safety Profile is a clear and concise summation of each entry's hazards. A chemical may present problems in occupational handling and use which is not reflected in the recommended air levels. Skin contact can cause irritation, skin corrosion and burns, allergenic reactions or skin penetration leading to toxic effects in the body. Other problems arise from fire and explosion potential. Many substances present storage problems because they are incompatible with other commonly encountered chemicals. The Safety Profiles disclose the various types and degrees of dangerous or harmful effects reported in the literature, condensed into a compact, understandable paragraph. The Safety Profiles are designed to quickly define and clarify the hazard of a given substance.

The final section of each entry contains a physical description of the material and gives useful physical and flammability properties. This information can aid in the identification of unknown materials and in the design and selection of proper storage and handling facilities.

This guide is designed to afford easy access to information on the adverse properties of commonly encountered industrial materials. A book of this size cannot hope to present all the data necessary to completely assess the proper use of these substances. The information provided should allow a quick assessment of the relative hazards of the material and the types and nature of the hazards likely to be encountered. Each entry is identified by a DPIM Number, an alphanumeric code in the form of three letters and three numbers, for example AAA123. These provide an easy pointer to the complete entry in *Dangerous Properties of Industrial Materials, Seventh Edition*. There can be found the detailed data and literature citations which form the basis for the Safety Profiles.

Many publications on hazardous materials attempt to provide information on all aspects of hazardous material control. It is our belief that such subjects as fire control, first aid, and the selection and use of personal protective equipment and respirators should not be treated as briefly as would be required by this format. Decisions on such crucial matters must be made with careful consideration of specific workplace conditions. I hope that the information in this book will stimulate actions to provide safer work environments.

The standards and recommended air concentrations are set by various mechanisms which vary in frequency of change. While the current values at time of publication are listed, the reader is cautioned to *verify* the data with the appropriate agency before undertaking major control efforts based on the data given here. Many substances are under test for carcinogenic activity. When a positive finding is reported, the recommended or mandatory control values can change rapidly. Transportation regulations change in detail as transport experience dictates.

I have strived for accuracy and completeness in this presentation, but recognize that perfection is rarely achieved. Please bring any errors or suggestions to my attention.

The Editor

# How to Use This Book

Each entry consists of four sections:

1. identifying information
2. standards and recommendations
3. Safety Profiles
4. physical properties.

## 1. *Identifying Information*

The first line of each entry contains the DPIM Number, The Chemical Abstracts Service (CAS:) number, the molecular formula, and, to the far right, the Hazard Rating (HR:). This rating varies from 3 indicating the highest hazard potential to 1 indicating the lowest hazard potential. Since the materials were selected for their importance in the design of a safe workplace, the majority, as expected, carry a rating of 3.

The second line of each entry contains the index name used to alphabetize the entry. This is usually the common name used by OSHA, ACGIH, the DFG, DOT, or NIOSH.

The third line contains the U.S. Department of Transportation (DOT:) hazard code. This code is recognized internationally and is in agreement with the United Nations coding system. The code is used on transport documents, labels, and placards. It is also used to determine the regulations for shipping the material. Appendix III contains a cross-index of the DOT hazard in numerical order.

Since chemicals are often known by several widely recognized names (synonyms), a few useful synonyms are included to aid in their identification. They are listed with each entry and alphabetized in Appendix I. If a name is not located in the entries, it may be a synonym for an entry and listed in Appendix I, with reference to the DPIM Number.

## 2. *Standards and Recommendations*

The five possible entries in this section are:

*OSHA PEL* which is followed by the Permissible Exposure Limit (PEL) as defined by the U.S. Occupational Safety and Health Administration (OSHA), Department of Labor. These standards are either time weighted average (TWA) concentrations for an 8-hour workday or ceiling levels (CL) indicating a value which must not be exceeded. Some entries have short term exposure limit (STEL) values which represent a 15-

minute concentration which should not be exceeded in an 8-hour workday. The notation "skin" indicates that the material penetrates intact skin, and skin contact should be avoided even though the PEL is not exceeded. These limits are found in 29 CFR (Code of Federal Regulations) 1910.1000. The CFR regulations also contain detailed requirements for control of some substances and special regulations for carcinogenic substances. Additional information is available from OSHA, Technical Data Center, U.S. Department of Labor, Washington, D.C. 20210.

*ACGIH TLV* which is followed by the Threshold Limit Value (TLV) of the American Conference of Governmental Industrial Hygienists (ACGIH). These standards are either time weighted average (TWA) concentrations for an 8-hour workday or ceiling levels (CL) indicating a value which must not be exceeded. Some entries have short term exposure limit (STEL) values which represent a 15-minute concentration which should not be exceeded in an 8-hour workday. The notation "skin" indicates that the material penetrates intact skin, and skin contact should be avoided even though the TLV concentration is not exceeded. Biological Exposure Indices (*BEI:*) are, according to the ACGIH, set to provide a warning level ". . . of biological response to the chemical, or warning levels of that chemical or its metabolic product(s) in tissues, fluids, or exhaled air of exposed workers . . ." The latest annual TLV list is contained in the publication *Threshold Limit Values and Biological Exposure Indices*. This publication should be consulted for future trends in recommendations. The ACGIH TLVs are adopted in whole or in part by many countries and local administrative agencies throughout the world. As a result, these recommendations have a major impact on the control of workplace contaminant concentrations. The ACGIH may be contacted for additional information at 6500 Glenway Ave., Cincinnati, Ohio 45211, USA.

*DFG MAK* which is followed by the German Research Society's Maximum Allowable Concentration values. Those materials which are classified as to workplace hazard potential by the German Research Society are noted on this line. The MAK values are also revised annually and discussions of materials under consideration for MAK assignment are included in the annual publication together with the current values. *BAT* indicates Biological Tolerance Value for a Working Material which is defined as, ". . . the maximum permissible quantity of a chemical compound, its metabolites, or any deviation from the norm of biological parameters induced by these substances in exposed humans." *TRK* values are Technical Guiding Concentrations for workplace control of carcinogens. For additional information, write to Deutsche Forschungsgemeinschaft (German Research Society), Kennedyallee 40, D-5300 Bonn 2, Federal Republic of Germany. The publication *Maximum Concentrations at the Workplace and Biological Tolerance Values for Working Materials* can be obtained from Verlag Chemie GmbH, Buchauslieferung, P.O. Box 1260/1280, D-6940 Weinheim, Federal Republic of Germany, or Verlag Chemie, Deerfield Beach, Florida.

*NIOSH REL* is followed by the recommended level contained in a NIOSH criteria document. These documents contain extensive data, analysis, and references. The more recent publications can be obtained from the National Institute for Occupational Safety and Health, U.S. Department of Health and Human Services, 4676 Columbia Parkway., Cincinnati, Ohio 45226.

*DOT Class* which is followed by the hazard classification according to the U.S. Department of Transportation (DOT) or the International Maritime Organization (IMO.) This classification gives an indication of the hazards expected in transportation, and serves as a guide to the development of proper labels, placards, and shipping instructions. Many materials are regulated under general headings such as "pesticides" or "combustible liquids" as defined in the regulations. These are not noted here as their specific concentration or properties must be known for proper classification. Special regulations may govern shipment by air. This information should serve *only as a guide* since the regulation of transported materials is carefully controlled in most countries by federal and local agencies. U.S. transportation regulations are found in 40 CFR, Parts 100 to 189. Contact the U.S. Department of Transportation, Materials Transportation Bureau, Washington, D.C. 20590.

### 3. *Safety Profiles*

This section contains a summary of the hazardous properties of the material. Acute immediate effects such as irritation, corrosion, or lethal action are reported in concise language. Chronic or delayed health effects are noted including cancer, reproductive, or allergenic effects. Reported human effects are specifically noted. The term "experimental" indicates that the effect was reported in experimental animals. The term "suspected carcinogen" indicates that a reviewing body such as the International Agency for Research on Cancer, has indicated that there is some evidence for carcinogenic activity, either in animals or humans. Where possible the specific human organ systems or body surfaces affected are listed. Toxic or hazardous decomposition products are identified. An assessment is given of flammable and explosive properties. Incompatible materials and instabilities are listed to guide in the safe storage and use of materials.

### 4. *Physical Properties*

This part gives a physical description of the material in terms of form, color, and odor to aid in positive identification. Here are listed the physical properties which are used for determination of hazard potential and assessment of correct storage and handling practices. When available, the boiling point, melting point, density, vapor pressure, vapor density, and refractive index are given. The flash point, autoignition temperature, and lower and upper explosive limits are included to aid in fire protection and control. An indication is given of the solubility or miscibility of the material in water and common solvents.

# Key to Abbreviations

abs - absolute
ACGIH - American Conference of Governmental Industrial Hygienists
alc - alcohol
alk - alkaline
amorph - amorphous
anhy - anhydrous
approx - approximately
aq - aqueous
atm - atmosphere
autoign - autoignition
aw - atomic weight
af - atomic formula
bp - boiling point
CAS - Chemical Abstracts Service
cc - cubic centimeter
CC - closed cup
CL - ceiling concentration
COC - Cleveland open cup
conc - concentrated
compd(s) - compound(s)
conc - concentration, concentrated
cryst, crys - crystal(s), crystalline
d - density
D - day(s)
decomp - decomposition
deliq - deliquescent
dil - dilute
DOT - U.S. Department of Transportation
EPA - U.S. Environmental Protection Agency
eth - ether
(F) - Fahrenheit
flash p - flash point
flam - flammable
fp - freezing point
g, gm - gram
gran - granular, granules
hygr - hygroscopic
H, hr - hour(s)
HR: - hazard rating
htd - heated
htg - heating

I - intermittent
IARC - International Agency for Research on Cancer
incomp - incompatible
insol - insoluble
IU - International Unit
kg - kilogram (one thousand grams)
L,l - liter
lel - lower explosive level
liq - liquid
M - minute(s)
$m^3$ - cubic meter
mg - milligram
misc - miscible
μ, u - micron
mL, ml - milliliter
mg - milligrams
mm - millimeter
mod - moderately
mp - melting point
mppcf - million particles per cubic foot
mf - molecular formula
mw - molecular weight
NIOSH - National Institute for Occupational Safety and Health
ng - nanogram
nonflam - nonflammable
OC - open cup
org - organic
OSHA - Occupational Safety and Health Administration
PEL - permissible exposure level
petr - petroleum
pg - picogram (one trillionth of a gram)
Pk - peak concentration
pmole - picomole
powd - powder
ppb - parts per billion (v/v)
pph - parts per hundred (v/v)(percent)
ppm - parts per million (v/v)
ppt - parts per trillion (v/v)
prep - preparation
PROP - properties
refr - refractive
rhomb - rhombic
S, sec - second(s)
sl, slt, sltly - slightly
sol - soluble
soln - solution
solv(s) - solvent(s)
spont - spontaneous(ly)
subl - sublimes

TCC - Tag closed cup
tech - technical
temp - temperature
TLV - Threshold Limit Value
TOC - Tag open cup
TWA - time weighted average
U, unk - unknown, unreported
μ, u - micron
uel - upper explosive limits
μg, ug - microgram
ULC, ulc - Underwriters Laboratory Classification
vac - vacuum
vap - vapor
vap d - vapor density
vap press - vapor pressure
vol - volume
visc - viscosity
W - week(s)
Y - year(s)
% - percent(age)
> - greater than
< - less than
<= - equal to or less than
=> - equal to or greater than
° - degrees of temperature in Celsius (Centigrade)
°(F),°F - temperature in Fahrenheit

**AAG250** CAS: 75-07-0 $C_2H_4O$ HR: 3
**ACETALDEHYDE**
DOT: 1089
SYNS: ACETIC ALDEHYDE ◇ ETHANAL ◇ ETHYL ALDEHYDE

OSHA PEL: (Transitional: TWA 200 ppm) TWA 100 ppm; STEL 150 ppm
ACGIH TLV: TWA 100 ppm; STEL 150 ppm
DFG MAK: 50 ppm (90 $mg/m^3$), Suspected Carcinogen.

SAFETY PROFILE: A human systemic irritant by inhalation. An experimental teratogen. An experimental tumorigen by inhalation. A skin and severe eye irritant. Highly flammable liquid. It can react violently with many substances.

PROP: Colorless, fuming liquid; pungent, fruity odor. Mp: −123.5°, bp: 20.8°, lel: 4.0%, uel: 57%, flash p: −36°F (CC), d: 0.804 @ 0°/20°, autoign temp: 347°F, vap d: 1.52. Misc in water, alc, and ether.

**AAI000** CAS: 60-35-5 $C_2H_5NO$ HR: 3
**ACETAMIDE**
SYNS: ACETIC ACID AMIDE ◇ ACETIMIDIC ACID ◇ ETHANAMIDE

DFG MAK: Suspected Carcinogen.
DOT Class: Flammable Liquid

SAFETY PROFILE: Experimental carcinogen and teratogen. When heated to decomposition it emits toxic fumes of $NO_x$.

PROP: Colorless crystals; mousey odor. Mp: 81°, bp: 221.2°, d: 1.159 @ 20°/4°, vap press: 1 mm @ 65°. Decomp in hot water.

**AAT250** CAS: 64-19-7 $C_2H_4O_2$ HR: 3
**ACETIC ACID**
DOT: 2789/2790
SYNS: ETHANOIC ACID ◇ GLACIAL ACETIC ACID ◇ VINEGAR ACID

OSHA PEL: TWA 10 ppm
ACGIH TLV: TWA 10 ppm; STEL 15 ppm
DFG MAK: 10 ppm (25 $mg/m^3$)

SAFETY PROFILE: A human poison. A severe eye and skin irritant. Experimental reproductive effects. A combustible liquid. Moderate fire and explosion hazard.

PROP: Clear, colorless liquid; pungent odor. Mp: 16.7°, bp: 118.1°, flash p: 109°F (CC), lel: 5.4%, uel: 16.0% @ 212°F, d: 1.049 @ 20°/4°, autoign temp: 869°F, vap press: 11.4 mm @ 20°, vap d: 2.07. Misc in water, alc and ether.

**AAU250** CAS: 18461-55-7 $C_9H_8N_2O_6$ HR: 3
**ACETIC ACID-4,6-DINITRO-o-CRESYL ESTER**

NIOSH REL: (Dinitro ortho-Cresyl) TWA 0.2 mg/m$^3$

SAFETY PROFILE: Poison by ingestion. A skin and severe eye irritant. When heated to decomposition it emits toxic fumes of $NO_x$.

**AAX500** CAS: 108-24-7 $C_4H_6O_3$ HR: 2
**ACETIC ANHYDRIDE**
DOT: 1715
SYNS: ACETIC ACID, ANHYDRIDE ◇ ACETIC OXIDE ◇ ACETYL ANHYDRIDE

OSHA PEL: CL 5 ppm
ACGIH TLV: CL 5 ppm
DFG MAK: 5 ppm (20 mg/m$^3$)
DOT Class: Corrosive, Flammable Liquid

SAFETY PROFILE: Moderately toxic by inhalation, ingestion, and skin contact. A skin and severe eye irritant. Moderate fire and explosion hazard. Reacts violently with many substances. Will react violently on contact with water or steam. To fight fire, use $CO_2$, dry chemical, water mist, alcohol foam.

PROP: Colorless, very mobile, strongly refractive liquid; very strong acetic odor. Mp: −73.1°, bp: 140°, flash p: 129°F (CC), d: 1.082 @ 20°/4°, lel: 2.9%, uel: 10.3%, autoign temp: 734°F, vap press: 10 mm @ 36.0°, vap d: 3.52. Somewhat sol in cold water; decomp in hot water and hot alc; misc in alc and ether.

**ABC750** CAS: 67-64-1 $C_3H_6O$ HR: 2
**ACETONE**
DOT: 1090
SYNS: DIMETHYL KETONE ◇ METHYL KETONE ◇ PROPANONE ◇ 2-PROPANONE

OSHA PEL: (Transitional: TWA 1000 ppm) TWA 750 ppm; STEL 1000 ppm
ACGIH TLV: TWA 750 ppm; STEL 1000 ppm
DFG MAK: 1000 ppm (2400 mg/m$^3$)
NIOSH REL: (Ketones) TWA 590 mg/m$^3$
DOT Class: IMO: Corrosive Material

SAFETY PROFILE: Moderately toxic A skin and severe eye irritant. Highly flammable liquid. A fire and explosion hazard. To fight fire, use $CO_2$, dry chemical, alcohol foam.

PROP: Colorless liquid; fragrant mint-like odor. Mp: −94.6°, bp: 56.48°, refr index: 1.356, flash p: 0°F (CC), lel: 2.6%, uel: 12.8%, d: 0.7972 @ 15°, autoign temp: (color) 869°F, vap press: 400 mm @ 39.5°, vap d: 2.00. Misc in water, alc, and ether.

**ABE500** CAS: 75-05-8 $C_2H_3N$ HR: 3
**ACETONITRILE**
DOT: 1648

SYNS: CYANOMETHANE ◇ ETHANENITRILE ◇ ETHYL NITRILE ◇ METHYL CYANIDE

OSHA PEL: TWA 40 ppm; STEL 60 ppm
ACGIH TLV: TWA 40 ppm; STEL 60 ppm (skin)
DFG MAK: 40 ppm (70 $mg/m^3$)
NIOSH REL: TWA 34 $mg/m^3$
DOT Class: Flammable Liquid

SAFETY PROFILE: Poison by ingestion. A skin and severe eye irritant. Human respiratory irritant. Experimental teratogenic and reproductive effects. Dangerous fire hazard. When heated to decomposition it emits highly toxic fumes of $CN^-$ and $NO_x$. To fight fire use foam, $CO_2$, dry chemical.

PROP: Colorless liquid, aromatic odor. Mp: −45°, bp: 81.1°, flash p: 42°F (COC), d: 0.7868 @ 20°/20°, vap d: 1.42, vap press: 100 mm @ 27°, lel: 4.4%, uel: 16%, autoign temp: 975°F. Misc in water, alc, and ether.

**ACI750** CAS: 74-86-2 $C_2H_2$ HR: 3
**ACETYLENE**
DOT: 1001
SYNS: ACETYLEN ◇ ETHINE ◇ ETHYNE

OSHA PEL: CL 2500 ppm
ACGIH TLV: Simple asphyxiant.
NIOSH REL: CL 2500 ppm
DOT Class: Flammable Liquid

SAFETY PROFILE: Mildly toxic by inhalation. Narcotic in high concentration. A very dangerous fire hazard. Moderate explosion hazard. To fight fire use $CO_2$, water spray, or dry chemical. Stop flow of gas.

PROP: Colorless gas, garlic-like odor. Flammable. Bp: −84.0° (sublimes), lel: 2.5%, uel: 82%, mp: −81.8°, flash p: 0°F (CC), d: 1.173 g/L @ 0°, autoign temp: 581°F, vap press: 40 atm @ 16.8°, vap d: 0.91; d: (liquid) 0.613 @ −80°. D: (solid) 0.730 @ −85°. Quite sol in water, very sol in alc, almost misc in ether.

**ACK000** CAS: 156-60-5 $C_2H_2Cl_2$ HR: 2
**trans-ACETYLENE DICHLORIDE**
SYN: trans-1,2-DICHLOROETHYLENE (MAK)

DFG MAK: 200 ppm (790 $mg/m^3$)
DOT Class: Forbidden; Flammable Gas

SAFETY PROFILE: Mildly toxic by inhalation. Dangerous fire hazard. When heated to decomposition it emits toxic fumes of $Cl^-$. To fight fire use water, foam, $CO_2$, dry chemical.

PROP: Colorless liquid, pleasant odor. Mp: −50°, bp: 48°, flash p: 36°F, autoign temp: 860°F, lel: 9.7%, uel: 12.8%, d: 1.2743 @ 25°/4°, vap press: 400 mm @ 30.8°, vap d: 3.34.

**ACK250** CAS: 79-27-6 $C_2H_2Br_4$ HR: 3
**ACETYLENE TETRABROMIDE**
DOT: 2504
SYNS: MUTHMANN'S LIQUID ◇ TBE ◇ TETRABROMOACETYLENE ◇ 1,1,2,2-TETRABROMOETHANE

OSHA PEL: TWA 1 ppm
ACGIH TLV: TWA 1 ppm
DFG MAK: 1 ppm (14 mg/m$^3$)

SAFETY PROFILE: Poison by inhalation and ingestion. An experimental neoplastigen. An eye and skin irritant. When heated to decomposition it emits highly toxic fumes of carbonyl bromide and $Br^-$.

PROP: Colorless to yellow liquid. Bp: 151° @ 54 mm, fp: −1°, d: 2.9638 @ 20°/4°, autoign temp: 635°F.

**ACV500** CAS: 110-22-5 $C_4H_6O_4$ HR: 3
**ACETYL PEROXIDE**
SYN: DIACETYL PEROXIDE (MAK)

DFG MAK: Strong Skin Effects.
DOT Class: ORM-A

SAFETY PROFILE: An experimental tumorigen. Severe skin and eye irritant. It may explode spontaneously in storage. It will react with water or steam to produce heat; can react vigorously with reducing materials; and emit toxic fumes on contact with acid. To fight fire use $CO_2$, dry chemical.

PROP: Solid or colorless crystals or liquid. Sltly sol in cold water, decomp. D: 1.18, mp: 30°, bp: 63° @ 21 mm.

**ADA725** CAS: 50-78-2 $C_9H_8O_4$ HR: 3
**ACETYLSALICYLIC ACID**
SYNS: 2-ACETOXYBENZOIC ACID ◇ ASPIRIN

OSHA PEL: TWA 5 mg/m$^3$
ACGIH TLV: TWA 5 mg/m$^3$
DOT Class: Forbidden.

SAFETY PROFILE: Poison by ingestion. A human teratogen. Human reproductive effects by ingestion. Combustible.

PROP: Colorless needles. Mp: 135°. Very sltly sol in alc, sol in benzene. Solubility in water = 1% @ 37°, in ether = 5% @ 20°.

**ADR000** CAS: 107-02-8 $C_3H_4O$ HR: 3
**ACROLEIN**
DOT: 1092
SYNS: ACRYLALDEHYDE ◇ 2-PROPENAL

OSHA PEL: (Transitional: TWA 0.1 ppm) TWA 0.1 ppm; STEL 0.3 ppm

ACGIH TLV: TWA 0.1 ppm; STEL 0.3 ppm
DFG MAK: 0.1 ppm (0.25 mg/m$^3$)

SAFETY PROFILE: Human poison by inhalation. An experimental carcinogen. Severe eye and skin irritant. Experimental reproductive effects. Dangerous fire and explosion hazard. To fight fire use $CO_2$, dry chemical or alcohol foam.

PROP: Colorless or yellowish liquid; disagreeable, choking odor. Sol in water, alc, and ether. Mp: −87.7°, bp: 52.5°, flash p: <0°F, d: 0.841 @ 20°/4°, autoign temp: unstable (455°F), lel: 2.8%, uel: 31%, vap d: 1.94.

**ADS250** CAS: 79-06-1 $C_3H_5NO$ HR: 3
**ACRYLAMIDE**
DOT: 2074
SYNS: ACRYLIC AMIDE ◇ PROPENAMIDE ◇ 2-PROPENAMIDE

OSHA PEL: (Transitional: TWA 0.3 mg/m$^3$ (skin)) TWA 0.03 mg/m$^3$ (skin)
ACGIH TLV: Suspected Human Carcinogen, TWA 0.03 mg/m$^3$ (skin)
DFG MAK: Animal Carcinogen, Suspected Human Carcinogen.
NIOSH REL: TWA 0.3 mg/m$^3$
DOT Class: Flammable Liquid

SAFETY PROFILE: Poison by ingestion and skin contact. An experimental carcinogen and neoplastigen. Experimental reproductive effects. A skin and eye irritant. Polymerizes violently at its melting point. When heated to decomposition it emits acrid fumes and $NO_x$.

PROP: White, crystalline solid. Very sol in water, alc and ether. Mp: 84.5 ± 0.3°, bp: 125° @ 25 mm, d: 1.122 @ 30°, vap press: 1.6 mm @ 84.5°, vap d: 2.45.

**ADS750** CAS: 79-10-7 $C_3H_4O_2$ HR: 3
**ACRYLIC ACID**
DOT: 2218
SYNS: ACROLEIC ACID ◇ ETHYLENECARBOXYLIC ACID ◇ 2-PROPENOIC ACID

OSHA PEL: TWA 10 ppm (skin)
ACGIH TLV: TWA 10 ppm (Proposed: 2 ppm (skin))
DFG MAK: 200 ppm (950 mg/m$^3$)
DOT Class: IMO: Poison B

SAFETY PROFILE: Poison by ingestion and skin contact. An experimental teratogen. A severe skin and eye irritant. Corrosive. A fire hazard.

PROP: Liquid, acrid odor. Misc in water, benzene, alc, chloroform, ether and acetone. Mp: 13°, bp: 141°, d: 1.062, vap press: 10 mm @ 39.9°, flash p: 130°F (OC), vap d: 2.45.

**ADX500** CAS: 107-13-1 $C_3H_3N$ HR: 3
**ACRYLONITRILE**
DOT: 1093
SYNS: CYANOETHYLENE ◇ 2-PROPENENITRILE ◇ VINYL CYANIDE

OSHA PEL: TWA 2 ppm; CL 10 ppm/15M; Cancer Hazard.
ACGIH TLV: Suspected Human Carcinogen, TWA 2 ppm (skin).
DFG TRK: 3 ppm (7 $mg/m^3$), Animal Carcinogen, Suspected Human Carcinogen.
NIOSH REL: TWA 1 ppm; CL 10 ppm/15M
DOT Class: IMO: Corrosive Material

SAFETY PROFILE: Poison by inhalation, ingestion, and skin contact. An experimental tumorigen and teratogen. Experimental reproductive effects. Dangerous fire hazard and highly reactive. When heated to decomposition it emits toxic fumes of $NO_x$ and $CN^-$.

PROP: Colorless, mobile liquid; mild odor. Sol in water. Mp: −82°, bp: 77.3°, fp: −83°, flash p: 30°F (TCC), lel: 3.1%, uel: 17%, d: 0.806 @ 20°/4°, autoign temp: 898°F, vap press: 100 mm @ 22.8°, vap d: 1.83, flash p: (of 5% aq sol): <50°F.

**AER250** CAS: 111-69-3 $C_6H_8N_2$ HR: 3
**ADIPONITRILE**
DOT: 2205
SYNS: ADIPIC ACID DINITRILE ◇ 1,4-DICYANOBUTANE ◇ TETRAMETHYLENE CYANIDE

NIOSH REL: TWA 18 $mg/m^3$
DOT Class: Flammable Liquid and Poison.

SAFETY PROFILE: Poison by inhalation and ingestion. Flammable. When heated to decomposition it emits toxic fumes of $CN^-$. To fight fire, use foam, $CO_2$, dry chemical.

PROP: Water-white liquid, practically odorless. Mp: 2.3°, bp: 295°, flash p: 199.4°F (OC), d: 0.965 @ 20°/4°, vap d: 3.73.

**AFK250** CAS: 309-00-2 $C_{12}H_8Cl_6$ HR: 3
**ALDRIN**
DOT: 2761
SYNS: ALTOX ◇ DRINOX ◇ OCTALENE

OSHA PEL: TWA 0.25 $mg/m^3$ (skin)
ACGIH TLV: TWA 0.25 $mg/m^3$/
DFG MAK: 0.25 $mg/m^3$
NIOSH REL: (Aldrin) Reduce to lowest detectable level.
DOT Class: IMO: Poison B

SAFETY PROFILE: Poison by ingestion and skin contact. An experimental carcinogen and teratogen. Experimental reproductive effects. Continued acute exposure causes liver damage. When heated to decomposition it emits toxic fumes of $Cl^-$.

PROP: Crystals. Mp: 104-105°. Insol in water; sol in aromatics, esters, ketones, paraffins, and halogenated solvents.

**AFV500** CAS: 107-18-6 $C_3H_6O$ HR: 3
**ALLYL ALCOHOL**
DOT: 1098
SYNS: 3-HYDROXYPROPENE ◇ PROPENOL ◇ VINYLCARBINOL

OSHA PEL: (Transitional: TWA 2 ppm (skin)) TWA 2 ppm; STEL 4 ppm (skin)
ACGIH TLV: TWA 2 ppm; STEL 4 ppm (skin)
DFG MAK 2 ppm (5 $mg/m^3$)
DOT Class: Poison B; ORM-A.

SAFETY PROFILE: Poison by inhalation, ingestion, skin contact. A skin and eye irritant. Dangerous fire and explosion hazard. To fight fire, use $CO_2$, alcohol foam, dry chemical.

PROP: Limpid liquid; pungent odor. Sol in water, alc and ether. Mp: −129°, bp: 96°-97°, lel: 2.5%, uel: 18%, flash p: 70°F (CC), d: 0.854 @ 20°/4°, autoign temp: 713°F, vap press: 10 mm @ 10.5°, vap d: 2.00. Misc in water, alc, and ether.

**AGB250** CAS: 107-05-1 $C_3H_5Cl$ HR: 3
**ALLYL CHLORIDE**
DOT: 1100
SYNS: CHLORALLYLENE ◇ 3-CHLOROPRENE ◇ 3-CHLORO-1-PROPYLENE

OSHA PEL: (Transitional: TWA 1 ppm) TWA 1 ppm; STEL 2 ppm
ACGIH TLV: TWA 1 ppm; STEL 2 ppm
DFG MAK: 1 ppm (3 $mg/m^3$), Suspected Carcinogen.
NIOSH REL: TWA 1 ppm; CL 3 ppm/15M
DOT Class: Flammable Liquid

SAFETY PROFILE: Poison by ingestion. Moderately toxic by inhalation and skin contact. An experimental tumorigen and teratogen. Experimental reproductive effects. A skin, eye, and mucous membrane irritant. Dangerous fire and explosion hazard. To fight fire, use $CO_2$, alcohol foam, dry chemical.

PROP: Colorless liquid. Mp: −136.4°, bp: 44.6°, d: 0.938 @ 20°/4°, flash p: −25°F, lel: 2.9%, uel: 11.2%, autoign temp: 905°F, vap d: 2.64. Solubility = <0.1 in water.

**AGH150** CAS: 106-92-3 $C_6H_{10}O_2$ HR: 3
**ALLYL GLYCIDYL ETHER**
DOT: 2219
SYNS: AGE ◇ ALLYL-2,3-EPOXYPROPYL ETHER

OSHA PEL: (Transitional: CL 10 ppm) TWA 5 ppm; STEL 10 ppm
ACGIH TLV: TWA 5 ppm; STEL 10 ppm (skin)
NIOSH REL: (Glycidyl Ethers) CL 45 $mg/m^3$/15M
DFG MAK: 10 ppm (45 $mg/m^3$)
DOT Class: Flammable Liquid

SAFETY PROFILE: Poison by ingestion. Moderately toxic by inhalation and skin contact. A severe skin and eye irritant. Flammable. To fight fire, use foam, $CO_2$, dry chemical.

PROP: Bp: 153.9°, fp: −100° (forms glass), flash p: 135°F (OC), d: 0.9698 @ 20°/4°, vap press: 21.59 mm @ 60°, vap d: 3.94.

**AGR500** CAS: 2179-59-1 $C_6H_{12}S_2$ HR: 1
**ALLYL PROPYL DISULFIDE**

OSHA PEL: (Transitional: TWA 2 PPM) TWA 2 ppm; STEL 3 ppm
ACGIH TLV: TWA 2 ppm; STEL 3 ppm
DFG MAK: 2 ppm (12 $mg/m^3$)

SAFETY PROFILE: A powerful irritant. Moderately flammable. When heated to decomposition it emits highly toxic $SO_x$. To fight fire, use foam, $CO_2$, dry chemical.

PROP: Liquid, pungent odor.

**AGX000** CAS: 7429-90-5 Al HR: 3
**ALUMINUM and COMPOUNDS**
DOT: 1309/1383/1396
SYN: ALUMINUM POWDER

OSHA PEL: Total Dust: TWA 15 $mg/m^3$; Respirable Fraction: TWA 5 $mg/m^3$; Pyro Powders and Welding Fumes: 5 $mg/m^3$; Soluble Salts and Alkyls: 2 $mg/m^3$.
ACGIH TLV: Metal and Oxide: TWA 10 $mg/m^3$ (dust); Pyro Powders and Welding Fumes: TWA 5 $mg/m^3$; Soluble Salts and Alkyls) TWA 2 $mg/m^3$
DFG MAK: 6 $mg/m^3$; BAT: 170 $\mu g/L$ in urine at end of shift.
DOT Class: Flammable or Combustible Liquid

SAFETY PROFILE: Although aluminum is not generally regarded as an industrial poison, inhalation of finely divided powder has been reported to cause pulmonary fibrosis. Many compounds are active chemically and thus exhibit dangerous toxic and reactive properties. The halides are generally irritants. Dust is moderately flammable/explosive and spontaneously combustible (pyrophoric). To fight fire, use special mixtures of dry chemical.

PROP: A silvery ductile metal. Mp: 660°, bp: 2450°, d: 2.702, vap press 1 mm @ 1284°. Sol in HCl, $H_2SO_4$, and alkalies.

**AHE250** CAS: 1344-28-1 $Al_2O_3$ HR: 2
**ALUMINUM OXIDE (2:3)**
SYN: $\alpha$-ALUMINA (OSHA)

OSHA PEL: Total Dust: (Transitional: TWA 5 $mg/m^3$) TWA 10 $mg/m^3$; Respirable Fraction: TWA 5 $mg/m^3$
ACGIH TLV: TWA (nuisance particulate) 10 $mg/m^3$ of total dust (when toxic impurities are not present, e.g., quartz $< 1\%$).
DFG MAK: 6 $mg/m^3$ (fume)
DOT Class: Flammable Solid

SAFETY PROFILE: Inhalation of finely divided particles may cause lung damage (Shaver's disease).

PROP: White powder. Mp: 2050°, bp: 2977°, d: 3.5-4.0, vap press: 1 mm @ 2158°.

**AIC250** CAS: 97-56-3 $C_{14}H_{15}N_3$ HR: 3
**2-AMINO-5-AZOTOLUENE**
SYNS: AAT ◇ o-AMINOAZOTOLUENE (MAK) ◇ BUTTER YELLOW ◇ C.I. 11160

DFG MAK: Animal Carcinogen, Suspected Human Carcinogen.

SAFETY PROFILE: An experimental carcinogen. Moderately toxic by ingestion. Experimental reproductive effects. When heated to decomposition it emits toxic fumes of $NO_x$.

**AJS100** CAS: 92-67-1 $C_{12}H_{11}N$ HR: 3
**4-AMINODIPHENYL**
SYNS: BIPHENYLAMINE ◇ 4-BIPHENYLAMINE

OSHA PEL: Carcinogen
ACGIH TLV: Confirmed Human carcinogen.
DFG MAK: Human Carcinogen.

SAFETY PROFILE: Poison by ingestion. A human carcinogen. When heated to decomposition it emits toxic fumes of $NO_x$.

PROP: Colorless crystals. Mp. 53°, bp: 302°, d: 1.160 @ 20°/20°, autoign temp: 842°F.

**AJV000** CAS: 132-32-1 $C_{14}H_{14}N_2$ HR: 3
**3-AMINO-9-ETHYLCARBAZOLE**
SYN: 3-AMINO-N-ETHYLCARBAZOLE

DFG MAK: Suspected Carcinogen.

SAFETY PROFILE: Poison by ingestion. An experimental carcinogen. When heated to decomposition it emits toxic fumes of $NO_x$.

**ALL750** CAS: 5307-14-2 $C_6H_7N_3O_2$ HR: 3
**4-AMINO-2-NITROANILINE**
SYNS: C.I. 76070 ◇ 2-NITRO-1,4-BENZENEDIAMINE

DFG MAK: Suspected Carcinogen.

SAFETY PROFILE: Moderately toxic by ingestion. An experimental carcinogen and teratogen. When heated to decomposition it emits toxic fumes of $NO_x$.

**AMI000** CAS: 504-29-0 $C_5H_6N_2$ HR: 3
**2-AMINOPYRIDINE**
DOT: 2671
SYN: α-AMINOPYRIDINE

OSHA PEL: TWA 0.5 ppm
ACGIH TLV: TWA 0.5 ppm
DFG MAK: 0.5 ppm (2 $mg/m^3$)

SAFETY PROFILE: Poison by inhalation. Toxic effects resemble strychnine poisoning. When heated to decomposition it emits highly toxic fumes of $NO_x$.

PROP: White powder or crystals. Mp: 58.1, bp: 210.6°. Sol in water and ether, very sol in alc, sltly sol in ligroin.

**AMY050** CAS: 61-82-5 $C_2H_4N_4$ HR: 3
**AMITROLE**
SYNS: 2-AMINOTRIAZOLE ◇ HERBIZOLE ◇ TRIAZOLAMINE

OSHA PEL: TWA 0.2 $mg/m^3$
ACGIH TLV: TWA 0.2 $mg/m^3$
DFG MAK: 0.2 $mg/m^3$
DOT Class: IMO: Poison B

SAFETY PROFILE: Moderately toxic by ingestion. An experimental carcinogen and teratogen. When heated to decomposition it emits toxic fumes of $NO_x$.

**AMY500** CAS: 7664-41-7 $H_3N$ HR: 3
**AMMONIA**
DOT: 1005
SYN: ANHYDROUS AMMONIA

OSHA PEL: TWA 35 ppm
ACGIH TLV: TWA 25 ppm; STEL 35 ppm
DFG MAK: 50 ppm (35 $mg/m^3$)
NIOSH REL: CL 50 ppm

SAFETY PROFILE: A human poison. An eye, mucous membrane, and systemic irritant by inhalation. Difficult to ignite. Emits toxic fumes of $NH_3$ and $NO_x$ when exposed to heat. To fight fire stop flow of gas.

PROP: Colorless gas, extremely pungent odor, liquefied by compression. Mp: −77.7°, bp: −33.35°, lel: 16%, uel: 25%, d: 0.771 g/liter @ 0°, 0.817 g/liter @ −79°, autoign temp: 1204°F, vap press: 10 atm @ 25.7°, vap d: 0.6. Very sol in water, moderately sol in alc.

**ANE500** CAS: 12125-02-9 $H_4N \cdot Cl$ HR: 2
**AMMONIUM CHLORIDE**
DOT: 9085
SYNS: AMMONIUM MURIATE ◇ SAL AMMONIAC

OSHA PEL: (Fume) TWA 10 $mg/m^3$; STEL 20 $mg/m^3$
ACGIH TLV: TWA 10 $mg/m^3$; STEL 20 $mg/m^3$
DOT Class: Nonflammable Gas

SAFETY PROFILE: Moderately toxic by ingestion. A severe eye irritant. When heated to decomposition it emits very toxic fumes of $NO_x$, $Cl^-$ and $NH_3$.

PROP: White crystals; salty taste. Bp: 520°, mp: 337.8°, d: 1.520, vap press: 1 mm @ 160.4° (sublimes). Sol in water, alc, and glycerin.

**ANK250** CAS: 1336-21-6 $H_4N \cdot HO$ HR: 3
**AMMONIUM HYDROXIDE**
DOT: 2672
SYN: AQUEOUS AMMONIA

NIOSH REL: CL 50 ppm
DOT Class: ORM-E.

SAFETY PROFILE: A human poison by ingestion. An experimental poison by inhalation. A severe eye irritant. Incompatible with many substances. When heated to decomposition it emits very toxic fumes of $NH_3$ and $NO_x$.

PROP: Clear, colorless liquid solution of ammonia; very pungent odor. D: 0.90, mp: −77°. Sol in water. Soln contains not more than 44% ammonia.

**ANP625** CAS: 3825-26-1 $C_8F_{15}O_2 \cdot H_4N$ HR: 3
**AMMONIUM PERFLUOROOCTANOATE**
SYNS: APFO ◇ PERFLUOROAMMONIUM OCTANOATE

ACGIH TLV: (Proposed: 0.1 mg/m$^3$)
DOT Class: Corrosive Material

SAFETY PROFILE: Poison by inhalation. Moderately toxic by ingestion. An eye and skin irritant. Experimental reproductive effects. When heated to decomposition it emits toxic fumes of $F^-$ and $NH_3$.

**ANU650** CAS: 7773-06-0 $H_2NO_3S \cdot H_4N$ HR: 2
**AMMONIUM SULFAMATE**
DOT: 9089
SYNS: AMS ◇ SULFAMIC ACID, MONOAMMONIUM SALT

OSHA PEL: (Transitional: TWA Total Dust: 15 mg/m$^3$; Respirable Fraction: 5 mg/m$^3$) TWA 10 mg/m$^3$; Respirable Fraction: 5 mg/m$^3$
ACGIH TLV: TWA 10 mg/m$^3$
DFG MAK: 15 mg/m$^3$

SAFETY PROFILE: Moderately toxic by ingestion. Somewhat explosive when heated: a powerful oxidizer. When heated to decomposition it emits very toxic fumes of $NH_3$, $NO_x$, and $SO_x$.

PROP: Deliquescent, crystalline material (white crystalline solid). Bp: 160° (decomp), mp: 131°.

**AOD725** CAS: 628-63-7 $C_7H_{14}O_2$ HR: 2
**n-AMYL ACETATE**
DOT: 1104
SYNS: AMYL ACETIC ESTER ◇ n-PENTYL ACETATE

OSHA PEL: TWA 100 ppm
ACGIH TLV: TWA 100 ppm
DOT Class: ORM-E

SAFETY PROFILE: A human eye irritant. Dangerous fire hazard; can react with oxidizing materials. Moderately explosive in the form of vapor when exposed to flame. To fight fire, use alcohol foam, dry chemical.

PROP: Colorless liquid; pear or banana-like odor. Mp: −78.5°, bp: 148° @ 737 mm, ULC: 55-60, lel: 1.1%, uel: 7.5%, flash p: 77°F (CC), d: 0.879 @ 20°/20°, autoign temp: 714°F, vap d: 4.5. Very sltly sol in water; misc in alc and ether.

**AOD735** CAS: 626-38-0 $C_7H_{14}O_2$ HR: 2
**sec-AMYL ACETATE**
DOT: 1104
SYNS: 2-ACETOXYPENTANE ◇ 2-PENTYL ACETATE

OSHA PEL: TWA 125 ppm
ACGIH TLV: TWA 125 ppm
DOT Class: Flammable or Combustible Liquid

SAFETY PROFILE: Mildly toxic by inhalation. Dangerous fire hazard; can react with oxidizing materials. Moderately explosive in the form of vapor. To fight fire, use alcohol foam, dry chemical.

PROP: Colorless liquid. Bp: 120°, flash p: 73.4°F (CC), d: 0.862-0.866 @ 20°/20°, vap d: 4.48, lel: 1.1%, uel: 7.5%. Sltly sol in water; misc in alc and ether.

**AOD750** $C_7H_{14}O_2$ HR: 2
**AMYL ACETATE (MIXED ISOMERS)**
SYN: ACETIC ACID, AMYL ESTER

DFG MAK: 100 ppm (525 mg/m$^3$)
DOT Class: Flammable Liquid

SAFETY PROFILE: A skin irritant. Mildly toxic by ingestion. Dangerous fire hazard; can react with oxidizing materials. Moderately explosive in the form of vapor. To fight fire, use alcohol foam, dry chemical.

PROP: Colorless liquid, pear-like odor. Mp: −78.5°, bp: 148° @ 737 mm, ULC: 55-60, lel: 1.1%, uel: 7.5%, flash p: 77°F (CC), d: 0.879 @ 20°/20°, autoign temp: 714°F, vap d: 4.5.

**AOQ000** CAS: 62-53-3 $C_6H_7N$ HR: 3
**ANILINE**
DOT: 1547
SYNS: AMINOBENZENE ◇ BENZENAMINE ◇ C.I. 76000 ◇ PHENYLAMINE

OSHA PEL: (Transitional: TWA 5 ppm (skin)) TWA 2 ppm (skin)
ACGIH TLV: TWA 2 ppm (skin) (Proposed: BEI: 50 mg/L total p-aminophenol in urine at end of shift.
DFG MAK: 2 ppm (8 mg/m$^3$), Suspected Carcinogen; BAT: 1 mg/L in urine at end of shift.

SAFETY PROFILE: A human poison. Poison experimentally by inhalation and ingestion. An experimental neoplastigen. A skin and severe eye irritant, and a mild sensitizer. Moderately flammable. Spontaneously explosive reactions with many substances. To fight fire, use alcohol foam, $CO_2$, dry chemical. When heated to decomposition it emits highly toxic fumes of $NO_x$.

PROP: Colorless, oily liquid; characteristic odor. Bp: 184.4°, lel: 1.3%, ULC: 20-25, flash p: 158°F (CC), fp: −6.2°, d: 1.02 @ 20°/4°, autoign temp: 1139°F, vap press: 1 mm @ 34.8°, vap d: 3.22.

**AOV900** CAS: 90-04-0 $C_7H_9NO$ HR: 2
**o-ANISIDINE**
DOT: 2431
SYNS: o-AMINOANISOLE ◇ o-METHOXYANILINE ◇ o-METHOXYPHENYLAMINE

OSHA PEL: TWA 0.5 $mg/m^3$
ACGIH TLV: TWA 0.5 $mg/m^3$ (skin)
DFG MAK: 0.1 ppm (0.5 $mg/m^3$)
DOT Class: Poison B.

SAFETY PROFILE: Moderately toxic by ingestion. When heated to decomposition it emits toxic fumes of $NO_x$.

**AOW000** CAS: 104-94-9 $C_7H_9NO$ HR: 2
**p-ANISIDINE**
SYNS: p-AMINOANISOLE ◇ 4-METHOXYBENZENAMINE

OSHA PEL: TWA 0.5 $mg/m^3$
ACGIH TLV: TWA 0.5 $mg/m^3$ (skin)
DFG MAK: 0.1 ppm (0.5 $mg/m^3$)
DOT Class: DOT-IMO: Poison B

SAFETY PROFILE: Moderately toxic. A mild sensitizer. May cause a contact dermatitis. When heated to decomposition it evolves toxic fumes of $NO_x$.

PROP: Plates from aq soln. D: 1.089 @ 55°/55°, mp: 57.2°, bp: 243°, vap d: 4.28. Sol in hot water, alc, and ether.

**APG500** CAS: 120-12-7 $C_{14}H_{10}$ HR: 3
**ANTHRACENE**
SYNS: ANTHRACIN ◇ PARANAPHTHALENE

OSHA PEL: TWA 0.2 $mg/m^3$

SAFETY PROFILE: An experimental tumorigen and neoplastigen. A skin irritant and allergen. Combustible and moderately explosive. To fight fire, use water, foam, $CO_2$, water spray or mist, dry chemical.

PROP: Colorless crystals, violet fluorescence. Mp: 217°, lel: 0.6%, flash p: 250°F (CC), d: 1.24 @ 27°/4°, autoign temp: 1004°F, vap press: 1 mm @ 145.0°, (sublimes), vap d: 6.15, bp: 339.9°. Insol in water. Solubility in alc @ 1.9/100 @ 20°; in ether 12.2/100 @ 20°.

**AQB750** CAS: 7440-36-0 Sb HR: 3
**ANTIMONY**
DOT: 2871
SYNS: C.I. 77050 ◇ STIBIUM

OSHA PEL: TWA 0.5 mg(Sb)/$m^3$
ACGIH TLV: TWA 0.5 mg(Sb)/$m^3$

DFG MAK: 0.5 mg(Sb)/$m^3$
NIOSH REL: TWA 0.5 mg(Sb)/$m^3$

SAFETY PROFILE: A human poison. Moderate fire and explosion hazard.

PROP: Silvery or gray, lustrous metal. Mp: 630°, bp: 1635°, d: 6.684 @ 25°, vap press: 1 mm @ 886°. Insol in water, sol in hot concentrated $H_2SO_4$.

**AQF000** CAS: 1327-33-9 $O_3Sb_2$ HR: 3
**ANTIMONY OXIDE**
SYN: ANTIMONY TRIOXIDE (MAK)

ACGIH TLV: TWA 500 μg/$m^3$
DFG MAK: Animal Carcinogen, Suspected Human Carcinogen.
NIOSH REL: TWA 0.5 mg(Sb)/$m^3$
DOT Class: Poison B

SAFETY PROFILE: Moderately toxic. An experimental carcinogen. When heated to decomposition it emits toxic Sb fumes.

PROP: White cubes. D: 5.2, mp: 650°, bp: 1550° subl. Very sltly sol in water; sol in KOH and HCl.

**AQN635** CAS: 86-88-4 $C_{11}H_{10}N_2S$ HR: 3
**ANTU**
DOT: 1651
SYNS: ALPHANAPHTHYL THIOUREA ◇ α-NAPHTHYLTHIOUREA (DOT)

OSHA PEL: TWA 0.3 mg/$m^3$
ACGIH TLV: TWA 0.3 mg/$m^3$
DFG MAK: 0.3 mg/$m^3$

SAFETY PROFILE: Poison by ingestion. An experimental tumorigen. Chronic toxicity has been known to cause dermatitis. When heated to decomposition it emits toxic fumes of $NO_x$ and $SO_x$.

PROP: Crystals; bitter taste. Mp: 198°. Sltly sol in hot alc.

**ARA750** CAS: 7440-38-2 As HR: 3
**ARSENIC and ARSENIC COMPOUNDS**
DOT: 1558
SYN: METALLIC ARSENIC

OSHA PEL: Inorganic: TWA 0.01 mg(As)/$m^3$; Cancer Hazard; Organic: TWA 0.5 mg(As)/$m^3$
ACGIH TLV: TWA 0.2 mg(As)/$m^3$
DFG TRK: 0.2 mg/$m^3$ calculated as arsenic in that portion of dust that can possibly be inhaled.
NIOSH REL: CL 2 μg(As)/$m^3$
DOT Class: Poison B

SAFETY PROFILE: Poisoning may be acute or chronic. Inorganic arsenicals are more toxic than organics. Trivalent is more toxic than pentavalent.

Arsenic can cause a variety of skin abnormalities including itching, pigmentation, and even cancerous changes. A recognized carcinogen of the skin, lungs, and liver. The metal is flammable in the form of dust. When heated to decomposition the compounds emit highly toxic fumes of As.

PROP: Metal: Silvery to black, brittle, crystalline and amorphous metalloid. Mp: 814° @ 36 atm, bp: subl @ 612°, d: black crystals 5.724 @ 14°; black amorphous 4.7, vap press: 1 mm @ 372° (sublimes). Insol in water; sol in $HNO_3$.

**ARI750** CAS: 1327-53-3 $As_4O_6$ HR: 3
**ARSENIC TRIOXIDE**
DOT: 1561
SYNS: ARSENIC SESQUIOXIDE ◇ ARSENIOUS ACID (MAK) ◇ ARSENIOUS OXIDE

OSHA PEL: TWA 0.01 mg(As)/$m^3$
ACGIH TLV: Production: Suspected Human Carcinogen
DFG MAK: Human Carcinogen.
NIOSH REL: CL 2 $\mu$g(As)/$m^3$/15M

SAFETY PROFILE: Poison by ingestion. A human carcinogen by inhalation. When heated to decomposition it emits highly toxic fumes of As.

PROP: Colorless, rhombic crystals (dimer, claudetite). D: 4.15, mp: 278°, bp: 460°. Solubility in water = 1.82/100 @ 20°; sol in alc. Colorless cubes. D: 3.865, mp: 309°. Solubility in water = 1.2/100 @ 20°.

**ARK250** CAS: 7784-42-1 $AsH_3$ HR: 3
**ARSINE**
DOT: 2188
SYNS: ARSENIC HYDRIDE ◇ ARSENIC TRIHYDRIDE ◇ HYDROGEN ARSENIDE

OSHA PEL: TWA 0.05 ppm
ACGIH TLV: TWA 0.05 ppm
DFG MAK: 0.05 ppm (0.2 mg/$m^3$)
NIOSH REL: CL 2 $\mu$g(As)/$m^3$/15M
DOT Class: Poison B

SAFETY PROFILE: Poison by inhalation. A human carcinogen. Flammable gas. Moderately explosive. When heated to decomposition it emits highly toxic fumes of As.

PROP: Colorless gas, mild garlic odor. D: 2.695 g/L; bp: −62.5°; vap d: 2.66; mp: −116°. Solubility in water = 28 mg/100 @ 20°. Sol in benzene and chloroform.

**ARM250** CAS: 1332-21-4 HR: 3
**ASBESTOS**
DOT: 2212/2590
SYNS: ACTINOLITE ASBESTOS ◇ AMOSITE ASBESTOS ◇ AMPHIBOLE ◇ ANTHOPHYLITE ◇ BLUE ASBESTOS (DOT) ◇ CHRYSOTILE ASBESTOS ◇ CROCIDOLITE ASBESTOS ◇ FERROANTHOPHYLLITE ◇ FIBROUS GRUNERITE ◇ SERPENTINE ◇ TREMOLITE ASBESTOS ◇ WHITE ASBESTOS

OSHA PEL: TWA 2 million fb/$m^3$; CL 10 million fb/$m^3$
ACGIH TLV: Human Carcinogen, TWA 2 fb/cc
DFG TRK: (Fine dust particles which are able to reach the alveolar area of the lung) crocidolite: $0.05 \times 10^6$ fibers/$m^3$ (0.025 mg/$m^3$) (definition of fiber: length greater than 5 μm; diameter less than 3 μm; length/diameter greater than 3:1, equivalent to 1 fiber/cc); chrysotile, amosite, anthophyllite, tremolite, actinolite: $1 \times 10^6$ fibers/$m^3$ (0.05 mg/$m^3$) applicable when there is more than 2.5% asbestos in the dust; 2.0 mg/$m^3$ applicable when there is less than or equal to 2.5 wt % asbestos in fine dust.
NIOSH REL: TWA 100,000 fb/$m^3$ over 5 μm in length
DOT Class: Poison A

SAFETY PROFILE: A human carcinogen.

**ARO500** CAS: 8052-42-4 HR: 3
**ASPHALT**
DOT: 1999
SYNS: ASPHALTUM ◇ BITUMEN (MAK) ◇ PETROLEUM PITCH ◇ ROAD TAR (DOT)

ACGIH TLV: TWA 5 mg/$m^3$
DFG MAK: Suspected Carcinogen.
NIOSH REL: CL 5 mg/$m^3$/15M

SAFETY PROFILE: A moderate irritant. May contain carcinogenic components. Combustible. To fight fire use foam, $CO_2$, dry chemical.

PROP: Black or dark brown mass. Bp: <470°, flash p: 400+°F (CC), d: 0.95−1.1, autoign temp: 905°F.

**ARQ725** CAS: 1912-24-9 $C_8H_{14}ClN_5$ HR: 3
**ATRAZINE**
SYNS: FENATROL ◇ PRIMATOL

OSHA PEL: TWA 5 mg/$m^3$
ACGIH TLV: TWA 5 mg/$m^3$
DFG MAK: 2 mg/$m^3$
DOT Class: ORM-C

SAFETY PROFILE: Moderately toxic by ingestion. Mildly toxic by inhalation and skin contact. An experimental tumorigen. Experimental reproductive effects. A skin and severe eye irritant. When heated to decomposition it emits toxic fumes of $Cl^-$ and $NO_x$.

PROP: Crystals. Mp: 171-174°. Solubility at 25°: in water, 70 ppm; ether, 12,000 ppm; chloroform, 52,000 ppm; methanol, 18,000 ppm.

**ASH500** CAS: 86-50-0 $C_{10}H_{12}N_3O_3PS_2$ HR: 3
**AZINPHOS METHYL**
DOT: 2783
SYNS: COTNION METHYL ◇ GUTHION (DOT) ◇ METILTRIAZOTION

OSHA PEL: TWA 0.2 mg/m$^3$ (skin)
ACGIH TLV: TWA 0.2 mg/m$^3$ (skin)
DFG MAK: 0.2 mg/m$^3$
DOT Class: Poison B

SAFETY PROFILE: Poison by inhalation, ingestion, and skin contact. An experimental tumorigen and teratogen. When heated to decomposition it emits very toxic fumes of $PO_x$, $SO_x$, and $NO_x$.

PROP: Crystals or brown, waxy solid. D: 1.44, mp: 74°. Sltly sol in water; sol in organic solvents.

**BAB750** CAS: 1395-21-7 HR: 3
**BACILLUS SUBTILIS BPN**
SYNS: BACILLOMYCIN (8CI, 9CI) ◇ SUBTILISINS (ACGIH)

OSHA PEL: CL 0.00006 mg/m$^3$
ACGIH TLV: CL 0.00006 mg/m$^3$

SAFETY PROFILE: A severe eye irritant. When heated to decomposition it emits toxic fumes of $NO_x$.

PROP: A commercial raw proteolytic enzyme used in laundry detergents.

**BAC000** CAS: 9014-01-1 HR: 3
**BACILLUS SUBTILIS CARLSBERG**
SYN: SUBTILISIN (9CI, ACGIH)

ACGIH TLV: CL 0.00006 mg/m$^3$

SAFETY PROFILE: Moderately toxic by ingestion. An eye irritant. When heated to decomposition it emits toxic fumes of $NO_x$.

PROP: A commercial raw proteolytic enzyme used in laundry detergents.

**BAK250** CAS: 10294-40-3 $Ba{\bullet}CrH_2O_4$ HR: 3
**BARIUM CHROMATE(VI)**
SYNS: C.I. 77103 ◇ ULTRAMARINE YELLOW

OSHA PEL: TWA 0.1 mg $(C_3O_3)m^3$; 0.5 mg(Ba)/m$^3$
ACGIH TLV: TWA 0.5 mg(Ba)/m$^3$; 0.05 mg(Cr)/m$^3$; Confirmed Human Carcinogen
NIOSH REL: TWA 0.001 mg(Cr(VI))/m$^3$

SAFETY PROFILE: A poison. A human carcinogen. It reacts vigorously with reducing materials.

PROP: Heavy, yellow, crystalline powder. D: 4.498 @ 15°.

**BAK500** HR: 3
**BARIUM COMPOUNDS (SOLUBLE)**

Barium and its compounds are on the Community Right-To-Know List.

OSHA PEL: Soluble Compounds: TWA 0.5 mg(Ba)/m$^3$

ACGIH TLV: Soluble Compounds:TWA 0.5 $mg/m^3$
DFG MAK: Soluble Compounds: 0.5 $mg/m^3$
DOT Class: Some barium compounds are flammable or explosive.

SAFETY PROFILE: The soluble barium salts, such as the chloride and sulfide, are poisonous when ingested. The insoluble sulfate used in radiography is not acutely toxic. The chromate is a human carcinogen. Some salts are skin, eye, and mucous membrane irritants producing dermatitis.

**BAP000** CAS: 7727-43-7 $O_4S \cdot Ba$ HR: 2
**BARIUM SULFATE**
SYNS: BARYTES ◇ C.I. 77120 ◇ PERMANENT WHITE

OSHA PEL: (Transitional: Total Dust: TWA 15 $mg/m^3$; Respirable Fraction: 5 $mg/m^3$) Total Dust: TWA 10 $mg/m^3$; Respirable Fraction: 5 $mg/m^3$
ACGIH TLV: TWA (nuisance particulate) 10 $mg/m^3$ of total dust (when toxic impurities are not present, e.g., quartz $< 1\%$).

SAFETY PROFILE: An experimental tumorigen. A relatively insoluble salt used as an opaque medium in radiography. When heated to decomposition it emits toxic fumes of $SO_x$.

PROP: White, heavy, odorless powder. D: 4.50 @ 15°, mp: 1580°. Insol in water or dilute acids.

**BAV575** CAS: 17804-35-2 $C_{14}H_{18}N_4O_3$ HR: 3
**BENOMYL**
SYNS: ARILATE ◇ FUNDASOL ◇ MBC

OSHA PEL: (Transitional: Total Dust: TWA 15 $mg/m^3$; Respirable Fraction: 5 $mg/m^3$) Total Dust: TWA 10 $mg/m^3$; Respirable Fraction: 5 $mg/m^3$
ACGIH TLV: TWA 10 $mg/m^3$

SAFETY PROFILE: Poison by ingestion. Mildly toxic by inhalation. An experimental teratogen. A human skin irritant. When heated to decomposition it emits toxic fumes of $NO_x$.

**BAY300** CAS: 98-87-3 $C_7H_6Cl_2$ HR: 3
**BENZAL CHLORIDE**
DOT: 1886
SYNS: BENZYLIDENE CHLORIDE (DOT) ◇ α,α-DICHLOROTOLUENE

DFG MAK: Suspected Carcinogen.
DOT Class: Poison B

SAFETY PROFILE: Poison by inhalation. Moderately toxic by ingestion. An experimental carcinogen. A suspected human carcinogen. A strong irritant and lachrymator. When heated to decomposition it emits toxic fumes of $Cl^-$.

PROP: Very refractive liquid. Mp: −16°, bp: 214°, d: 1.29.

**BBL250** CAS: 71-43-2 $C_6H_6$ HR: 3
**BENZENE**
DOT: 1114
SYNS: BENZOL (DOT) ◇ COAL NAPHTHA ◇ MINERAL NAPHTHA

OSHA PEL: (Transitional: TWA 10 ppm; CL 25 ppm; Pk 50 ppm/10M) TWA 1 ppm; STEL 5 ppm; Pk 5 ppm/15M/8H; Cancer Hazard
ACGIH TLV: TWA 10 ppm; Suspected Human Carcinogen; BEI: 50 mg(total phenol)/L in urine at end of shift recommended as a mean value.
DFG TRK: 5 ppm (16 $mg/m^3$) Human Carcinogen.
NIOSH REL: TWA 0.32 $mg/m^3$; CL 3.2 $mg/m^3$/15M
DOT Class: Flammable Liquid

SAFETY PROFILE: A human poison by inhalation. A severe eye and moderate skin irritant. A human carcinogen. An experimental carcinogen and teratogen. A dangerous fire hazard. Moderate explosion hazard when exposed to heat or flame. To fight fire, use foam, $CO_2$, dry chemical.

PROP: Clear, colorless liquid. Mp: 5.51°, bp: 80.093°-80.094°, flash p: 12°F (CC), d: 0.8794 @ 20°, autoign temp: 1044°F, lel: 1.4%, uel: 8.0%, vap press: 100 mm @ 26.1°, vap d: 2.77, ULC: 95-100.

**BBP000** CAS: 123-61-5 $C_8H_4N_2O_2$ HR: 3
**BENZENE-1,3-DIISOCYANATE**
SYN: m-PHENYLENE ISOCYANATE

NIOSH REL: TWA (Diisocyanates) 0.005 ppm; CL 0.02 ppm/10M

SAFETY PROFILE: An allergic sensitizer. An eye and severe skin irritant. When heated to decomposition it emits toxic fumes of $NO_x$ and $CN^-$.

**BBQ000** CAS: 319-84-6 $C_6H_6Cl_6$ HR: 3
**BENZENE HEXACHLORIDE-α-isomer**
SYNS: α-BHC ◇ α-LINDANE

DFG MAK: 0.5 $mg/m^3$

SAFETY PROFILE: Poison by ingestion. An experimental carcinogen. When heated to decomposition it emits toxic fumes of $Cl^-$.

**BBQ500** CAS: 58-89-9 $C_6H_6Cl_6$ HR: 3
**BENZENE HEXACHLORIDE-γ isomer**
DOT: 2761
SYNS: γ-BHC ◇ LINDANE (ACGIH, DOT, USDA)



OSHA PEL: TWA 0.5 $mg/m^3$ (skin)
ACGIH TLV: TWA 0.5 $mg/m^3$ (skin)
DFG MAK: 0.5 $mg/m^3$
DOT Class: ORM-A

SAFETY PROFILE: A human poison by ingestion. An experimental neoplastigen and teratogen. When heated to decomposition it emits toxic fumes of $Cl^-$, HCl, and phosgene.

**BBR000** CAS: 319-85-7 $C_6H_6Cl_6$ HR: 3
**trans-α-BENZENEHEXACHLORIDE**
SYNS: β-BHC ◇ β-LINDANE

DFG MAK: 0.5 mg/m$^3$

SAFETY PROFILE: An experimental neoplastigen. Mildly toxic by ingestion. When heated to decomposition it emits very toxic fumes of $Cl^-$, HCl, and phosgene.

**BBX000** CAS: 92-87-5 $C_{12}H_{12}N_2$ HR: 3
**BENZIDINE**
DOT: 1885
SYNS: 4,4′-BIPHENYLDIAMINE ◇ C.I. 37225

OSHA: Carcinogen
ACGIH TLV: Confirmed Human Carcinogen
DFG MAK: Human Carcinogen.
DOT Class: Poison B

SAFETY PROFILE: Poison by ingestion. A human carcinogen. An experimental carcinogen and tumorigen. When heated to decomposition it emits highly toxic fumes of $NO_x$.

PROP: Grayish-yellow, crystalline powder; white or sltly reddish crystals, powder, or leaf. Mp: 127.5-128.7° @ 740 mm, bp: 401.7°, d: 1.250 @ 20°/4°.

**BBY000** CAS: 531-86-2 $C_{12}H_{12}N_2 \cdot H_2O_4S$ HR: 3
**BENZIDINE SULFATE**
SYN: (1,1′-BIPHENYL)-4,4′-DIAMINE SULFATE (1:1)

OSHA: Carcinogen

SAFETY PROFILE: An experimental carcinogen. When heated to decomposition it emits toxic fumes of $SO_x$ and $NO_x$.

**BCS750** CAS: 50-32-8 $C_{20}H_{12}$ HR: 3
**BENZO(a)PYRENE**
SYNS: BENZO(d,e,f)CHRYSENE ◇ 3,4-BENZOPYRENE

OSHA PEL: TWA 0.2 mg/m$^3$

SAFETY PROFILE: An experimental carcinogen,,s25tumorigen, and teratogen.

PROP: Yellow crystals. Mp: 179°, bp: 312° @ 10 mm. Insol in water; sol in benzene, toluene, and xylene.

**BDS000** CAS: 94-36-0 $C_{14}H_{10}O_4$ HR: 3
**BENZOYL PEROXIDE**
DOT: 2085
SYNS: BENZOIC ACID, PEROXIDE ◇ DIBENZOYL PEROXIDE (MAK) ◇ VANOXIDE

OSHA PEL: TWA 5 $mg/m^3$
ACGIH TLV: TWA 5 $mg/m^3$
DFG MAK: 5 $mg/m^3$
NIOSH REL: TWA 5 $mg/m^3$
DOT Class: Organic Peroxide

SAFETY PROFILE: Poison by ingestion. An experimental tumorigen. Can cause dermatitis, asthmatic effects, testicular atrophy. An allergen and eye irritant. Moderate fire hazard. To fight fire, use water spray, foam.

PROP: White, granular, tasteless, odorless powder. Mp: 103-106° (decomp), bp: decomposes explosively, autoign temp: 176°F. Sol in benzene, acetone, chloroform; sltly sol in alc; insol in water.

**BEE375** CAS: 100-44-7 $C_7H_7Cl$ HR: 3
**BENZYL CHLORIDE**
DOT: 1738
SYNS: CHLOROMETHYLBENZENE ◇ α-CHLOROTOLUENE ◇ TOLYL CHLORIDE

OSHA PEL: TWA 1 ppm
ACGIH TLV: TWA 1 ppm
DFG MAK: 1 ppm (5 $mg/m^3$); Suspected Carcinogen.
NIOSH REL: (Benzyl Chloride) CL 5 $mg/m^3$/15M
DOT Class: Corrosive Material

SAFETY PROFILE: Poison by inhalation. Moderately toxic by ingestion. An experimental carcinogen and tumorigen. Experimental reproductive effects. A corrosive irritant to skin, eyes, and mucous membranes. Flammable and moderately explosive. When heated to decomposition it emits toxic fumes of $Cl^-$.

PROP: Colorless liquid, very refractive; irritating, unpleasant odor. Mp: −43°, bp: 179°, lel: 1.1%, flash p: 153°F, d: 1.1026 @ 18/4°, autoign temp: 1085°F, vap d: 4.36.

**BFL250** CAS: 98-07-7 $C_7H_5Cl_3$ HR: 3
**BENZYL TRICHLORIDE**
DOT: 2226
SYNS: BENZENYL CHLORIDE ◇ BENZOTRICHLORIDE (DOT, MAK) ◇ TRICHLOROMETHYLBENZENE

DFG MAK: Suspected Carcinogen.
DOT Class: IMO: Corrosive Material

SAFETY PROFILE: Poison by inhalation. Moderately toxic by ingestion. An experimental carcinogen by skin contact. Corrosive to the skin, eyes, and mucous membranes. When heated to decomposition it emits toxic fumes of $Cl^-$.

PROP: Clear, colorless to yellowish liquid; penetrating odor. Mp: −5°, bp: 221°, d: 1.38 @ 15.5°/15.5°, vap d: 6.77.

**BFO750** CAS: 7440-41-7 Be HR: 3

**BERYLLIUM and COMPOUNDS**

DOT: 1567

SYN: GLUCINUM

OSHA PEL: (Transitional: TWA 0.002 mg(Be)/$m^3$; CL 0.005; Pk 0.025/30M/8H) TWA 0.002 mg(Be)/$m^3$; STEL 0.005 mg(Be)/$m^3$/30M; CL 0.025 mg(Be)/$m^3$

ACGIH TLV: TWA 0.002 mg/$m^3$, Suspected Human Carcinogen.

DFG TRK: Animal Carcinogen, Suspected Human Carcinogen. Grinding of beryllium metal and alloys: 0.005 mg/$m^3$ calculated as Be in that portion of dust that can possibly be inhaled; other Be compounds: 0.002 mg/$m^3$ calculated as Be in that portion of dust that can possibly be inhaled

NIOSH REL: CL not to exceed 0.0005 mg(Be)/$m^3$

DOT Class: Poison B, Flammable Solid Powder and Poison (metal).

SAFETY PROFILE: A suspected human carcinogen. An experimental carcinogen. Beryllium and its compounds can enter the body through inhalation of dusts and fumes, and may act locally on the skin. Even alloys of low beryllium content have been shown to be dangerous. A moderate fire hazard in the form of dust or powder.

PROP: A grayish-white, hard, light metal. Mp: 1278°, bp: 2970°, d: 1.85.

**BFW750** CAS: 128-37-0 $C_{15}H_{24}O$ HR: 2

**BHT (FOOD GRADE)**

SYNS: BUTYLATED HYDROXYTOLUENE ◇ DIBUTYLATED HYDROXYTOLUENE ◇ 2,6-DI-tert-BUTYL-p-CRESOL (OSHA, ACGIH)

OSHA PEL: TLV 10 mg/$m^3$

ACGIH TLV: TLV 10 mg/$m^3$

SAFETY PROFILE: Moderately toxic by ingestion. An experimental carcinogen. Experimental reproductive effects. A skin and eye irritant. Combustible. To fight fire, use $CO_2$, dry chemical.

PROP: White, crystalline solid; faint characteristic odor. Bp: 265°, fp: 68°, flash p: 260°F (TOC), d: 1.048 @ 20°/4°, vap d: 7.6. Sol in alc; insol in water and propylene glycol.

**BGE000** CAS: 92-52-4 $C_{12}H_{10}$ HR: 3

**BIPHENYL**

SYNS: DIPHENYL ◇ PHENYLBENZENE

OSHA PEL: TWA 0.2 ppm

ACGIH TLV: TWA 0.2 ppm

DFG MAK: 0.2 ppm (1 mg/$m^3$)

SAFETY PROFILE: Poison by intravenous route. Moderately toxic by ingestion. An experimental tumorigen and neoplastigen. A powerful irritant by inhalation. Combustible. To fight fire, use $CO_2$, dry chemical, water spray, mist, fog.

PROP: White scales, pleasant odor. Mp: 70°, bp: 255°, flash p: 235°F (CC), d: 0.991 @ 75°/4°, autoign temp: 1004°F, vap d: 5.31, lel: 0.6% @ 232°, uel: 5.8% @ 331°F.

**BIE250** CAS: 51-75-2 $C_5H_{11}Cl_2N$ HR: 3
**BIS(β-CHLOROETHYL)METHYLAMINE**
SYNS: CHLORMETHINE ◇ N-METHYL-BIS(2-CHLOROETHYL)AMINE (MAK) ◇ NITROGEN MUSTARD

DFG MAK: Human Carcinogen.

SAFETY PROFILE: A deadly poison by inhalation, ingestion, and skin contact. A human carcinogen which produces skin tumors by skin contact. A powerful irritant to skin and eyes. When heated to decomposition it emits very toxic fumes of $Cl^-$ and $NO_x$.

PROP: Dark liquid. Mp: 1° @ 10 mm, d: 1.09 @ 25°, vap press: 0.17 mm @ 25°, vap d: 5.9.

**BIH250** CAS: 505-60-2 $C_4H_8Cl_2S$ HR: 3
**BIS(2-CHLOROETHYL)SULFIDE**
SYNS: 2,2′-DICHLOROETHYL SULPHIDE (MAK) ◇ DISTILLED MUSTARD ◇ MUSTARD GAS

DFG MAK: Human Carcinogen.

SAFETY PROFILE: A human poison by inhalation. An experimental poison by skin contact. A severe human skin and eye irritant. An experimental carcinogen. To fight fire, use water, foam, $CO_2$, dry chemical. Dangerous; when heated to decomposition or on contact with acid or acid fumes it emits highly toxic fumes of $SO_x$ and $Cl^-$.

PROP: Colorless (if pure), to light yellow, oily liquid. Bp: 228°, fp: 14.4°, flash p: 221°F, d: 1.2741 @ 20°/4°, vap d: 5.4, vap press: 0.09 mm @ 30°.

**BIK000** CAS: 542-88-1 $C_2H_4Cl_2O$ HR: 3
**BIS(CHLOROMETHYL) ETHER**
DOT: 2249
SYNS: BCME ◇ BIS-CME ◇ sym-DICHLOROMETHYL ETHER

OSHA: Carcinogen
ACGIH TLV: TWA 0.001 ppm; Confirmed Human Carcinogen
DFG MAK: Human Carcinogen.

SAFETY PROFILE: Poison by inhalation, ingestion, and skin contact. A human carcinogen. A dangerous fire hazard. When heated to decomposition it emits very toxic fumes of $Cl^-$.

PROP: Volatile liquid. Bp: 105°, d: 1.315 @ 20°, vap d: 4.0. flash p: $<$19°.

**BMG000** CAS: 1303-86-2 $B_2O_3$ HR: 2
**BORON OXIDE**
SYNS: BORON SESQUIOXIDE ◇ BORON TRIOXIDE

OSHA PEL: (Transitional: Total Dust: TWA 15 $mg/m^3$; Respirable Fraction: TWA 5 $mg/m^3$) Total Dust: TWA 10 $mg/m^3$; Respirable Fraction: TWA 5 $mg/m^3$
ACGIH TLV: TWA 10 $mg/m^3$
DFG MAK: 15 $mg/m^3$

SAFETY PROFILE: Moderately toxic by ingestion. An eye and skin irritant.

PROP: Vitreous, colorless crystals. Mp: 450° (approx), bp: 1860°, d: 2.46.

**BMG400** CAS: 10294-33-4 $BBr_3$ HR: 3
**BORON TRIBROMIDE**
DOT: 2692
SYNS: BORON BROMIDE ◇ TRONA

OSHA PEL: CL 1 ppm
ACGIH TLV: CL 1 ppm
DOT Class: Corrosive Material

SAFETY PROFILE: A poison. Corrosive. Dangerous; may explode when heated. When heated to decomposition it emits toxic fumes of $Br^-$.

PROP: Colorless, fuming liquid. Mp: −45°, bp: 91.7°, d: 2.650 @ 0°, vap press: 40 mm @ 14.0°, 100 mm @ 33.5°.

**BMG700** CAS: 7637-07-2 $BF_3$ HR: 3
**BORON TRIFLUORIDE**
DOT: 1008
SYN: BORON FLUORIDE

OSHA PEL: CL 1 ppm
ACGIH TLV: CL 1 ppm
DFG MAK: 1 ppm (3 $mg/m^3$)
NIOSH REL: (Boron Trifluoride) No Exposure Limit
DOT Class: Nonflammable Gas

SAFETY PROFILE: A poison by inhalation. A strong irritant. Dangerous; when heated to decomposition or upon contact with water or steam, will produce toxic and corrosive fumes of $F^-$.

PROP: Colorless gas; pungent, irritating odor. Mp: −126.8°, bp: −99.9°, d: 2.99 g/L.

**BMM650** CAS: 314-40-9 $C_9H_{13}BrN_2O_2$ HR: 2
**BROMACIL**
SYNS: CYNOGAN ◇ HYVAR ◇ URAGAN

OSHA PEL: TWA 1 ppm
ACGIH TLV: TWA 1 ppm

SAFETY PROFILE: Moderately toxic by ingestion. An experimental teratogen. When heated to decomposition it emits very toxic fumes of $Br^-$ and $NO_x$.

**BMP000** CAS: 7726-95-6 $Br_2$ HR: 3
**BROMINE**
DOT: 1744

OSHA PEL: (Transitional: TWA 0.1 ppm) TWA 0.1 ppm; STEL 0.3 ppm
ACGIH TLV: TWA 0.1 ppm; STEL 0.3 ppm
DFG MAK: 0.1 ppm (0.7 $mg/m^3$)
DOT Class: Corrosive Material

SAFETY PROFILE: A human poison by ingestion and moderately toxic by inhalation. Corrosive. A very powerful oxidizer.

PROP: Rhombic crystals or dark red liquid. Fp: −7.3°, bp: 58.73°, d: 2.928 @ 59°, 3.12 @ 20°, vap press: 175 mm @ 21°, 1 atm @ 58.2°, vap d: 5.5.

**BNL000** CAS: 75-25-2 $CHBr_3$ HR: 3
**BROMOFORM**
DOT: 2515
SYNS: METHENYL TRIBROMIDE ◇ TRIBROMOMETHANE

OSHA PEL: TWA 0.5 ppm (skin)
ACGIH TLV: TWA 0.5 ppm (skin)
DOT Class: Poison B

SAFETY PROFILE: A human poison by ingestion. An experimental neoplastigen. A lachrymator. When heated to decomposition it emits highly toxic fumes of $Br^-$.

PROP: Colorless liquid or hexagonal crystals. Mp: 6-7°, bp: 149.5°, flash p: none, d: 2.890 @ 20°/4°.

**BOP500** CAS: 106-99-0 $C_4H_6$ HR: 3
**1,3-BUTADIENE**
SYNS: DIVINYL ◇ VINYLETHYLENE

OSHA PEL: TWA 1000 ppm
ACGIH TLV: TWA 10 ppm; Suspected Human Carcinogen
DFG MAK: Processing after polymerization and loading: 15 ppm; Others: 5 ppm; Animal Carcinogen, Suspected Human Carcinogen.
NIOSH REL: Reduce to lowest feasible level
DOT Class: Flammable Gas

SAFETY PROFILE: An experimental carcinogen and teratogen. An eye irritant. If spilled on skin or clothing, it can cause burns or frost bite (due to rapid vaporization). Dangerous fire hazard. To fight fire, stop flow of gas.

PROP: Colorless gas, mild aromatic odor. Very reactive. Bp: −4.5°, mp: −113°, fp: −108.9°, flash p: −105°F, lel: 2.0%, uel: 11.5%, d: 0.621 @ 2 0°/4°, autoign temp: 788°F, vap d: 1.87, vap press: 1840 mm @ 21°.

**BOR500** CAS: 106-97-8 $C_4H_{10}$ HR: 1
**BUTANE**
DOT: 1011
SYNS: DIETHYL ◇ METHYLETHYLMETHANE

OSHA PEL: TWA 800 ppm
ACGIH TLV: TWA 800 ppm
DFG MAK: 1000 ppm (2350 mg/m$^3$)
DOT Class: Flammable Gas

SAFETY PROFILE: Mildly toxic via inhalation. Very dangerous fire hazard. Highly explosive when exposed to flame. To fight fire, stop flow of gas.

PROP: Colorless gas; faint disagreeable odor. Bp: −0.5°, fp: −138°, lel: 1.9%, uel: 8.5%, flash p: −76°F (CC), d: 0.599, autoign temp: 761°F, vap press: 2 atm @ 18.8°, vap d: 2.046.

**BOU250** CAS: 1633-83-6 $C_4H_8O_3S$ HR: 3
**BUTANE SULTONE**
SYNS: 1,4-BUTANESULTONE (MAK) ◇ Δ-VALEROSULTONE

DFG MAK: Suspected Carcinogen.

SAFETY PROFILE: Moderately toxic by ingestion. An experimental tumorigen. When heated to decomposition it emits toxic fumes of $SO_x$.

**BPJ850** CAS: 111-76-2 $C_6H_{14}O_2$ HR: 3
**2-BUTOXYETHANOL**
DOT: 2369
SYNS: BUTYL CELLOSOLVE ◇ o-BUTYL ETHYLENE GLYCOL ◇ ETHYLENE GLYCOL MONOBUTYL ETHER (MAK, DOT)

OSHA PEL: (Transitional: TWA 50 ppm (skin)) TWA 25 ppm (skin)
ACGIH TLV: TWA 25 ppm (skin)
DFG MAK: 20 ppm (100 mg/m$^3$)
DOT Class: Poison B

SAFETY PROFILE: Poison via ingestion and skin contact. Moderately toxic via inhalation. An experimental teratogen. A skin and eye irritant. Flammable liquid. To fight fire, use foam, $CO_2$, dry chemical.

PROP: Clear, mobile liquid; pleasant odor. Bp: 168.4-170.2°, fp: −74.8°, flash p: 160°F (COC), d: 0.9012 @ 20°/20°, vap press: 300 mm @ 140°.

**BPM000** CAS: 112-07-2 $C_8H_{16}O_3$ HR: 2
**2-BUTOXYETHYL ACETATE**
SYNS: BUTYL CELLOSOLVE ACETATE ◇ ETHYLENE GLYCOL MONOBUTYL ETHER ACETATE (MAK)

DFG MAK: 20 ppm (135 mg/m$^3$)

SAFETY PROFILE: Moderately toxic by ingestion and skin contact. Mild skin irritant. Flammable. To fight fire, use alcohol foam.

PROP: Colorless liquid, fruity odor. Bp: 192.3°, d: 0.9424 @ 20°/20°, fp: −63.5°, flash p: 190°F. Sol in hydrocarbons and organic solvents, insol in water.

**BPU750** CAS: 123-86-4 $C_6H_{12}O_2$ HR: 2
**n-BUTYL ACETATE**
DOT: 1123
SYNS: ACETIC ACID n-BUTYL ESTER ◇ BUTYL ETHANOATE

OSHA PEL: (Transitional: TWA 150 ppm) TWA 150 ppm; STEL 200 ppm
ACGIH TLV: TWA 150 ppm; STEL 200 ppm
DFG MAK: 200 ppm (950 $mg/m^3$)
DOT Class: Flammable Liquid

SAFETY PROFILE: Mildly toxic by inhalation and ingestion. An experimental teratogen. A skin and severe eye irritant. A mild allergen. Flammable. Moderately explosive. To fight fire, use alcohol foam, $CO_2$, dry chemical.

PROP: Colorless liquid; strong fruity odor. Bp: 126°, fp: −73.5°, ULC: 50-60, lel: 1.4%, uel: 7.5°, flash p: 72°F, d: 0.88 @ 20°/20°, refr index: 1.393-1.396, autoign temp: 797°F, vap press: 15 mm @ 25°. Misc with alc, ether, propylene glycol; sltly sol in water.

**BPV000** CAS: 105-46-4 $C_6H_{12}O_2$ HR: 2
**sec-BUTYL ACETATE**
DOT: 1123
SYNS: ACETIC ACID-2-BUTOXY ESTER ◇ 2-BUTYL ACETATE

OSHA PEL: TWA 200 ppm
ACGIH TLV: TWA 200 ppm
DFG MAK: 200 ppm (950 $mg/m^3$)
DOT Class: Flammable Liquid

SAFETY PROFILE: An irritant and allergen. Flammable. To fight fire, use alcohol foam, $CO_2$, dry chemical.

PROP: Colorless liquid, mild odor. Bp: 112°, flash p: 18°, d: 0.862-0.866 @ 20°/20°, vap d: 4.00. lel: 1.3%, uel: 7.5%.

**BPV100** CAS: 540-88-5 $C_6H_{12}O_2$ HR: 3
**tert-BUTYL ACETATE**
DOT: 1123
SYNS: ACETIC ACID-tert-BUTYL ESTER ◇ ACETIC ACID-1,1-DIMETHYLETHYL ESTER

OSHA PEL: TWA 200 ppm
ACGIH TLV: TWA 200 ppm
DFG MAK: 200 ppm (950 $mg/m^3$)
DOT Class: IMO: Flammable Liquid.

SAFETY PROFILE: Poison by inhalation and ingestion. Flammable. To fight fire, use alcohol foam, $CO_2$, dry chemical.

**BPW100** CAS: 141-32-2 $C_7H_{12}O_2$ HR: 2

**n-BUTYL ACRYLATE**

DOT: 2348

SYNS: ACRYLIC ACID n-BUTYL ESTER (MAK) ◇ BUTYL-2-PROPENOATE

OSHA PEL: TWA 10 ppm
ACGIH TLV: TWA 10 ppm
DFG MAK: 10 ppm (55 mg/$m^3$)
DOT Class: Flammable or Combustible Liquid

SAFETY PROFILE: Moderately toxic by ingestion, inhalation, and skin contact. Experimental reproductive effects. A skin and eye irritant. Flammable. To fight fire, use foam, $CO_2$, dry chemical.

PROP: Water-white, extremely reactive monomer. Bp: 69° @ 50 mm, fp: −64.6°, flash p: 120°F (OC), d: 0.89 @ 25°/25°, vap press: 10 mm @ 35.5°, vap d: 4.42.

**BPW500** CAS: 71-36-3 $C_4H_{10}O$ HR: 3

**n-BUTYL ALCOHOL**

DOT: 1120

SYNS: BUTANOL (DOT) ◇ PROPYLCARBINOL

OSHA PEL: (Transitional: TWA 100 ppm) CL 50 ppm (skin)
ACGIH TLV: CL 50 ppm (skin)
DFG MAK: 100 ppm (300 mg/$m^3$)
DOT Class: Flammable or Combustible Liquid

SAFETY PROFILE: Moderately toxic by skin contact, ingestion. A skin and severe eye irritant. Flammable liquid and moderately explosive. To fight fire, use water spray, alcohol foam, $CO_2$, dry chemical.

PROP: Colorless liquid; vinous odor. Bp: 117.5°, ULC: 40, lel: 1.4%, uel: 11.2%, fp: −88.9°, flash p: 95-100°F, d: 0.80978 @ 20°/4°, autoign temp: 689°F, vap press: 5.5 mm @ 20°, vap d: 2.55. Misc in alc, ether, and organic solvents; sltly sol in water.

**BPW750** CAS: 78-92-2 $C_4H_{10}O$ HR: 3

**sec-BUTYL ALCOHOL**

DOT: 1120

SYNS: sec-BUTANOL (DOT) ◇ 2-HYDROXYBUTANE

OSHA PEL: (Transitional: TWA 150 ppm) TWA 100 ppm
ACGIH TLV: TWA 100 ppm
DFG MAK: 100 ppm (300 mg/$m^3$)
DOT Class: Flammable or Combustible Liquid

SAFETY PROFILE: Mildly toxic by ingestion. An eye irritant. Dangerous fire hazard. Auto-oxidizes to an explosive peroxide. To fight fire, use water spray, alcohol foam, $CO_2$, dry chemical.

PROP: Colorless liquid. Mp: −89°, bp: 99.5°, flash p: 14°, d: 0.808 @ 20°/4°, autoign temp: 763°F, vap press: 10 mm @ 20°, vap d: 2.55, lel: 1.7% @ 212°F, uel: 9.8% @ 212°F.

**BPX000** CAS: 75-65-0 $C_4H_{10}O$ HR: 2
**tert-BUTYL ALCOHOL**
DOT: 1120
SYNS: tert-BUTANOL ◇ TRIMETHYLCARBINOL

OSHA PEL: (Transitional: TWA 100 ppm) TWA 100 ppm; STEL 150 ppm
ACGIH TLV: TWA 100 ppm; STEL 150 ppm
DFG MAK: 100 ppm (300 mg/m$^3$)
DOT Class: Flammable Liquid

SAFETY PROFILE: Moderately toxic by ingestion. Experimental reproductive effects. Dangerous fire hazard. Moderately explosive. To fight fire, use alcohol foam, $CO_2$, dry chemical.

PROP: Colorless liquid or rhombic prisms or plates. Mp: 25.3°, bp: 82.8°, flash p: 50°F (CC), d: 0.7887 @ 20°/4°, autoign temp: 896°F, vap press: 40 mm @ 24.5°, vap d: 2.55, lel: 2.4%, uel: 8.0%.

**BPX750** CAS: 109-73-9 $C_4H_{11}N$ HR: 3
**n-BUTYLAMINE**
DOT: 1125
SYNS: 1-AMINOBUTANE ◇ NORVALAMINE

OSHA PEL: CL 5 ppm (skin)
ACGIH TLV: CL 5 ppm
DFG MAK: 5 ppm (15 mg/m$^3$)
DOT Class: Flammable Liquid

SAFETY PROFILE: Poison by ingestion and skin contact. Moderately toxic by inhalation. An experimental tumorigen. A severe skin irritant. Dangerous fire hazard. To fight fire, use alcohol foam, $CO_2$, dry chemical. When heated to decomposition it emits toxic fumes of $NO_x$.

PROP: Liquid, ammonia-like odor. Mp: −50°, bp: 77°, flash p: 10°F (OC), 10°F (CC), d: 0.74-0.76 @ 20°/20°, autoign temp: 594°F, vap d: 2.52, lel: 1.7%, uel: 9.8%.

**BPY000** CAS: 13952-84-6 $C_4H_{11}N$ HR: 3
**sec-BUTYLAMINE**
DOT: 1125
SYNS: 2-AMINOBUTANE ◇ 1-METHYLPROPYLAMINE

DFG MAK: 5 ppm (15 mg/m$^3$)
DOT Class: Flammable Liquid

SAFETY PROFILE: A poison by ingestion. A powerful irritant. Moderately toxic by skin contact. Dangerous fire hazard. To fight fire, use alcohol foam, water spray or mist, dry chemical. When heated to decomposition it emits toxic fumes of $NO_x$.

PROP: Liquid. Mp: −104°, bp: 63°, flash p: 15°F, d: 0.724 @ 20°.

**BPY250** CAS: 75-64-9 $C_4H_{11}N$ HR: 3
**tert-BUTYLAMINE**
DOT: 1125
SYNS: 2-AMINOISOBUTANE ◇ 1,1-DIMETHYLETHYLAMINE ◇ TRIMETHYL-AMINOMETHANE

DFG MAK: 5 ppm (15 mg/m$^3$)
DOT Class: Flammable Liquid

SAFETY PROFILE: Poison by ingestion. Very dangerous fire hazard. To fight fire, use alcohol foam. When heated to decomposition it emits toxic fumes of $NO_x$.

PROP: Colorless liquid. Mp: −67.5°, bp: 44-46°, d: 0.700 @ 15°, lel: 1.7% @ 212°F, uel: 8.9% @ 212°F, vap d: 2.5, autoign temp: 716°F.

**BRK750** CAS: 2426-08-6 $C_7H_{14}O_2$ HR: 3
**n-BUTYL GLYCIDYL ETHER**
SYNS: BGE ◇ 2,3-EPOXYPROPYL BUTYL ETHER

OSHA PEL: (Transitional: TWA 50 ppm) TWA 25 ppm
ACGIH TLV: TWA 25 ppm
DFG MAK: Suspected Carcinogen.
NIOSH REL: (Glycidyl Ethers) CL 30 mg/m$^3$/15M

SAFETY PROFILE: Moderately toxic by ingestion and skin contact. Mildly toxic by inhalation. An experimental teratogen. A skin and severe eye irritant.

**BRM250** CAS: 75-91-2 $C_4H_{10}O_2$ HR: 3
**tert-BUTYLHYDROPEROXIDE**
DOT: 2093/2094
SYNS: 1,1-DIMETHYLETHYL HYDROPEROXIDE ◇ 2-HYDROPEROXY-2-METHYLPROPANE

DFG MAK: Moderate skin effects.
DOT Class: Organic Peroxide.

SAFETY PROFILE: A poison by ingestion and inhalation. A severe skin and eye irritant. Very dangerous fire hazard. Moderately explosive. To fight fire, use alcohol foam, $CO_2$, dry chemical.

PROP: Water-white liquid, sltly sol in water, very sol in esters and alcohols. Flash p: 80°F or above, fp: −35°, d: 0.860, vap d: 2.07.

**BRR600** CAS: 138-22-7 $C_7H_{14}O_3$ HR: 3
**n-BUTYL LACTATE**
SYNS: 2-HYDROXYPROPANOIC ACID, BUTYL ESTER ◇ LACTIC ACID, BUTYL ESTER

OSHA PEL: TWA 5 ppm
ACGIH TLV: TWA 5 ppm

SAFETY PROFILE: A skin irritant. Flammable. To fight fire, use alcohol foam, foam, $CO_2$, dry chemical.

PROP: Liquid. Sltly sol in water; misc in alc and ether. Mp: −43°, bp: 188°, flash p: 160°F (OC), d: 0.968, autoign temp: 720°F, vap d: 5.04, vap press: 0.4 mm @ 20°.

**BRR900** CAS: 109-79-5 $C_4H_{10}S$ HR: 3
**n-BUTYL MERCAPTAN**
DOT: 2347
SYN: BUTANETHIOL

OSHA PEL: (Transitional: TWA 10 ppm) TWA 0.5 ppm
ACGIH TLV: TWA 0.5 ppm
DFG MAK: 0.5 ppm (1.5 mg/m$^3$)
NIOSH REL: (n-Alkane Mono Thiols) CL 0.5 ppm/15M
DOT Class: Flammable Liquid

SAFETY PROFILE: Moderately toxic by inhalation and ingestion. An eye irritant. Dangerous fire hazard. To fight fire, use alcohol foam. When heated to decomposition it emits toxic $SO_x$.

PROP: Colorless liquid, skunk-like odor. Mp: −116°, bp: 98°, d: 0.8365 @ 25°/4°, flash p: 35°F, vap d: 3.1.

**BRY500** CAS: 924-16-3 $C_8H_{18}N_2O$ HR: 3
**n-BUTYL-N-NITROSO-1-BUTAMINE**
SYNS: DIBUTYLNITROSOAMINE ◇ N-NITROSODI-n-BUTYLAMINE (MAK)

DFG MAK: Animal Carcinogen, Suspected Human Carcinogen.

SAFETY PROFILE: Moderately toxic by ingestion. An experimental carcinogen and teratogen. When heated to decomposition it emits toxic fumes of $NO_x$.

PROP: Pale yellow liquid. Bp: 235°.

**BSC250** CAS: 107-71-1 $C_6H_{12}O_3$ HR: 3
**tert-BUTYL PERACETATE**
SYN: tert-BUTYL PEROXYACETATE

DFG MAK: Moderate skin irritant.
DOT Class: Forbidden.

SAFETY PROFILE: Moderately toxic by ingestion. Mildly toxic by inhalation. Moderate skin and eye irritant. A shock- and heat- sensitive explosive. Dangerous fire hazard. To fight fire, use dry chemical, alcohol foam, spray and mist.

PROP: Clear, colorless, benzene solution; insol in water, sol in organic solvents. D: 0.923, vap press: 50 mm @ 26°, flash p: <80°F (COC).

**BSC750** CAS: 110-05-4 $C_8H_{18}O_2$ HR: 2
**tert-BUTYL PEROXIDE**
DOT: 2102
SYNS: DI-tert-BUTYL PEROXIDE (MAK) ◇ DTBP

DFG MAK: Mild skin irritant.
DOT Class: Organic Peroxide

SAFETY PROFILE: An experimental tumorigen. A powerful irritant by ingestion and inhalation. A mild skin and eye irritant. Warning: Water may not work to fight fire.

PROP: Clear, water white liquid. Mp: −40°, bp: 80° @ 284 mm, flash p: 65°F (OC), d: 0.79, vap press: 19.51 mm @ 20°, vap d: 5.03.

**BSE000** CAS: 89-72-5 $C_{10}H_{14}O$ HR: 3
**o-sec-BUTYLPHENOL**

OSHA PEL: TWA 5 ppm (skin)
ACGIH TLV: TWA 5 ppm (skin)

SAFETY PROFILE: Moderately toxic by ingestion and skin contact. A severe skin and eye irritant. Combustible. To fight fire, use foam, spray, $CO_2$, dry chemical.

PROP: Colorless liquid. Bp: 226-228° @ 25 mm, fp: 12°, flash p: 225°F, d: 0.981 @ 25°/25°.

**BSE500** CAS: 98-54-4 $C_{10}H_{14}O$ HR: 3
**4-tert-BUTYLPHENOL**
SYNS: p-tert-BUTYLPHENOL (MAK) ◇ 1-HYDROXY-4-tert-BUTYLBENZENE

DFG MAK: 0.08 ppm (0.5 $mg/m^3$)

SAFETY PROFILE: Moderately toxic by skin contact and ingestion. A skin and severe eye irritant. Combustible. To fight fire, use foam, $CO_2$, dry chemical.

PROP: Crystals or practically white flakes. Bp: 238°, fp: 97°, d: 0.9081 @ 114°/4°, vap press: 1 mm @ 70.0°, vap d: 5.1.

**BSP500** CAS: 98-51-1 $C_{11}H_{16}$ HR: 2
**p-tert-BUTYLTOLUENE**
SYNS: 1-METHYL-4-tert-BUTYLBENZENE ◇ TBT

OSHA PEL: (Transitional: TWA 10 ppm) TWA 10 ppm; STEL 20 ppm
ACGIH TLV: TWA 10 ppm; STEL 20 ppm, Suspected Carcinogen.
DFG MAK: 10 ppm (60 $mg/m^3$)

SAFETY PROFILE: Moderately toxic by inhalation and ingestion. A skin and human eye irritant. Flammable.

PROP: Colorless liquid.

**BSX250** CAS: 109-74-0 $C_4H_7N$ HR: 3
**BUTYRONITRILE**
DOT: 2411
SYNS: BUTYRIC ACID NITRILE ◇ 1-CYANOPROPANE ◇ PROPYL CYANIDE

NIOSH REL: TWA 22 $mg/m^3$
DOT Class: Flammable Liquid

SAFETY PROFILE: A poison by ingestion and skin contact. Moderately toxic by inhalation. A skin irritant. Dangerous fire hazard. To fight fire, use alcohol foam. When heated to decomposition it emits toxic fumes of $NO_x$ and $CN^-$.

PROP: Colorless liquid, sltly sol in water, sol in alc and ether. D: 0.796 @ 15°, mp: −112.6°, bp: 117°, flash p: 79°F (OC).

**CAD000** CAS: 7440-43-9 Cd HR: 3
**CADMIUM and COMPOUNDS**
SYN: C.I. 77180

OSHA PEL: Fume: TWA 0.1 $mg(Cd)/m^3$; CL 0.6 $mg(Cd)/m^3$; Dust: TWA 0.2 $mg(Cd)/m^3$; CL 0.6 $mg(Cd)/m^3$
ACGIH TLV: Dust and Salts: TWA 0.05 $mg(Cd)/m^3$ (Proposed: TWA 0.01 $mg(Cd)/m^3$ (dust), Human carcinogen); BEI: 10 μg/g creatinine in urine; 10 μg/L in blood.)
DFG BAT: Blood 1.5 μg/dL; Urine 15 μg/dL. MAK: Suspected Carcinogen.
NIOSH REL: (Cadmium) Reduce to lowest feasible level

SAFETY PROFILE: A human poison by inhalation. Inhalation causes lung cancer in humans. An experimental carcinogen and teratogen. Cadmium oxide fumes can cause metal fume fever resembling that caused by zinc oxide fumes. The dust ignites spontaneously in air and is flammable and explosive. When heated to decomposition they emit toxic fumes of Cd.

PROP: Hexagonal crystals, silver-white, malleable metal. Mp: 320.9°, bp: 767 ± 2°, d: 8.642, vap press: 1 mm @ 394°.

**CAO000** CAS: 1317-65-3 $CO_3$•Ca HR: 2
**CALCIUM CARBONATE**
SYNS: LIMESTONE (FCC) ◇ MARBLE

OSHA PEL: Total Dust: 15 $mg/m^3$; Respirable Fraction: 5 $mg/m^3$
ACGIH TLV: TWA (nuisance particulate) 10 $mg/m^3$ of total dust (when toxic impurities are not present, e.g., quartz $< 1\%$).

SAFETY PROFILE: A severe eye and moderate skin irritant.

PROP: White microcrystalline powder. Mp: 825° (α), 1339° (β) @ 102.5 atm; d: 2.7-2.95. Found in nature as the minerals limestone, marble, aragonite, calcite, and vaterite. Odorless, tasteless powder or crystals. Practically insol in water and alc; sol in dilute acids.

**CAQ250** CAS: 156-62-7 $CN_2$•Ca HR: 3
**CALCIUM CYANAMIDE**
DOT: 1403
SYNS: CALCIUM CARBIMIDE ◇ LIME-NITROGEN (DOT)

OSHA PEL: TWA 0.5 $mg/m^3$
ACGIH TLV: TWA 0.5 $mg/m^3$
DFG MAK: 1 $mg/m^3$
DOT Class: ORM-C

SAFETY PROFILE: Poison by ingestion, inhalation, and skin contact. Moderately toxic to humans by ingestion. An experimental tumorigen. Flammable solid. Reaction with water forms the explosive acetylene gas. When heated to decomposition it emits toxic fumes of $NO_x$ and $CN^-$.

PROP: Hexagonal, rhombohedral, colorless crystals. Mp: 1300°, subl > 1500°.

**CAQ500** CAS: 592-01-8 $C_2CaN_2$ HR: 3
**CALCIUM CYANIDE**
DOT: 1575
SYNS: CALCYANIDE ◇ CYANOGAS

OSHA PEL: TWA 5 $mg(CN)/m^3$
ACGIH TLV: TWA 5 $mg(CN)/m^3$ (skin)
DFG MAK: 5 $mg/m^3$
NIOSH REL: (Cyanide) CL 5 $mg(CN)/m^3$/10M
DOT Class: Poison B

SAFETY PROFILE: A deadly poison by ingestion. When heated to decomposition it emits toxic fumes of $NO_x$ and $CN^-$.

PROP: Rhombohedral crystals or white powder. Mp: decomp > 350°.

**CAT250** CAS: 1305-62-0 $CaH_2O_2$ HR: 2
**CALCIUM HYDROXIDE**
SYNS: HYDRATED LIME ◇ SLAKED LIME

OSHA PEL: TWA 5 $mg/m^3$
ACGIH TLV: TWA 5 $mg/m^3$

SAFETY PROFILE: Mildly toxic by ingestion. A skin and severe eye irritant. Causes dermatitis.

PROP: Rhombic, trigonal, colorless crystals or white power; sltly bitter taste. Mp: loses $H_2O$ @ 580°, bp: decomp, d: 2.343. Sol in water and glycerin; insol in alc.

**CAU500** CAS: 1305-78-8 CaO HR: 3
**CALCIUM OXIDE**
DOT: 1910
SYNS: BURNT LIME ◇ CALX ◇ LIME ◇ QUICKLIME (DOT)

OSHA PEL: TWA 5 $mg/m^3$
ACGIH TLV: TWA 2 $mg/m^3$
DFG MAK: 5 $mg/m^3$
DOT Class: ORM-B

SAFETY PROFILE: A caustic and irritating material. The powdered oxide may react explosively with water and other substances.

PROP: Cubic, white crystals. Mp: 2580°, d: 3.37, bp: 2850°. Sol in water and glycerin; insol in alc.

**CAW850** CAS: 1344-95-2 HR: 1
**CALCIUM SILICATE**

OSHA PEL: Total Dust: 15 $mg/m^3$; Respirable Fraction: 5 $mg/m^3$
ACGIH TLV: TWA (nuisance particulate) 10 $mg/m^3$ of total dust (when toxic impurities are not present, e.g., quartz < 1%).

SAFETY PROFILE: A nuisance dust.

PROP: Varying proportions of CaO and $SiO_2$. White powder. Insol in water.

**CAX500** CAS: 7778-18-9 $CaSO_4$ HR: 3
**CALCIUM SULFATE**
SYNS: GYPSUM ◇ PLASTER of PARIS

OSHA PEL: Total Dust: 15 $mg/m^3$; Respirable Fraction: 5 $mg/m^3$
ACGIH TLV: TWA (nuisance particulate) 10 $mg/m^3$ of total dust (when toxic impurities are not present, e.g., quartz < 1%)7

SAFETY PROFILE: A nuisance dust. When heated to decomposition it emits toxic fumes of $SO_x$.

PROP: Pure anhydrous, white powder or odorless crystals. D: 2.964; mp: 1450°.

**CAX750** CAS: 10101-41-4 $O_4S{\cdot}Ca{\cdot}2H_2O$ HR: 1
**CALCIUM(II) SULFATE DIHYDRATE (1:1:2)**
SYNS: GYPSUM ◇ MAGNESIA WHITE

OSHA PEL: Total Dust: 15 $mg/m^3$; Respirable Fraction: 5 $mg/m^3$
ACGIH TLV: TWA (nuisance particulate) 10 $mg/m^3$ of total dust (when toxic impurities are not present, e.g., quartz < 1%)

SAFETY PROFILE: Long considered a nuisance dust (depending on silica content). When heated to decomposition it emits toxic fumes of $SO_x$.

PROP: Colorless crystals. D: 2.32, mp: 128° (loses $1.5H_2O$), bp: 163° (loses $2H_2O$).

**CBA750** CAS: 76-22-2 $C_{10}H_{16}O$ HR: 3
**CAMPHOR**
DOT: 2717
SYNS: 2-BORNANONE ◇ 2-OXOBORNANE

OSHA PEL: TWA 2 $mg/m^3$
ACGIH TLV: TWA 2 ppm; STEL 3 ppm
DFG MAK: 2 ppm (13 $mg/m^3$)
DOT Class: Flammable Solid

SAFETY PROFILE: A human poison by ingestion. An experimental poison by inhalation. Flammable. Vapor is explosive. To fight fire, use foam, carbon dioxide, dry chemical.

PROP: White, transparent, crystalline masses; penetrating odor; pungent, aromatic taste. Mp: 180°, bp: 204°, lel: 0.6%, uel: 3.5%, flash p: 150°F (CC), d: 0.992 @ 25°/4°, autoign temp: 871°F, vap d: 5.24.

**CBF700** CAS: 105-60-2 $C_6H_{11}NO$ HR: 3
**CAPROLACTAM**

SYNS: omega-CAPROLACTAM (MAK) ◇ CYCLOHEXANONE ISO-OXIME ◇ HEXANONE ISOXIME

OSHA PEL: Dust: 1 $mg/m^3$; STEL 3 $mg/m^3$; Vapor: 5 ppm; STEL 10 ppm
ACGIH TLV: Dust: 1 $mg/m^3$; STEL 3 $mg/m^3$; Vapor: 5 ppm; STEL 10 ppm; (Proposed: TWA Dust: 1 $mg/m^3$; 5 ppm (vapor and aerosol)
DFG MAK: 25 $mg/m^3$

SAFETY PROFILE: Moderately toxic by ingestion and skin contact. Experimental reproductive effects. A skin and eye irritant. When heated to decomposition it emits toxic fumes of $NO_x$.

PROP: White crystals. Mp: 69°, vap press: 6 mm @ 120°.

**CBF800** CAS: 2425-06-1 $C_{10}H_9Cl_4NO_2S$ HR: 3
**CAPTAFOL**

SYNS: DIFOLATAN ◇ SULPHEIMIDE

OSHA PEL: TWA 0.1 $mg/m^3$
ACGIH TLV: TWA 0.1 $mg/m^3$

SAFETY PROFILE: Moderately toxic by ingestion. An experimental carcinogen and teratogen. When heated to decomposition it emits very toxic fumes of $Cl^-$, $NO_x$ and $SO_x$.

**CBG000** CAS: 133-06-2 $C_9H_8Cl_3NO_2S$ HR: 3
**CAPTAN**

DOT: 9099

SYNS: MERPAN ◇ ORTHOCIDE ◇ N-TRICHLOROMETHYLTHIO-3A,4,7,7A-TETRAHYDROPHTHALIMIDE ◇ VANICIDE

OSHA PEL: TWA 5 $mg/m^3$
ACGIH TLV: TWA 5 $mg/m^3$
DOT Class: ORM-E

SAFETY PROFILE: Moderately toxic to humans by ingestion. Moderately toxic experimentally by inhalation. An experimental tumorigen and teratogen. A suspected carcinogen. When heated to decomposition it emits toxic fumes of $Cl^-$, $SO_x$, and $NO_x$.

PROP: Odorless crystals. Insol in water; sol in benzene and chloroform.

**CBM750** CAS: 63-25-2 $C_{12}H_{11}NO_2$ HR: 3
**CARBARYL**

SYNS: CRAG SEVIN ◇ METHYLCARBAMATE-1-NAPHTHALENOL ◇ SEVIN

OSHA PEL: TWA 5 $mg/m^3$

ACGIH TLV: TWA 5 $mg/m^3$
DFG MAK: 5 $mg/m^3$
NIOSH REL: TWA 5 $mg/m^3$
DOT Class: ORM-A

SAFETY PROFILE: Poison by ingestion. An experimental carcinogen and teratogen. Experimental reproductive effects. An eye and severe skin irritant. When heated to decomposition it emits toxic fumes of $NO_x$.

PROP: White crystals. Mp: 142°, d: 1.232 @ 20°/20°.

**CBS275** CAS: 1563-66-2 $C_{12}H_{15}NO_3$ HR: 3
**CARBOFURAN**
DOT: 2757
SYNS: CURATERR ◇ FURADAN ◇ YALTOX

OSHA PEL: TWA 0.1 $mg/m^3$
ACGIH TLV: TWA 0.1 $mg/m^3$
DOT Class: Poison B

SAFETY PROFILE: Poison by inhalation, ingestion, and skin contact. An experimental teratogen. Experimental reproductive effects. When heated to decomposition it emits toxic fumes of $NO_x$.

PROP: White, crystalline solid; odorless. Mp: 105-152°, d: 1.180 @ 20°/20°, vap press: $2 \times 10^{-5}$ mm @ 33°. Sltly sol in sol.

**CBT500** CAS: 7440-44-0 C HR: 1
**CARBON**
SYNS: CHARCOAL BLACK ◇ C.I. 77266 ◇ GRAPHITE (MAK)

OSHA PEL: (Natural graphite) (Transitional: TWA 50 mppcf) TWA 2.5 $mg/m^3$; (Synthetic graphite) (Transitional: TWA Total Dust: 15 $mg/m^3$; Respirable Fraction: 5 $mg/m^3$) TWA Total Dust: 10 $mg/m^3$; Respirable Fraction: 5 $mg/m^3$
ACGIH TLV: (Proposed: 2.5 $mg/m^3$ (respirable))
DFG MAK: 6 $mg/m^3$

SAFETY PROFILE: Experimental reproductive effects. It can cause a dust irritation, particularly to the eyes and mucous membranes. Combustible. Dust is explosive.

PROP: Black crystals, powder or diamond form. Mp: 3652-3697° (sublimes), bp: approx 4200°, d (amorphous): 1.8-2.1, d (graphite): 2.25, d (diamond): 3.51, vap press: 1 mm @ 3586°.

**CBT750** CAS: 1333-86-4 HR: 1
**CARBON BLACK**
SYNS: ACETYLENE BLACK ◇ CHANNEL BLACK ◇ FURNACE BLACK ◇ LAMP BLACK

OSHA PEL: TWA 3.5 $mg/m^3$
ACGIH TLV: TWA 3.5 $mg/m^3$

NIOSH REL: TWA 3.5 mg/m$^3$

SAFETY PROFILE: Mildly toxic by ingestion, inhalation, and skin contact. A nuisance dust in high concentrations.

PROP: A generic term applied to a family of high-purity colloidal carbons commercially produced by carefully controlled pyrolysis of gaseous or liquid hydrocarbons. Carbon blacks, including commercial colloidal carbons such as furnace blacks, lamp blacks and acetylene blacks, usually contain less than several tenths percent of extractable organic matter and less than one percent ash.

**CBU250** CAS: 124-38-9 $CO_2$ HR: 1
**CARBON DIOXIDE**
DOT: 1013/1845/2187
SYNS: CARBONIC ACID GAS ◇ CARBONIC ANHYDRIDE

OSHA PEL: (Transitional: TWA 5000 ppm) TWA 10,000 ppm; STEL 30,000 ppm
ACGIH TLV: TWA 5000 ppm; STEL 30,000 ppm
DFG MAK: 5000 ppm (9000 mg/m$^3$)
NIOSH REL: TWA 10000 ppm; CL 30000 ppm/10M
DOT Class: Nonflammable Gas

SAFETY PROFILE: An asphyxiant. An experimental teratogen. Other experimental reproductive effects. Contact of carbon dioxide snow with the skin can cause burns.

PROP: Colorless, odorless gas. Mp: subl @ −78.5° (−56.6° @ 5.2 atm), vap d: 1.53.

**CBV000** CAS: 53569-62-3 $CO_2 \cdot N_2O$ HR: 2
**CARBON DIOXIDE mixed with NITROUS OXIDE**
DOT: 1015

NIOSH REL: (Carbon Dioxide) TWA 10000 ppm; CL 30000 ppm/10M
NIOSH REL: ($N_2O$ as Anesthetic Agent) TWA 25 ppm/1H
DOT Class: Nonflammable Gas

SAFETY PROFILE: An anesthetic mixture. Combustible. Can react with reducing materials.

**CBV500** CAS: 75-15-0 $CS_2$ HR: 3
**CARBON DISULFIDE**
DOT: 1131
SYNS: CARBON BISULFIDE (DOT) ◇ DITHIOCARBONIC ANHYDRIDE

OSHA PEL: (Transitional: TWA 20 ppm; CL 30 ppm; PK 100 ppm/30 min) TWA 4 ppm; STEL 12 (skin)
ACGIH TLV: TWA 10 ppm (skin); BEI: 5 mg(2-thiothiazolidine-4-carboxylic acid (TTCA))/g creatinine in urine.
DFG MAK: 10 ppm (30 mg/m$^3$); BAT: 8 mg/L of 4-thio-4-thiazolidine carboxylic acid (TTCA) at end of shift.

NIOSH REL: TWA 1 ppm; CL 10 ppm/15M
DOT Class: Flammable Liquid

SAFETY PROFILE: A human poison by ingestion. Human reproductive effects by inhalation. An experimental teratogen. Flammable. A dangerous fire and explosion hazard. To fight fire, use water, $CO_2$, dry chemical, fog, mist. When heated to decomposition it emits highly toxic fumes of $SO_x$.

PROP: Clear, colorless liquid; nearly odorless when pure. Mp: −110.8°, bp: 46.5°, lel: 1.3%, uel: 50%, flash p: −22°F (CC), d: 1.261 @ 20°/20°, autoign temp: 257°F, vap press: 400 mm @ 28°, vap d: 2.64.

**CBW750** CAS: 630-08-0 CO HR: 3
**CARBON MONOXIDE**
DOT: 1016/9202
SYNS: CARBONIC OXIDE ◇ FLUE GAS

OSHA PEL: (Transitional: TWA 50 ppm) TWA 35; CL 200 ppm
ACGIH TLV: TWA 50 ppm; STEL 400 ppm; BEI: less than 8% carboxyhemoglobin in blood at end of shift; less than 40 ppm CO in end-exhaled air at end of shift.
DFG MAK: 30 ppm (33 mg/m$^3$); BAT: 5% in blood at end of shift.
NIOSH REL: TWA 35 ppm; CL 200 ppm
DOT Class: Flammable Gas

SAFETY PROFILE: Mildly toxic by inhalation in humans but has caused many fatalities. An experimental teratogen. Experimental reproductive effects. A dangerous fire hazard and severe explosion hazard. To fight fire, stop flow of gas.

PROP: Colorless, odorless gas. Mp: −207°, bp: −191.3°, lel: 12.5%, uel: 74.2%, d: (gas) 1.250 g/L @ 0°, (liquid) 0.793, autoign temp: 1128°F.

**CBX750** CAS: 558-13-4 $CBr_4$ HR: 3
**CARBON TETRABROMIDE**
DOT: 2516
SYNS: CARBON BROMIDE ◇ TETRABROMOMETHANE

OSHA PEL: TWA 0.1 ppm; STEL 0.3 ppm
ACGIH TLV: TWA 0.1 ppm; STEL 0.3 ppm
DOT Class: Poison B

SAFETY PROFILE: Moderately toxic by ingestion. Narcotic in high concentration. When heated to decomposition it emits toxic fumes of $Br^-$.

PROP: Colorless, monoclinic tablets. Mp: (α) 48.4°, (β) 90.1°, bp: 189.5°, d: 3.42, vap press: 40 mm @ 96.3°.

**CBY000** CAS: 56-23-5 $CCl_4$ HR: 3
**CARBON TETRACHLORIDE**
DOT: 1846
SYNS: CARBONA ◇ CARBON TET ◇ METHANE TETRACHLORIDE

OSHA PEL: (Transitional: TWA 10 ppm; CL 25 ppm; PK 200 ppm/5 min) TWA 2 ppm
ACGIH TLV: TWA 5 ppm; STEL 30 (skin); Suspected Human Carcinogen
DFG MAK: 10 ppm (65 mg/$m^3$); BEI: 1.6 mL/$m^3$ in alveolar air 1 hour after exposure; Suspected Carcinogen.
NIOSH REL: CL 2 ppm/60M
DOT Class: ORM-A

SAFETY PROFILE: A human poison by ingestion. Mildly toxic by inhalation. An experimental carcinogen and teratogen. An eye and skin irritant. When heated to decomposition it emits toxic fumes of $Cl^-$ and phosgene.

PROP: Colorless liquid; heavy, ethereal odor. Mp: −22.6°, bp: 76.8°, fp: −22.9°, flash p: none, d: 1.597 @ 20°, vap press: 100 mm @ 23.0°.

**CCA500** CAS: 353-50-4 $CF_2O$ HR: 3
**CARBONYL FLUORIDE**
DOT: 2417
SYNS: CARBON DIFLUORIDE OXIDE ◇ FLUOROPHOSGENE

OSHA PEL: TWA 2 ppm; STEL 5 ppm
ACGIH TLV: TWA 2 ppm; STEL 5 ppm
DOT Class: Poison A

SAFETY PROFILE: A poison. Moderately toxic by inhalation. A powerful irritant. When heated to decomposition it emits toxic fumes of CO and $F^-$.

PROP: Colorless gas; pungent. Hygroscopic, mp: −114°, bp: −83°, d: 1.139 @ −114°.

**CCP850** CAS: 120-80-9 $C_6H_6O_2$ HR: 3
**CATECHOL**
SYNS: o-BENZENEDIOL ◇ C.I. 76500 ◇ o-HYDROQUINONE ◇ PYROCATECHOL

OSHA PEL: TWA 5 ppm (skin)
ACGIH TLV: TWA 5 ppm

SAFETY PROFILE: Poison by ingestion. Moderately toxic by skin contact. Experimental reproductive effects. Can cause dermatitis on skin contact. An allergen. Combustible. To fight fire, use water, $CO_2$, dry chemical.

PROP: Colorless crystals. Mp: 105°, bp: 246°, flash p: 261°F (CC), d: 1.341 @ 15°, vap press: 10 mm @ 118.3°, vap d: 3.79. Sol in water, chloroform and benzene; very sol in alc and ether.

**CCU150** CAS: 9004-34-6 HR: 1
**CELLULOSE, POWDERED**

OSHA PEL: Total Dust: 15 mg/$m^3$; Respirable Fraction: 5 mg/$m^3$

ACGIH TLV: TWA (nuisance particulate) 10 mg/m$^3$ of total dust (when toxic impurities are not present, e.g., quartz $< 1\%$).

SAFETY PROFILE: A nuisance dust.

PROP: Fine white fibrous particles from treatment of bleached cellulose from wood or cotton. Insol in water and most organic solvents.

**CDD750** CAS: 21351-79-1 CsHO HR: 3
**CESIUM HYDROXIDE**
DOT: 2681/2682
SYN: CESIUM HYDRATE

OSHA PEL: TWA 2 mg/m$^3$
ACGIH TLV: TWA 2 mg/m$^3$
DOT Class: Corrosive; IMO: Corrosive Material

SAFETY PROFILE: Moderately toxic by ingestion. A powerful caustic. A corrosive skin and eye irritant.

PROP: Colorless to yellowish, very deliquescent crystals. Mp: 272.3°, d: 3.675.

**CDN200** CAS: 78-95-5 $C_3H_5ClO$ HR: 3
**CHLORACETONE**
DOT: 1695
SYNS: ACETONYL CHLORIDE ◇ CHLOROACETONE ◇ CHLOROPROPANONE

ACGIH TLV: CL 1 ppm (skin)
DOT Class: Forbidden.

SAFETY PROFILE: Poison by inhalation, ingestion, and skin contact. An experimental tumorigen. A lachrymator poison gas. Flammable.

PROP: Colorless liquid, pungent odor. Mp: −44.5°, bp: 119°, d: 1.162.

**CDR750** CAS: 57-74-9 $C_{10}H_6Cl_8$ HR: 3
**CHLORDANE**
DOT: 2762
SYNS: CHLORODANE ◇ NIRAN ◇ TOXICHLOR

OSHA PEL: TWA 0.5 mg/m$^3$ (skin)
ACGIH TLV: TWA 0.5 mg/m$^3$ (skin)
DFG MAK: Suspected Carcinogen.
DOT Class: Combustible Liquid

SAFETY PROFILE: Poison to humans by ingestion. An experimental poison by inhalation. Moderately toxic by skin contact. A suspected human carcinogen. An experimental carcinogen and teratogen. Combustible liquid. When heated to decomposition it emits toxic fumes of $Cl^-$.

PROP: Colorless to amber; odorless, viscous liquid. Bp: 175°, d: 1.57-1.63 @ 15.5°/15.5°.

**CDV100** CAS: 8001-35-2 $C_{10}H_{10}Cl_8$ HR: 3
**CHLORINATED CAMPHENE**
DOT: 2761
SYNS: CAMPHECHLOR ◇ CHLOROCAMPHENE ◇ OCTACHLOROCAMPHENE ◇ POLYCHLORINATED CAMPHENES ◇ TOXAPHENE

OSHA PEL: (Transitional: TWA 0.5 $mg/m^3$ (skin)) TWA 0.5 $mg/m^3$; STEL 1 $mg/m^3$ (skin)
ACGIH TLV: TWA 0.5 $mg/m^3$; STEL 1 $mg/m^3$ (skin)
DFG MAK: 0.5 $mg/m^3$
DOT Class: ORM-A

SAFETY PROFILE: Human poison by ingestion. An experimental carcinogen and teratogen. A skin irritant. When heated to decomposition it emits toxic fumes of $Cl^-$.

PROP: Yellow, waxy solid; pleasant piney odor. Mp: 65-90°. Almost insol in water; very sol in aromatic hydrocarbons.

**CDV175** CAS: 55720-99-5 $C_{12}H_4Cl_6O$ HR: 3
**CHLORINATED DIPHENYL OXIDE**
SYNS: HEXACHLOROPHENYL ETHER ◇ PHENYL ETHER HEXACHLORO

OSHA PEL: TWA 0.5 $mg/m^3$
ACGIH TLV: TWA 0.5 $mg/m^3$
DFG MAK: 0.5 $mg/m^3$

SAFETY PROFILE: Poison by ingestion and probably by inhalation. Combustible. To fight fire, use water spray, fog, foam, dry chemical, $CO_2$. When heated to decomposition it emits toxic fumes of $Cl^-$.

PROP: Light yellow, very viscous liquid. Bp: 230-260° @ 8 mm, d: 1.60 @ 20°/60°, autoign temp: 1148°F, vap d: 13.0.

**CDV750** CAS: 7782-50-5 $Cl_2$ HR: 3
**CHLORINE**
DOT: 1017

OSHA PEL: (Transitional: TWA CL 1 ppm) TWA 0.5 ppm; STEL 1 ppm
ACGIH TLV: TWA 0.5 ppm; STEL 1 ppm
DFG MAK: 0.5 ppm (1.5 $mg/m^3$)
NIOSH REL: CL 0.5 ppm/15M
DOT Class: Nonflammable Gas

SAFETY PROFILE: Moderately toxic and very irritating to humans by inhalation. Chlorine is extremely irritating to the mucous membranes of the eyes and the respiratory tract at 3 ppm. Can react, upon contact with many substances, to cause fires or explosions.

PROP: Greenish-yellow gas, liquid, or rhombic crystals. Mp: −101°, bp: −34.5°, d: (liquid) 1.47 @ 0° (3.65 atm), vap press: 4800 mm @ 20°, vap d: 2.49. Sol in water.

**CDW450** CAS: 10049-04-4 $ClO_2$ HR: 3
**CHLORINE DIOXIDE**
DOT: 9191
SYNS: CHLORINE PEROXIDE ◇ CHLORINE OXIDE

OSHA PEL: (Transitional: TWA 0.1 ppm) TWA 0.1 ppm; STEL 0.3 ppm
ACGIH TLV: TWA 0.1 ppm; STEL 0.3 ppm
DFG MAK: 0.1 ppm (0.3 $mg/m^3$)
DOT Class: Forbidden (not hydrated); Oxidizer

SAFETY PROFILE: Moderately toxic by inhalation. Experimental reproductive effects. An eye irritant. A powerful explosive and oxidizer. Reacts with water or steam to produce toxic and corrosive fumes of HCl. When heated to decomposition it emits toxic fumes of $Cl^-$.

PROP: Red-yellow gas or orange-red crystals. Mp: −59°, bp: 9.9° @ 731 mm explodes, d: 3.09 g/L @ 11°.

**CDX750** CAS: 7790-91-2 $ClF_3$ HR: 3
**CHLORINE TRIFLUORIDE**
DOT: 1749
SYN: CHLORINE FLUORIDE

OSHA PEL: CL 0.1 ppm
ACGIH TLV: CL 0.1 ppm
DFG MAK: 0.1 ppm (0.4 $mg/m^3$)
DOT Class: Oxidizer

SAFETY PROFILE: Human poison by inhalation. An eye irritant. Spontaneously flammable. A powerful oxidant which may react violently with oxidizable materials. When heated to decomposition or in reaction with water or steam it emits toxic fumes of $F^-$ and $Cl^-$.

PROP: Colorless gas to yellow liquid, sweet odor, mp: −83°, bp: 11.8°, d: 1.77 @ 13°.

**CDY500** CAS: 107-20-0 $C_2H_3ClO$ HR: 3
**CHLOROACETALDEHYDE**
DOT: 2232
SYNS: 2-CHLOROACETALDEHYDE ◇ 2-CHLORO-1-ETHANAL

OSHA PEL: CL 1 ppm
ACGIH TLV: CL 1 ppm
DFG MAK: 1 ppm (3 $mg/m^3$)
DOT Class: Poison B

SAFETY PROFILE: Poison by ingestion, skin contact, and intraperitoneal routes. Combustible. To fight fire, use water, foam, $CO_2$, dry chemical. When heated to decomposition it emits toxic fumes of $Cl^-$.

PROP: Clear, colorless liquid; pungent odor. Bp: 90.0°-100.1° (40% soln), fp: −16.3° (40% soln), flash p: 190°F, d: 1.19 @ 25°/25° (40% soln), vap press: 100 mm @ 45° (40% soln).

**CEA750** CAS: 532-27-4 $C_8H_7ClO$ HR: 3
**α-CHLOROACETOPHENONE**
DOT: 1697
SYNS: CHEMICAL MACE ◇ omega-CHLOROACETOPHENONE ◇ MACE (lacrimator) ◇ PHENYLCHLOROMETHYLKETONE

OSHA PEL: TWA 0.05 ppm
ACGIH TLV: TWA 0.05 ppm
DOT Class: Irritating Material

SAFETY PROFILE: A human poison by inhalation. An experimental poison by ingestion. An experimental neoplastigen by skin contact. A severe eye and moderate skin irritant. When heated to decomposition it emits toxic fumes of $Cl^-$.

**CEC250** CAS: 79-04-9 $C_2H_2Cl_2O$ HR: 3
**CHLOROACETYL CHLORIDE**
DOT: 1752
SYNS: CHLOROACETIC ACID CHLORIDE ◇ MONOCHLOROACETYL CHLORIDE

OSHA PEL: TWA 0.05 ppm
ACGIH TLV: TWA 0.05 ppm
DOT Class: Corrosive Material

SAFETY PROFILE: Poison by ingestion. Mildly toxic by inhalation. Corrosive. A lacrymator. When heated to decomposition it emits toxic fumes of $Cl^-$.

PROP: Water-white or sltly yellow liquid. Bp: 105-106°, fp: −22.5°, flash p: none, d: 1.495 @ 0°.

**CEJ125** CAS: 108-90-7 $C_6H_5Cl$ HR: 2
**CHLOROBENZENE**
DOT: 1134
SYNS: BENZENE CHLORIDE ◇ CHLOROBENZOL (DOT) ◇ MCB ◇ MONOCHLOROBENZENE

OSHA PEL: TWA 75 ppm
ACGIH TLV: TWA 75 ppm; (Proposed: 10 ppm)
DFG MAK: 50 ppm (230 $mg/m^3$)
DOT Class: Flammable or Combustible Liquid

SAFETY PROFILE: Poison by ingestion. An experimental teratogen. Other experimental reproductive effects by inhalation. Strong narcotic with slight irritant qualities. Dangerous fire hazard. Moderate explosion hazard. To fight fire, use foam, $CO_2$, dry chemical, water to blanket fire.

PROP: Clear, colorless liquid. Bp: 131.7°, lel: 1.3%, uel: 7.1%, @ 150°, mp: −45°, flash p: 85°F (CC), d: 1.113 @ 15.5°/15.5°, autoign temp: 1180°F, vap press: 10 mm @ 22.2°, vap d: 3.88.

**CEQ600** CAS: 2698-41-1 $C_{10}H_5ClN_2$ HR: 3

**o-CHLOROBENZYLIDENE MALONONITRILE**

SYNS: 2-CHLOROBENZAL MALONONITRILE ◇ CS

OSHA PEL: (Transitional: TWA 0.05 ppm) CL 0.05 ppm (skin)
ACGIH TLV: CL 0.05 ppm (skin)

OSHA PEL: TWA 0.05 ppm

SAFETY PROFILE: Poison by ingestion. Moderately toxic by inhalation. A human skin and eye irritant. When heated to decomposition it emits very toxic fumes of Cl , $NO_x$, and CN .

PROP: White crystals. Mp: 95°, bp: 313°.

**CES650** CAS: 74-97-5 $CH_2BrCl$ HR: 3

**CHLOROBROMOMETHANE**

DOT: 1887

SYNS: BROMOCHLOROMETHANE ◇ HALON 1011 ◇ METHYLENE CHLOROBROMIDE

OSHA PEL: TWA 200 ppm
ACGIH TLV: TWA 200 ppm
DFG MAK: 200 ppm (1050 mg/m$^3$)
DOT Class: Poison B

SAFETY PROFILE: A poison. Mildly toxic by ingestion and inhalation. Dangerous; when heated to decomposition it emits highly toxic fumes of $Br^-$ and $Cl^-$.

PROP: Clear, colorless liquid; sweet odor. Bp: 67.8°, fp: −88°, flash p: none, d: 1.930 @ 25°/25°, vap d: 4.46.

**CFX500** CAS: 75-45-6 $CHClF_2$ HR: 1

**CHLORODIFLUOROMETHANE**

DOT: 1018

SYNS: FLUOROCARBON-22 ◇ FREON 22 ◇ MONOCHLORODIFLUOROMETHANE

OSHA PEL: TWA 1000 ppm
ACGIH TLV: TWA 1000 ppm
DFG MAK: 500 ppm (1800 mg/m$^3$)
DOT Class: Nonflammable Gas

SAFETY PROFILE: Mildly toxic by inhalation. Experimental reproductive effects. An asphyxiant in high concentrations. At elevated pressures, 50% mixtures with air are combustible although ignition is difficult. When heated to decomposition it emits toxic fumes of $F^-$ and $Cl^-$.

PROP: Gas. D: 3.87 air @ 0°, mp: −146°, bp: −40.8°, autoign temp: 1170°F.

**CHI900** CAS: 593-70-4 $CH_2ClF$ HR: 3

**CHLOROFLUOROMETHANE**

SYNS: MONOCHLOROMONOFLUOROMETHANE ◇ FREON 31

DFG MAK: Animal Carcinogen, Suspected Human Carcinogen.

SAFETY PROFILE: Moderately toxic by inhalation. An experimental carcinogen. When heated to decomposition it emits very toxic fumes of $Cl^-$ and $F^-$.

**CHJ500** CAS: 67-66-3 $CHCl_3$ HR: 3
**CHLOROFORM**
DOT: 1888
SYNS: FORMYL TRICHLORIDE ◇ METHANE TRICHLORIDE ◇ TCM ◇ TRICHLOROMETHANE

OSHA PEL: (Transitional: CL 50 ppm) TWA 2 ppm
ACGIH TLV: TWA 10 ppm; Suspected Human Carcinogen
DFG MAK: Suspected Carcinogen.
NIOSH REL: (Waste Anesthetic Gases and Vapors) CL 2 ppm/1H; (Chloroform) CL 2 ppm/60M
DOT Class: ORM-A

SAFETY PROFILE: A human poison by ingestion and inhalation. An experimental carcinogen and teratogen. Nonflammable. When heated to decomposition it emits toxic fumes of $Cl^-$.

PROP: Colorless liquid; heavy, ethereal odor. Mp: −63.5°, bp: 61.26°, fp: −63.5°, flash p: none, d: 1.49845 @ 15°, vap press: 100 mm @ 10.4°, vap d: 4.12.

**CIO250** CAS: 107-30-2 $C_2H_5ClO$ HR: 3
**CHLOROMETHYL METHYL ETHER**
DOT: 1239
SYNS: CMME ◇ DIMETHYLCHLOROETHER ◇ METHYLCHLOROMETHYL ETHER (DOT) ◇ MONOCHLORODIMETHYL ETHER (MAK)

OSHA: Carcinogen
ACGIH TLV: Suspected Human Carcinogen.
DFG MAK: Human Carcinogen.
DOT Class: Flammable Liquid

SAFETY PROFILE: Poison by inhalation. Moderately toxic by ingestion. A suspected human carcinogen. An experimental carcinogen. A very dangerous fire hazard. To fight fire use alcohol foam, water, $CO_2$, dry chemical. When heated to decomposition it emits toxic fumes of $Cl^-$.

PROP: Flash p: <73.4°F.

**CJE000** CAS: 600-25-9 $C_3H_6ClNO_2$ HR: 3
**1-CHLORO-1-NITROPROPANE**
SYNS: KORAX ◇ LANSTAN

OSHA PEL: (Transitional: TWA 20 ppm) TWA 2 ppm
ACGIH TLV: TWA 2 ppm
DFG MAK: 20 ppm (100 $mg/m^3$)

SAFETY PROFILE: Poison by ingestion. Moderately toxic by inhalation. Combustible. Moderately explosive. To fight fire use alcohol foam, water, $CO_2$, dry chemical. When heated to decomposition it emits toxic fumes of $Cl^-$ and $NO_x$.

PROP: Liquid. Bp: 139.5°, flash p: 144°F (OC), d: 1.209 @ 20°/20°, vap d: 4.26.

**CJI500** CAS: 76-15-3 $C_2ClF_5$ HR: 1
**CHLOROPENTAFLUOROETHANE**
DOT: 1020
SYNS: FLUOROCARBON-115 ◇ FREON 115 ◇ MONOCHLOROPENTAFLUOROETHANE (DOT)

OSHA PEL: TWA 1000 ppm
ACGIH TLV: TWA 1000 ppm
DOT Class: Nonflammable Gas

SAFETY PROFILE: Mildly toxic by inhalation. A nonflammable gas. When heated to decomposition it emits toxic fumes of $F^-$ and $Cl^-$.

PROP: Colorless gas. Bp: −39.3°, mp: −38°, d: 1.5678 @ -42°. Insol in water; sol in alc and ether.

**CKN500** CAS: 76-06-2 $CCl_3NO_2$ HR: 3
**CHLOROPICRIN**
DOT: 1580/1583
SYNS: DOLOCHLOR ◇ NITROTRICHLOROMETHANE ◇ TRICHLORONITROMETHANE

OSHA PEL: TWA 0.1 ppm
ACGIH TLV: TWA 0.1 ppm
DFG MAK: 0.1 ppm (0.7 $mg/m^3$)
DOT Class: Poison B

SAFETY PROFILE: Poison by ingestion. Moderately toxic by inhalation. An experimental tumorigen. A powerful irritant that affects all body surfaces. When heated to decomposition it emits very toxic fumes of $Cl^-$ and $NO_x$.

PROP: Sltly oily, colorless liquid. D: 1.651 @ 22.8°/4°, mp: −64°, bp: 112.3 @ 766 mm, vap press: 40 mm @ 33.80, vap d: 6.69. Sol in water, alc, and ether.

**CLE750** CAS: 2039-87-4 $C_8H_7Cl$ HR: 1
**o-CHLOROSTYRENE**

OSHA PEL: TWA 50 ppm; STEL: 75 ppm
ACGIH TLV: TWA 50 ppm; STEL: 75 ppm

SAFETY PROFILE: Mildly toxic by ingestion and skin contact. A skin and eye irritant. When heated to decomposition it emits toxic fumes of $Cl^-$.

**CLK100** CAS: 95-49-8 $C_7H_7Cl$ HR: 2
**o-CHLOROTOLUENE**
DOT: 2238
SYNS: 2-CHLORO-1-METHYLBENZENE (9CI) ◇ 2-METHYLCHLOROBENZENE ◇ o-TOLYL CHLORIDE

OSHA PEL: TWA 50 ppm
ACGIH TLV: TWA 50 ppm
DOT Class: Flammable or Combustible Liquid

SAFETY PROFILE: Moderately toxic. Flammable. When heated to decomposition it emits toxic fumes of $Cl^-$.

PROP: Liquid. Bp: 158.97°, d: (20/4) 1.0826, mp: -35.59°. Volatile with steam. Sltly sol in water; freely sol in alc, benzene, chloroform, and ether.

**CLK220** CAS: 95-69-2 $C_7H_8ClN$ HR: 3
**4-CHLORO-o-TOLUIDINE**
SYNS: 5-CHLORO-2-AMINOTOLUENE ◇ RED TR BASE

DFG MAK: Human Carcinogen.

SAFETY PROFILE: Poison by ingestion. An experimental carcinogen. When heated to decomposition it emits toxic fumes of $Cl^-$ and $NO_x$.

**CLK225** CAS: 95-79-4 $C_7H_8ClN$ HR: 3
**5-CHLORO-o-TOLUIDINE**
SYN: 2-AMINO-4-CHLOROTOLUENE ◇ FAST RED KB BASE

DFG MAK: Suspected Carcinogen.

SAFETY PROFILE: An experimental carcinogen and tumorigen. Moderately toxic by ingestion. When heated to decomposition it emits very toxic fumes of $Cl^-$ and $NO_x$.

PROP: Solid. Bp: 241°, mp: 29°.

**CLP750** CAS: 1929-82-4 $C_6H_3Cl_4N$ HR: 3
**2-CHLORO-6-(TRICHLOROMETHYL)PYRIDINE**
SYNS: NITRAPYRIN (ACGIH) ◇ N-SERVE NITROGEN STABILIZER

OSHA PEL: Total Dust: 15 $mg/m^3$; Respirable Fraction: 5 $mg/m^3$
ACGIH TLV: TWA 10 $mg/m^3$; STEL 20 $mg/m^3$

SAFETY PROFILE: Poison by ingestion. Moderately toxic by skin contact. When heated to decomposition it emits very toxic fumes of $Cl^-$ and $NO_x$.

**CMA100** CAS: 2921-88-2 $C_9H_{11}Cl_3NO_3PS$ HR: 3
**CHLORPYRIFOS**
DOT: 2783
SYNS: DURSBAN ◇ PYRINEX

OSHA PEL: TWA 0.2 $mg/m^3$ (skin)
ACGIH TLV: TWA 0.2 $mg/m^3$ (skin)
DOT Class: ORM-A

SAFETY PROFILE: Poison by ingestion, skin contact, and inhalation. An experimental teratogen. When heated to decomposition it emits very toxic fumes of $Cl^-$, $NO_x$, $PO_x$, and $SO_x$.

**CMH250** CAS: 7738-94-5 $CrH_2O_4$ HR: 3
**CHROMIC ACID**
SYN: CHROMIC(VI) ACID

OSHA PEL: (Transitional: CL 1 mg/10m$^3$) CL 0.1 mg($CrO_3$)/m$^3$
ACGIH TLV: TWA 0.05 mg(Cr)/m$^3$, Confirmed Human Carcinogen.
DFG MAK: Animal Carcinogen, Suspected Human Carcinogen.
NIOSH REL: TWA 0.025 mg(Cr(VI))/m$^3$; CL 0.05/15M

SAFETY PROFILE: A carcinogen. A powerful oxidizer. A storage hazard; it may burst a sealed container due to carbon dioxide release. May ignite on contact with acetone or alcohols.

**CMI750** CAS: 7440-47-3 Cr HR: 3
**CHROMIUM**
SYN: CHROME

OSHA PEL: TWA 1 mg/m$^3$
ACGIH TLV: TWA 0.5 mg/m$^3$

SAFETY PROFILE: Human poison by ingestion. An experimental tumorigen and suspected carcinogen. Powder will explode spontaneously in air. Incompatible with oxidants.

**CML810** CAS: 218-01-9 $C_{18}H_{12}$ HR: 3
**CHRYSENE**
SYNS: 1,2-BENZOPHENANTHRENE ◇ BENZO(a)PHENANTHRENE

OSHA PEL: 0.2 mg/m$^3$
ACGIH TLV: Suspected Human Carcinogen
DFG MAK: Animal Carcinogen, Suspected Human Carcinogen.
NIOSH REL: (Chrysene) To be controlled as a carcinogen.

SAFETY PROFILE: An experimental carcinogen by skin contact.

PROP: Occurs in coal tar. Orthorhombic bipyramidal plates from benzene. D: 1.274, mp: 254°, bp: 448°. Sltly sol in alc, ether, carbon bisulfide, and glacial acetic acid; moderately sol in boiling benzene; insol in water.

**CMX850** CAS: 2971-90-6 $C_7H_7Cl_2NO$ HR: 1
**CLOPIDOL**
SYNS: COYDEN ◇ 3,5-DICHLORO-2,6-DIMETHYL-4-PYRIDINOL ◇ METHYLCHLOROPINDOL

OSHA PEL: Total Dust: 15 mg/m$^3$; Respirable Fraction: 5 mg/m$^3$
ACGIH TLV: TWA 10 mg/m$^3$

SAFETY PROFILE: A nuisance dust. When heated to decomposition it emits very toxic fumes of $Cl^-$ and $NO_x$.

**CMY635** HR: 2
**COAL DUST**
DOT: 1361
SYNS: ANTHRACITE PARTICLES ◇ COAL, GROUND BITUMINOUS (DOT)

OSHA PEL: (Transitional: Respirable Quartz Fraction less than 5% $SiO_2$: TWA 2.4 mg/m$^3$; Respirable Quartz Fraction greater than or equal to 5% $SiO_2$: 10 mg/m$^3$) Respirable Quartz Fraction less than 5% $SiO_2$: TWA 2 mg/m$^3$; Respirable Quartz Fraction greater than or equal to 5% $SiO_2$: 0.1 mg/m$^3$
ACGIH TLV: TWA 2 mg/m$^3$
DOT Class: Flammable Solid

SAFETY PROFILE: Variable toxicity depending upon $SiO_2$ content. Moderately flammable.

PROP: Black powder or dust.

**CMY800** CAS: 8007-45-2 HR: 3
**COAL TAR**
DOT: 1999
SYNS: TAR, COAL ◇ TAR, LIQUID (DOT)

OSHA PEL: TWA 0.2 mg/m$^3$
DFG MAK: Human Carcinogen.
NIOSH REL: TWA 0.1 mg/m$^3$
DOT Class: Flammable or Combustible Liquid

SAFETY PROFILE: An experimental carcinogen. A human and experimental skin irritant.

**CMY825** CAS: 8001-58-9 HR: 2
**COAL TAR CREOSOTE**
DOT: 1136
SYNS: COAL TAR OIL (DOT) ◇ CREOSOTE ◇ NAPHTHALENE OIL

NIOSH REL: TWA 0.1 mg/m$^3$
DOT Class: Flammable or Combustible Liquid

SAFETY PROFILE: Moderately toxic by ingestion. Experimental reproductive effects.

**CMZ100** CAS: 65996-93-2 HR: 3
**COAL TAR PITCH VOLATILES**
SYNS: PITCH ◇ PITCH, COAL TAR

OSHA PEL: TWA 0.2 mg/m$^3$
ACGIH TLV: TWA 0.2 mg/m$^3$ (volatile), Confirmed Human Carcinogen
NIOSH REL: TWA 0.1 mg/m$^3$

SAFETY PROFILE: An experimental carcinogen by skin contact.

**CNA250** CAS: 7440-48-4 Co HR: 3
**COBALT and COMPOUNDS**
SYN: C.I. 77320

OSHA PEL: (Transitional: TWA 0.1 mg/m$^3$) TWA 0.05 mg/m$^3$
ACGIH TLV: (metal, dust, and fume) TWA 0.05 mg(Co)/m$^3$
DFG TRK: 0.5 mg/m$^3$ calculated as cobalt in that portion of dust that can possibly be inhaled in the production of cobalt powder and catalysts;

hard metal (tungsten carbide) and magnet production (processing of powder, machine pressing, and mechanical processing of unsintered articles); others 0.1 mg/m$^3$ calculated as cobalt in that portion of dust that can possibly be inhaled. Animal Carcinogen, Suspected Human Carcinogen.

SAFETY PROFILE: Moderately toxic by ingestion. An experimental neoplastigen. Inhalation of the dust may cause pulmonary damage. The powder may cause dermatitis. Powdered cobalt ignites spontaneously in air. Flammable.

PROP: Gray, hard, magnetic, ductile, somewhat malleable metal. D 8.92, mp 1493°, bp about 3100°, Brinell hardness: 125, specific heat (15-100°): 0.1056 cal/g/°C. Readily sol in dil $HNO_3$; very slowly attacked by HCl or cold $H_2SO_4$. The hydrated salts of cobalt are red, and the sol salts form red solns which become blue on adding concentrated HCl.

**CNB500** CAS: 10210-68-1 $C_8Co_2O_8$ HR: 3
**COBALT CARBONYL**
SYNS: COBALT OCTACARBONYL ◇ COBALT TETRACARBONYL ◇ DICOBALT OCTACARBONYL

OSHA PEL: TWA 0.1 mg(Co)/m$^3$
ACGIH TLV: TWA 0.1 mg(Co)/m$^3$

SAFETY PROFILE: Poison by ingestion and inhalation. Decomposes in air to form a product which ignites spontaneously in air.

PROP: Orange platelets. D: 1.87, mp: 51°, decomp above 52°. Decomp on exposure to air. Insol in water; sol in organic solvents.

**CNC230** CAS: 16842-03-8 $C_4HCoO_4$ HR: 3
**COBALT HYDROCARBONYL**

OSHA PEL: TWA 0.1 mg(Co)/m$^3$
ACGIH TLV: TWA 0.1 mg(Co)/m$^3$

SAFETY PROFILE: Poison by inhalation.

**CNI000** CAS: 7440-50-8 Cu HR: 3
**COPPER**
SYNS: C.I. 77400 ◇ RANEY COPPER

OSHA PEL: TWA (dust, mist) 1 mg(Cu)/m$^3$; (fume) 0.1 mg/m$^3$
ACGIH TLV: TWA (dust, mist) 1 mg(Cu)/m$^3$; (fume) 0.2 mg/m$^3$
DFG MAK: (dust) 1 mg/m$^3$; (fume) 0.1 mg/m$^3$

SAFETY PROFILE: An experimental tumorigen and teratogen. Other experimental reproductive effects.

PROP: A metal with a distinct reddish color. Mp: 1083°, bp: 2324°, d: 8.92, vap press: 1 mm @ 1628°.

**CNT750** HR: 2
**COTTON DUST**

OSHA PEL: TWA 1 $mg/m^3$ (raw dust); 0.2 $mg/m^3$ (yarn manufacturing); 0.75 $mg/m^3$ (slashing and weaving); 0.5 $mg/m^3$ (other operations)
ACGIH TLV: TWA 0.2 $mg/m^3$ (raw dust)
DFG MAK: 1.5 $mg/m^3$ (raw cotton)
NIOSH REL: CL 0.200 $mg/m^3$ lint-free

SAFETY PROFILE: Human pulmonary effects. Causes a mild febrile condition of the lungs resembling metal fume fever. It is a mild allergen. Inhalation may produce bronchial asthma, sneezing and eczema in sensitized persons. Moderate fire and explosion hazard.

**CNW000** CAS: 136-78-7 $C_8H_7Cl_2O_5S{\cdot}Na$ HR: 2
**CRAG HERBICIDE**
SYNS: CRAG SESONE ◇ SESONE (ACGIH) ◇ SODIUM-2,4-DICHLOROPHENOXYETHYL SULPHATE

OSHA PEL: (Transitional: TWA Total Dust: 15 $mg/m^3$; Respirable Fraction: 5 $mg/m^3$) TWA Total Dust: 10 $mg/m^3$; Respirable Fraction: 5 $mg/m^3$
ACGIH TLV: TWA 10 $mg/m^3$

SAFETY PROFILE: Moderately toxic by ingestion. Strong solutions are skin irritants. When heated to decomposition it emits very toxic fumes of $Cl^-$, $Na_2O$, and $SO_x$.

**CNW500** CAS: 1319-77-3 $C_7H_8O$ HR: 3
**CRESOL**
DOT: 2022/2076
SYN: CRESYLIC ACID

OSHA PEL: TWA 5 ppm (skin)
ACGIH TLV: TWA 5 ppm
DFG MAK: (all isomers) 5 ppm (22 $mg/m^3$)
NIOSH REL: TWA 10 $mg/m^3$
DOT Class: Corrosive Material

SAFETY PROFILE: Moderately toxic by ingestion and skin contact. Corrosive to skin and mucous membranes. Flammable. When heated to decomposition it emits highly toxic and irritating fumes. To fight fire, use foam, $CO_2$, dry chemical.

PROP: Description (U.S.P. XVI): mixture of isomeric cresols obtained from coal tar, colorless or yellowish to brown-yellow or pinkish liquid, phenolic odor. Mp: 10.9-35.5°, bp: 191-203°, flash p: 178°F, d: 1.030-1.038 @ 25°/25°, vap press: 1 mm @ 38-53°, vap d: 3.72.

**CNW750** CAS: 108-39-4 $C_7H_8O$ HR: 3
**m-CRESOL**
DOT: 2076
SYNS: m-CRESYLIC ACID ◇ m-HYDROXYTOLUENE

OSHA PEL: TWA 5 ppm (skin)
ACGIH TLV: TWA 5 ppm
NIOSH REL: TWA 10 $mg/m^3$
DOT Class: Poison B

SAFETY PROFILE: Poison by ingestion. Moderately toxic by skin contact. An experimental neoplastigen. Severe eye and skin irritant. Combustible. Moderately explosive.

PROP: Colorless to yellowish liquid, phenolic odor. Mp: 10.9° bp: 202.8°, lel: 1.1% @ 302°F, flash p: 202°F, d: 1.034 @ 20°/4°, autoign temp: 1038°F, vap press: 1 mm @ 52.0°, vap d: 3.72.

**CNX000** CAS: 95-48-7 $C_7H_8O$ HR: 3
**o-CRESOL**
DOT: 2076
SYNS: o-CRESYLIC ACID ◇ o-HYDROXYTOLUENE ◇ ORTHOCRESOL ◇ o-OXYTOLUENE

OSHA PEL: TWA 5 ppm (skin)
ACGIH TLV: TWA 5 ppm
NIOSH REL: TWA 10 $mg/m^3$
DOT Class: Poison B

SAFETY PROFILE: Poison by ingestion, inhalation. Moderately toxic by skin contact. A severe eye and skin irritant. An experimental neoplastigen. Combustible. To fight fire, water may be used to blanket fire; foam, fog, mist, dry chemical.

PROP: Crystals or liquid darkening with exposure to air and light. Mp: 30.8°, bp: 190.8°, flash p: 178°F, d: 1.047 @ 20°/4°, autoign temp: 1110°F, vap press: 1 mm @ 38.2°, vap d: 3.72, lel: 1.4% @ 300°F.

**CNX250** CAS: 106-44-5 $C_7H_8O$ HR: 3
**p-CRESOL**
DOT: 2076
SYNS: p-CRESYLIC ACID ◇ p-HYDROXYTOLUENE ◇ p-OXYTOLUENE ◇ p-TOLUOL

OSHA PEL: TWA 5 ppm (skin)
ACGIH TLV: TWA 5 ppm
NIOSH REL: TWA 10 $mg/m^3$
DOT Class: Poison B

SAFETY PROFILE: Poison by ingestion and skin contact. An experimental neoplastigen A severe skin and eye irritant. Combustible. Moderately explosive in the form of vapor. To fight fire, use $CO_2$, dry chemical, alcohol foam.

PROP: Crystals; phenolic odor. Mp: 35.5°, bp: 201.8°, lel: 1.1% @ 302°F, flash p: 202°F, d: 1.0341 @ 20°/4°, autoign temp: 1038°F, vap press 1 mm @ 53.0°, vap d: 3.72. Found in a score of essential oils, including ylang-ylang and oil of jasmine.

**COB250** CAS: 4170-30-3 $C_4H_6O$ HR: 3
**CROTONALDEHYDE**
DOT: 1143
SYNS: 2-BUTENAL ◇ CROTONIC ALDEHYDE ◇ β-METHYL ACROLEIN

OSHA PEL: TWA 2 ppm
DFG MAK: Suspected Carcinogen.
DOT Class: Flammable Liquid

SAFETY PROFILE: Poison by ingestion and inhalation. An experimental carcinogen by ingestion. An eye, skin, and mucous membrane irritant. Dangerous fire hazard. To fight fire, use alcohol foam, $CO_2$, dry chemical.

PROP: Water-white, mobile liquid; pungent suffocating odor. Bp: 104°, fp: −76.0°, lel: 2.1%, uel: 15.5%, flash p: 55°F, d: 0.853 @ 20°/20°, vap d: 2.41, autoign temp: 405°F.

**COD850** CAS: 299-86-5 $C_{12}H_{19}ClNO_3P$ HR: 3
**CRUFORMATE**
SYN: RUELENE

OSHA PEL: TWA 5 mg/m$^3$
ACGIH TLV: TWA 5 mg/m$^3$

SAFETY PROFILE: A poison by ingestion and inhalation. Moderately toxic via skin contact. Experimental reproductive effects. When heated to decomposition it emits very toxic fumes of $PO_x$, $NO_x$, and $Cl^-$.

**COE750** CAS: 98-82-8 $C_9H_{12}$ HR: 2
**CUMENE**
DOT: 1918
SYNS: BENZENE ISOPROPYL ◇ ISOPROPYL BENZENE

OSHA PEL: TWA 50 ppm (skin)
ACGIH TLV: TWA 50 ppm (skin)
DFG MAK: 50 ppm (245 mg/m$^3$)
DOT Class: Flammable or Combustible Liquid

SAFETY PROFILE: Moderately toxic by ingestion. Mildly toxic by inhalation and skin contact. An eye and skin irritant. Flammable. To fight fire use foam $CO_2$, dry chemical.

PROP: Colorless liquid. Mp: −96.0°, bp: 152°, flash p: 111°F, d: 0.864 @ 20°/4°, vap press: 10 mm @ 38.3°, autoign temp: 795°F, lel: 0.9%, uel: 6.5%, vap d: 4.1.

**COH500** CAS: 420-04-2 $CH_2N_2$ HR: 3
**CYANAMIDE**
SYNS: CARBAMONITRILE ◇ CYANOAMINE ◇ CYANOGEN NITRIDE ◇ HYDROGEN CYANAMIDE

OSHA PEL: TWA 2 mg/m$^3$
ACGIH TLV: TWA 2 mg/m$^3$

SAFETY PROFILE: Poison by ingestion and inhalation. Moderately toxic by skin contact. A severe eye irritant. Combustible. To fight fire, use $CO_2$, dry chemical. When heated to decomposition or on contact with acid or acid fumes, it emits toxic fumes of $CN^-$, and $NO_x$.

PROP: Deliquescent crystals. Mp: 45°, bp: 260°, flash p: 285°F, d: 1.282, vap d: 1.45.

**COI500** CAS: 57-12-5 $CN^-$ HR: 3
**CYANIDE**
SYN: ISOCYANIDE

OSHA PEL: TWA 5 mg(CN)/$m^3$
ACGIH TLV: TWA 5 mg/$m^3$ (skin)
DFG MAK: 5 mg/$m^3$
NIOSH REL: TWA CL 5 mg/$m^3$/10M

SAFETY PROFILE: Very poisonous. Flammable. When heated to decomposition or on contact with acid, acid fumes, water or steam, it emits toxic and flammable vapors of $CN^-$.

**COO000** CAS: 460-19-5 $C_2N_2$ HR: 3
**CYANOGEN**
DOT: 1026
SYNS: CARBON NITRIDE ◇ DICYANOGEN ◇ ETHANEDINITRILE ◇ PRUSSITE

OSHA PEL: TWA 10 ppm
ACGIH TLV: TWA 10 ppm
DFG MAK: 10 ppm (22 mg/$m^3$)
DOT Class: Poison A

SAFETY PROFILE: Moderately toxic by inhalation. A human eye irritant. Very dangerous fire hazard. To fight fire, stop flow of gas.

PROP: Colorless gas, pungent odor. Mp: −34.4°, bp: −21.0°, d: 0.866 @ 17°/4°, lel: 6.6%, uel: 32%, vap d: 1.8.

**COO750** CAS: 506-77-4 CClN HR: 3
**CYANOGEN CHLORIDE**
DOT: 1589
SYN: CHLORINE CYANIDE

OSHA PEL: CL 0.3 ppm
ACGIH TLV: CL 0.3 ppm
DOT Class: Poison A

SAFETY PROFILE: Poison by ingestion and inhalation. A severe human eye irritant. Flammable. When heated to decomposition or on contact with water or steam, it will react to produce highly toxic and corrosive fumes of $Cl^-$, $CN^-$ and $NO_x$.

PROP: Colorless liquid or gas; lachrymatory and irritating odor. Mp: −6.5°, bp: 13.1°, d: 1.218 @ 4°/4°, vap press: 1010 mm @ 20°, vap d: 1.98.

**CPB000** CAS: 110-82-7 $C_6H_{12}$ HR: 3
**CYCLOHEXANE**
DOT: 1145
SYNS: HEXAHYDROBENZENE ◇ HEXAMETHYLENE ◇ HEXANAPHTHENE

OSHA PEL: TWA 300 ppm
ACGIH TLV: TWA 300 ppm
DFG MAK: 300 ppm (1050 mg/m$^3$)
DOT Class: Flammable Liquid

SAFETY PROFILE: Moderately toxic by ingestion. A systemic irritant by inhalation and ingestion. A skin irritant. Dangerous fire hazard. Moderate explosion hazard in the form of vapor. To fight fire, use foam, $CO_2$, dry chemical, spray, fog.

PROP: Colorless, mobile liquid; pungent odor. Mp: 6.5°, bp: 80.7°, fp: 4.6°, flash: p: 1.4°F, ULC: 90-95, lel: 1.3%, uel: 8.4%, d: 0.7791 @ 20°/4°, autoign temp: 473°F, vap press: 100 mm @ 60.8°, vap d: 2.90.

**CPB625** CAS: 1569-69-3 $C_6H_{12}S$ HR: 2
**CYCLOHEXANETHIOL**

NIOSH REL: CL 0.5 ppm/15M

SAFETY PROFILE: An eye and severe skin irritant. When heated to decomposition it emits toxic fumes of $SO_x$.

**CPB750** CAS: 108-93-0 $C_6H_{12}O$ HR: 3
**CYCLOHEXANOL**
SYNS: CYCLOHEXYL ALCOHOL ◇ HYDROXYCYCLOHEXANE ◇ NAXOL

OSHA PEL: TWA 50 ppm (skin)
ACGIH TLV: TWA 50 ppm (skin)
DFG MAK: 50 ppm (200 mg/m$^3$)

SAFETY PROFILE: Moderately toxic by ingestion. Mildly toxic by skin contact. Experimental reproductive effects. A severe eye irritant. Flammable. To fight fire, use alcohol foam, foam, $CO_2$, dry chemical.

PROP: Colorless needles or viscous liquid; hygroscopic, camphor-like odor. Mp: 24°, bp: 161.5°, flash p: 154°F (CC), d: 0.9449 @ 25°/4°, vap press: 1 mm @ 21.0°, vap d: 3.45, autoign temp: 572°F.

**CPC000** CAS: 108-94-1 $C_6H_{10}O$ HR: 2
**CYCLOHEXANONE**
DOT: 1915
SYN: SEXTONE

OSHA PEL: (Transitional: TWA 50 ppm) TWA 25 ppm (skin)
ACGIH TLV: TWA 25 ppm (skin)
DFG MAK: 50 ppm (200 mg/m$^3$)

NIOSH REL: TWA 100 mg/m$^3$
DOT Class: Flammable or Combustible Liquid

SAFETY PROFILE: Moderately toxic by ingestion and inhalation. A skin and severe eye irritant. Experimental reproductive effects. Flammable. To fight fire, use alcohol foam, dry chemical or $CO_2$.

PROP: Colorless liquid, acetone-like odor. Mp: −45.0°, bp: 115.6°, ULC: 35-40, lel: 1.1% @ 100°, flash p: 111°F, d: 0.9478 @ 20°/4°, autoign temp: 788°F, vap press: 10 mm @ 38.7°, vap d: 3.4.

**CPC579** CAS: 110-83-8 $C_6H_{10}$ HR: 2
**CYCLOHEXENE**
DOT: 2256
SYNS: BENZENETETRAHYDRIDE ◇ 1,2,3,4-TETRAHYDROBENZENE

OSHA PEL: 300 ppm
ACGIH TLV: 300 ppm
DFG MAK: 300 ppm (1015 mg/m$^3$)
DOT Class: Flammable Liquid.

SAFETY PROFILE: Moderately toxic by inhalation and ingestion. A very dangerous fire hazard when exposed to flame; can react with oxidizers. To fight fire, use foam, $CO_2$, dry chemical.

PROP: Colorless liquid. Bp: 83°, fp: −103.7°, flash p: < 21.2°F, d: 0.8102 @ 20°/4°, vap press: 160 mm @ 38°, autoign temp: 590°F, vap d: 2.8, lel: 1.2%.

**CPF500** CAS: 108-91-8 $C_6H_{13}N$ HR: 3
**CYCLOHEXYLAMINE**
DOT: 2357
SYNS: CHA ◇ HEXAHYDROBENZENAMINE

OSHA PEL: TWA 10 ppm
ACGIH TLV: TWA 10 ppm
DFG MAK: 10 ppm (40 mg/m$^3$)
DOT Class: Flammable Liquid

SAFETY PROFILE: A poison by ingestion and skin contact. An experimental teratogen. Other experimental reproductive effects. Severe human skin irritant. Flammable. To fight fire, use alcohol foam, $CO_2$, dry chemical. When heated to decomposition it emits toxic fumes of $NO_x$.

PROP: Liquid; strong, fishy odor. Mp: −17.7°, bp: 134.5°, flash p: 69.8°F, d: 0.865 @ 25°/25°, autoign temp: 560°F, vap d. 3.42.

**CPR800** CAS: 121-82-4 $C_3H_6N_6O_6$ HR: 3
**CYCLONITE**
DOT: 0072/0118
SYNS: CYCLOTRIMETHYLENETRINITRAMINE ◇ HEXOGEN (explosive) ◇ HEXOLITE ◇ RDX ◇ 1,3,5-TRINITRO-1,3,5-TRIAZACYCLOHEXANE

OSHA PEL: TWA 1.5 mg/m$^3$ (skin)
ACGIH TLV: TWA 1.5 mg/m$^3$ (skin)
DOT Class: Class A Explosive

SAFETY PROFILE: Poison by ingestion. Experimental reproductive effects. A corrosive irritant to skin, eyes, and mucous membranes. One of the most powerful high explosives. When heated to decomposition it emits toxic fumes of $NO_x$.

PROP: White, crystalline powder. Mp: 202°.

**CPU500** CAS: 542-92-7 $C_5H_6$ HR: 3
**1,3-CYCLOPENTADIENE**
SYNS: PENTOLE ◇ PYROPENTYLENE

OSHA PEL: TWA 75 ppm
ACGIH TLV: TWA 75 ppm
DFG MAK: 75 ppm (200 mg/m$^3$)

SAFETY PROFILE: Probably moderately toxic by inhalation. A dangerous fire hazard. Moderate explosion hazard.

PROP: Colorless liquid. Mp: −85°, bp: 42.5°, d: 0.80475 @ 19°/4°. flash p: 77°F.

**CPV000** CAS: 12079-65-1 $C_8H_5MnO_3$ HR: 3
**CYCLOPENTADIENYLMANGANESE TRICARBONYL**
SYNS: MANGANESE CYCLOPENTADIENYL TRICARBONYL ◇ MCT

OSHA PEL: TWA 0.1 mg(Mn)/m$^3$ (skin)
ACGIH TLV: TWA 0.1 mg(Mn)/m$^3$

SAFETY PROFILE: A poison by ingestion and inhalation. A mild narcotic.

**CPV750** CAS: 287-92-3 $C_5H_{10}$ HR: 1
**CYCLOPENTANE**
DOT: 1146
SYN: PENTAMETHYLENE

OSHA PEL: TWA 600 ppm
ACGIH TLV: TWA 600 ppm
DOT Class: Flammable Liquid

SAFETY PROFILE: Mildly toxic by ingestion and inhalation. A very dangerous fire hazard. To fight fire, use foam, $CO_2$, dry chemical.

PROP: Colorless liquid. Bp: 49.3°, fp: −93.7°, flash p: 19.4°F, autoign temp: 716°F, d: 0.745 @ 20°/4°, vap press: 400 mm @ 31.0°, vap d: 2.42.

**DAA800** CAS: 94-75-7 $C_8H_6Cl_2O_3$ HR: 3
**2,4-D**
DOT: 2765
SYNS: CHLOROXONE ◇ 2,4-DICHLOROPHENOXYACETIC ACID (DOT) ◇ SALVO

OSHA PEL: TWA 10 $mg/m^3$
ACGIH TLV: TWA 10 $mg/m^3$
DFG MAK: 10 $mg/m^3$
DOT Class: ORM-A

SAFETY PROFILE: Poison by ingestion. Moderately toxic by skin contact. An experimental carcinogen and teratogen. A suspected human carcinogen. A skin and severe eye irritant. When heated to decomposition it emits toxic fumes of $Cl^-$.

PROP: White powder. Mp: 141°, bp: 160° @ 0.4 mm, vap d: 7.63.

**DAD200** CAS: 50-29-3 $C_{14}H_9Cl_5$ HR: 3
**DDT**
DOT: 2761
SYNS: DICHLORODIPHENYLTRICHLOROETHANE (DOT) ◇ MUTOXIN ◇ PARACHLOROCIDUM

OSHA PEL: TWA 1 $mg/m^3$ (skin)
ACGIH TLV: TWA 1 $mg/m^3$
NIOSH REL: (DDT) TWA 0.5 $mg/m^3$; avoid skin contact
DFG MAK: 1 $mg/m^3$
DOT Class: ORM-A

SAFETY PROFILE: Human poison by ingestion. Experimental poison by skin contact. A suspected human carcinogen. An experimental carcinogen and teratogen. When heated to decomposition it emits toxic fumes of $Cl^-$.

PROP: Colorless crystals or white to slightly off-white powder. Odorless or with slight aromatic odor. Mp: 108.5-109°.

**DAE400** CAS: 17702-41-9 $B_{10}H_{14}$ HR: 3
**DECABORANE**
DOT: 1868

OSHA PEL: (Transitional: TWA 0.05 ppm (skin)) TWA 0.05 ppm; STEL 0.15 ppm (skin)
ACGIH TLV: TWA 0.05 ppm; STEL 0.15 ppm (skin)
DFG MAK: 0.05 ppm (0.3 $mg/m^3$)
DOT Class: Flammable Solid

SAFETY PROFILE: Poison by inhalation, ingestion, and skin contact. Ignites in $O_2$ at 100°C. Forms impact-sensitive explosive mixtures with ethers and halocarbons. When heated to decomposition it emits toxic fumes of boron oxides.

PROP: Colorless needles. Mp: 99.7°, d: 0.94. (solid), d: 0.78 (liquid @ 100°), vap press: 19 mm @ 100°.

**DAO600** CAS: 8065-48-3 $C_8H_{19}O_3PS_2 \cdot C_8H_{19}O_3PS_2$ HR: 3
**DEMETON-O + DEMETON-S**
SYNS: DEMOX ◇ MERCAPTOPHOS ◇ SYSTOX ◇ ULV

OSHA PEL: TWA 0.1 mg/m$^3$ (skin)
ACGIH TLV: TWA 0.01 ppm (skin)
DFG MAK: 0.01 ppm (0.1 mg/m$^3$)

SAFETY PROFILE: A deadly human poison by ingestion. Poison experimentally by ingestion, inhalation, and skin contact. An experimental teratogen. When heated to decomposition it emits very toxic fumes of $PO_x$ and $SO_x$.

PROP: A light brown liquid, sulfur compound odor.

**DBF750** CAS: 123-42-2 $C_6H_{12}O_2$ HR: 3
**DIACETONE ALCOHOL**
DOT: 1148
SYNS: 4-HYDROXY-4-METHYLPENTANONE-2 ◇ PYRANTON

OSHA PEL: TWA 50 ppm
ACGIH TLV: TWA 50 ppm
DFG MAK: 50 ppm (240 mg/m$^3$)
NIOSH REL: (Ketones) TWA 240 mg/m$^3$
DOT Class: Flammable Liquid

SAFETY PROFILE: Moderately toxic by ingestion. Mildly toxic by skin contact. A skin, mucous membrane and severe eye irritant. Flammable and explosive in the form of vapor. To fight fire, use alcohol foam, foam, $CO_2$, dry chemical.

PROP: Liquid; faint, pleasant odor. Mp: −47 to −54°, bp: 167.9°, flash p: 148°F, d: 0.9306 @ 25°/4°, autoign temp: 1118°F, vap d: 4.00, vap press: 1.1 mm @ 20°, lel: 1.8%, uel: 6.9%, flash p: (acetone free): 136°F.

**DBO000** CAS: 615-05-4 $C_7H_{10}N_2O$ HR: 3
**2,4-DIAMINOANISOLE**
SYNS: C.I. 76050 ◇ 4-METHOXY-m-PHENYLENEDIAMINE

DFG MAK: Animal Carcinogen, Suspected Human Carcinogen.
NIOSH REL: (2,4-diaminoanisole) Reduce to lowest feasible level

SAFETY PROFILE: Moderately toxic by ingestion. A suspected human carcinogen. A skin irritant. When heated to decomposition it emits toxic fumes of $NO_x$.

**DCJ200** CAS: 119-90-4 $C_{14}H_{16}N_2O_2$ HR: 3
**o-DIANISIDINE**
SYNS: C.I. 24110 ◇ DIATO BLUE BASE B ◇ LAKE BLUE B BASE

DFG MAK: Animal Carcinogen, Suspected Human Carcinogen.
NIOSH REL: (Benzidine-based dye) Reduce to lowest feasible level

SAFETY PROFILE: Moderately toxic by ingestion. An experimental carcinogen and tumorigen. Combustible. When heated to decomposition it emits toxic fumes of $NO_x$.

PROP: Colorless crystals. Mp: 137-138°, flash p: 403°F, vap d: 8.5.

**DCJ400** CAS: 91-93-0 $C_{16}H_{12}N_2O_4$ HR: 3
**DIANISIDINE DIISOCYANATE**
SYNS: 3,3′-DIMETHOXYBENZIDINE-4,4′-DIISOCYANATE ◇ 3,3′-DIMETHOXY-4,4′-BIPHENYLENE DIISOCYANATE

NIOSH REL: (Diisocyanates) TWA 0.005 ppm; CL 0.02 ppm/10M

SAFETY PROFILE: An inhalation sensitizer. An experimental carcinogen. When heated to decomposition it emits toxic fumes of $NO_x$.

**DCJ800** CAS: 68855-54-9 HR: 1
**DIATOMACEOUS EARTH**
SYNS: DIATOMITE ◇ INFUSORIAL EARTH ◇ KIESELGUHR

OSHA PEL: (Transitional: TWA 80 $mg/m^3/\%SiO_2$) TWA 6 $mg/m^3$
ACGIH TLV: TWA (nuisance particulate) 10 $mg/m^3$ of total dust (when toxic impurities are not present, e.g., quartz $< 1\%$).

SAFETY PROFILE: The dust may cause fibrosis of the lungs. Roasting or calcining at high temperatures produces cristobalite and tridymite, thus increasing the fibrogenicity of the material.

PROP: Composed of skeletons of small aquatic plants related to algae and contains as much as 88% amorphous silica. White to buff colored solid. Insol in water; sol in hydrofluoric acid.

**DCM750** CAS: 333-41-5 $C_{12}H_{21}N_2O_3PS$ HR: 3
**DIAZINON**
DOT: 2783
SYNS: DIETHYL 4-(2-ISOPROPYL-6-METHYLPYRIMIDINYL)PHOSPHOROTHIONATE ◇ SPECTRACIDE

OSHA PEL: TWA 0.1 $mg/m^3$ (skin)
ACGIH TLV: TWA 0.1 $mg/m^3$
DFG MAK: 1 $mg/m^3$
DOT Class: ORM-A

SAFETY PROFILE: Poison by ingestion and skin contact. Mildly toxic by inhalation. An experimental teratogen. Experimental reproductive effects. A skin and severe eye irritant. When heated to decomposition it emits very toxic fumes of $NO_x$, $PO_x$, and $SO_x$.

PROP: Liquid with faint ester-like odor. Bp: 84° @ 0.002 mm, d: 1.116 @ 20°/4°. Miscible in organic solvents.

**DCP800** CAS: 334-88-3 $CH_2N_2$ HR: 3
**DIAZOMETHANE**
SYN: AZIMETHYLENE

OSHA PEL: TWA 0.2 ppm
ACGIH TLV: TWA 0.2 ppm
DFG MAK: Animal Carcinogen, Suspected Human Carcinogen.

SAFETY PROFILE: An experimental tumorigen and carcinogen. A poison irritant by inhalation. A powerful allergen. Highly explosive. When heated to decomposition or on contact with acid or acid fumes it emits highly toxic fumes of $NO_x$.

PROP: Yellow gas at ordinary temperature. Mp: −145°, bp: −23°, d: 1.45.

**DDI450** CAS: 19287-45-7 $B_2H_6$ HR: 3
**DIBORANE**
DOT: 1911
SYNS: BOROETHANE ◇ BORON HYDRIDE ◇ DIBORON HEXAHYDRIDE

OSHA PEL: TWA 0.1 ppm
ACGIH TLV: TWA 0.1 ppm
DFG MAK: 0.1 ppm (0.1 $mg/m^3$)
DOT Class: Flammable Gas

SAFETY PROFILE: Poison by inhalation. An irritant to skin, eyes, and mucous membranes. Dangerously flammable. Highly explosive. Explosive reaction with air.

PROP: Colorless gas, sickly sweet odor. Mp: −165.5°, bp: −92.5°, d: 0.447 (liquid @ −112°), 0.577 (solid @ −183°), vap press: 224 mm @ −112°, autoign temp: 38-52°, lel: 0.9%, uel: 98%, flash p: −90°F.

**DDL800** CAS: 96-12-8 $C_3H_5Br_2Cl$ HR: 3
**1,2-DIBROMO-3-CHLOROPROPANE**
DOT: 2872
SYNS: DBCP ◇ NEMATOCIDE

OSHA PEL: TWA 0.001 ppm; Cancer Hazard.
DFG MAK: Animal Carcinogen, Suspected Human Carcinogen.
NIOSH REL: CL 0.01 ppm/30M

SAFETY PROFILE: Poison by ingestion and inhalation. Moderately toxic by skin contact. An eye and severe skin irritant. An experimental carcinogen and teratogen. A suspected human carcinogen. Flammable. When heated to decomposition it emits toxic fumes of $Cl^-$ and $Br^-$.

PROP: Bp: 196°, flash p: 170°F (TOC).

**DDU600** CAS: 102-81-8 $C_{10}H_{23}NO$ HR: 3
**2-N-DIBUTYLAMINOETHANOL**
DOT: 2873
SYNS: N,N-DI-n-BUTYLAMINOETHANOL (DOT) ◇ β-N-DIBUTYLAMINOETHYL ALCOHOL

OSHA PEL: TWA 2 ppm
ACGIH TLV: TWA 2 ppm (skin)
DOT Class: Poison B

SAFETY PROFILE: Moderately toxic by ingestion and skin contact. A severe eye and skin irritant. Flammable. To fight fire, use $CO_2$, dry chemical. When heated to decomposition it emits toxic fumes of $NO_x$.

PROP: Liquid. Bp: 222°, flash p: 220°F (OC), d: 0.85, vap d: 6.0.

**DEG600** CAS: 2528-36-1 $C_{14}H_{23}O_4P$ HR: 2
**DIBUTYL PHENYL PHOSPHATE**
SYN: PHOSPHORIC ACID, DIBUTYL PHENYL ESTER

ACGIH TLV: (Proposed: TWA 0.3 ppm (skin))

SAFETY PROFILE: Moderately toxic by ingestion. When heated to decomposition it emits toxic fumes of $PO_x$.

**DEG700** CAS: 107-66-4 $C_8H_{19}PO_4$ HR: 2
**DIBUTYL PHOSPHATE**

OSHA PEL: TWA 1 ppm; STEL 2 ppm
ACGIH TLV: TWA 1 ppm; STEL 2 ppm

SAFETY PROFILE: Moderately toxic by ingestion. When heated to decomposition it emits toxic fumes of $PO_x$.

PROP: Pale amber liquid. Bp: decomp > 100°.

**DEH200** CAS: 84-74-2 $C_{16}H_{22}O_4$ HR: 3
**DIBUTYL PHTHALATE**
DOT: 9095
SYNS: n-BUTYL PHTHALATE (DOT) ◇ DBP

OSHA PEL: TWA 5 $mg/m^3$
ACGIH TLV: TWA 5 $mg/m^3$
DOT Class: ORM-E

SAFETY PROFILE: Mildly toxic by ingestion. An experimental teratogen. Experimental reproductive effects. Combustible. To fight fire, use $CO_2$, dry chemical.

PROP: Oily liquid, mild odor. Bp: 340°, fp: −35°, flash p: 315°F (CC), d: 1.047-1.049 @ 20°/20°, autoign temp: 757°F, vap d: 9.58.

**DEN600** CAS: 7572-29-4 $C_2Cl_2$ HR: 3
**DICHLOROACETYLENE**
SYN: DICHLOROETHYNE

OSHA PEL: CL 0.1 ppm
ACGIH TLV: CL 0.1 ppm
DFG MAK: Animal Carcinogen, Suspected Human Carcinogen.
DOT Class: Forbidden.

SAFETY PROFILE: Poison by inhalation. An experimental carcinogen. Strong explosive when shocked or exposed to heat or air. When heated to decomposition or on contact with acid or acid fumes it emits highly toxic fumes of $Cl^-$.

**DEP600** CAS: 95-50-1 $C_6H_4Cl_2$ HR: 3
**o-DICHLOROBENZENE**
DOT: 1591
SYNS: DCB ◇ 1,2-DICHLOROBENZENE (MAK) ◇ DOWTHERM E

OSHA PEL: CL 50 ppm
ACGIH TLV: CL 50 ppm

DFG MAK: 50 ppm (300 mg/m$^3$)
DOT Class: ORM-A

SAFETY PROFILE: Poison by ingestion. Moderately toxic by inhalation. An eye, skin, and mucous membrane irritant. An experimental teratogen and suspected carcinogen. Experimental reproductive effects. Flammable. To fight fire, use water, foam, $CO_2$, dry chemical. When heated to decomposition it emits toxic fumes of $Cl^-$.

PROP: Clear liquid. Mp: −17.5°, bp: 180-183°, fp: −22°, flash p: 151°F, d: 1.307 @ 20°/20°, vap d: 5.05, autoign temp: 1198°F, lel: 2.2%, uel: 9.2%.

**DEP800** CAS: 106-46-7 $C_6H_4Cl_2$ HR: 3
**p-DICHLOROBENZENE**
DOT: 1592
SYNS: p-CHLOROPHENYL CHLORIDE ◇ 1,4-DICHLOROBENZENE (MAK) ◇ PARADICHLOROBENZENE ◇ PDCB

OSHA PEL: (Transitional: TWA 75 ppm) TWA 75 ppm; STEL 110 ppm
ACGIH TLV: TWA 75 ppm; STEL 110 ppm
DFG MAK: 75 ppm (450 mg/m$^3$)
DOT Class: ORM-A

SAFETY PROFILE: A human poison. An experimental carcinogen and teratogen. Other experimental reproductive effects. A human eye irritant. Flammable. To fight fire, use water, foam, $CO_2$, dry chemical. When heated to decomposition it emits toxic fumes of $Cl^-$.

PROP: White crystals, penetrating odor. Mp: 53°, bp: 173.4°, flash p: 150°F (CC), d: 1.4581 @ 20.5°/4°, vap press: 10 mm @ 54.8°, vap d: 5.08.

**DEQ600** CAS: 91-94-1 $C_{12}H_{10}Cl_2N_2$ HR: 3
**3′,3′-DICHLOROBENZIDINE**
SYN: C.I. 23060

OSHA PEL: Carcinogen
ACGIH TLV: Suspected Human Carcinogen.
DFG TRK: 0.1 mg/m$^3$, Animal Carcinogen, Suspected Human Carcinogen.
NIOSH REL: (Benzidine-based Dye) Reduce to lowest feasible level.

SAFETY PROFILE: Mildly toxic by ingestion. An experimental carcinogen and tumorigen. A suspected human carcinogen. When heated to decomposition it emits very toxic fumes of $Cl^-$ and $NO_x$.

PROP: Crystals. Mp: 133°. Insol in water; sol in alc, benzene, and glacial acetic acid.

**DEQ800** CAS: 612-83-9 $C_{12}H_{10}Cl_2N_2 \cdot 2ClH$ HR: 3
**3,3′-DICHLOROBENZIDINE DIHYDROCHLORIDE**
SYN: 3,3′-DICHLORO-(1,1′-BIPHENYL)-4,4′-DIAMINE DIHYDROCHLORIDE

OSHA PEL: Carcinogen

SAFETY PROFILE: Moderately toxic by ingestion. A carcinogen. When heated to decomposition it emits very toxic fumes of $Cl^-$ and $NO_x$.

**DEV000** CAS: 764-41-0 $C_4H_6Cl_2$ HR: 3
**1,4-DICHLORO-2-BUTENE**
SYNS: DCB ◇ 1,4-DICHLOROBUTENE-2 (MAK)

DFG MAK: Animal Carcinogen, Suspected Human Carcinogen.

SAFETY PROFILE: Poison by ingestion and inhalation. Moderately toxic by skin contact. An experimental teratogen. A severe skin and eye irritant. When heated to decomposition it emits toxic fumes of $Cl^-$.

PROP: Colorless liquid. Mp: 1°-3°; bp: 156°; d: 1.183 @ 25°/4°.

**DFA400** CAS: 76-38-0 $C_3H_4Cl_2F_2O$ HR: 3
**2,2-DICHLORO-1,1-DIFLUOROETHYL METHYL ETHER**
SYNS: INGALAN ◇ METHOXYFLURANE

NIOSH REL: (Waste Anesthetic Gases and Vapors) CL 2 ppm/1H

SAFETY PROFILE: A human poison by ingestion. Mildly toxic by inhalation. An experimental teratogen. An eye irritant. When heated to decomposition it emits very toxic fumes of $Cl^-$ and $F^-$.

**DFA600** CAS: 75-71-8 $CCl_2F_2$ HR: 1
**DICHLORODIFLUOROMETHANE**
DOT: 1028
SYNS: FLUOROCARBON-12 ◇ FREON F-12



OSHA PEL: TWA 1000 ppm
ACGIH TLV: TWA 1000 ppm
DFG MAK: 1000 ppm (5000 mg/m$^3$)
DOT Class: Nonflammable Gas

SAFETY PROFILE: Narcotic in high concentrations. Nonflammable gas. When heated to decomposition it emits highly toxic fumes of phosgene, $Cl^-$, and $F^-$.

PROP: Colorless, almost odorless gas. Mp: −158°, bp: −29°, vap press: 5 atm @ 16.1°.

**DFE200** CAS: 118-52-5 $C_5H_6Cl_2N_2O_2$ HR: 2
**1,3-DICHLORO-5,5-DIMETHYL HYDANTOIN**
SYNS: DCA ◇ DICHLORANTIN ◇ HALANE

OSHA PEL: TWA 0.2 mg/m$^3$; STEL 0.4 mg/m$^3$
ACGIH TLV: TWA 0.2 mg/m$^3$; STEL 0.4 mg/m$^3$

SAFETY PROFILE: Moderately toxic by ingestion. Mildly toxic by inhalation. A severe skin irritant. Will react with water or steam to produce toxic and corrosive fumes. When heated to decomposition it emits toxic fumes of $Cl^-$ and $NO_x$.

PROP: Crystals. Liberates chlorine on contact with hot water. Mp: 132°, subl @ 100°, conflagrates @ 212°, d: 1.5 @ 20°, vap d: 6.8.

**DFE800** CAS: 28675-08-3 $C_{12}H_8Cl_2O$ HR: 2

**DICHLORO DIPHENYL OXIDE**

SYN: DICHLOROPHENYL ETHER

OSHA PEL: TWA 0.5 mg/m$^3$

SAFETY PROFILE: Moderately toxic by ingestion. When heated to decomposition it emits toxic fumes of $Cl^-$.

PROP: Liquid. Vap d: 8.2.

**DFF809** CAS: 75-34-3 $C_2H_4Cl_2$ HR: 3

**1,1-DICHLOROETHANE**

DOT: 2362

SYNS: CHLORINATED HYDROCHLORIC ETHER ◇ ETHYLIDENE DICHLORIDE

OSHA PEL: TWA 100 ppm
ACGIH TLV: TWA 200 ppm; STEL 250 ppm
DFG MAK: 100 ppm (400 mg/m$^3$)
DOT Class: Flammable Liquid

SAFETY PROFILE: Moderately toxic by ingestion. An experimental tumorigen and teratogen. Experimental reproductive effects. A very dangerous fire hazard and moderate explosion hazard. To fight fire use alcohol foam, water, foam, $CO_2$, dry chemical. When heated to decomposition it emits highly toxic fumes of phosgene and $Cl^-$.

PROP: Colorless liquid; aromatic, ethereal odor; hot, saccharine taste. Mp: −97.7°, lel: 5.6%, bp: 57.3°, flash p: 22°F (TOC), d: 1.174 @ 20°/4°, vap press: 230 mm @ 25°, vap d: 3.44, autoign temp: 856°F.

**DFI100** CAS: 540-59-0 $C_2H_2Cl_2$ HR: 3

**1,2-DICHLOROETHYLENE**

SYNS: ACETYLENE DICHLORIDE ◇ DIOFORM

OSHA PEL: TWA 200 ppm
ACGIH TLV: TWA 200 ppm
DFG MAK: 200 ppm (790 mg/m$^3$)

SAFETY PROFILE: Poison by inhalation. Moderately toxic by ingestion. When heated to decomposition it emits highly toxic fumes of $Cl^-$.

**DFI200** CAS: 156-59-2 $C_2H_2Cl_2$ HR: 1

**cis-DICHLOROETHYLENE**

SYN: 1,2-DICHLOROETHYLENE

DFG MAK: 200 ppm (790 mg/m$^3$)

SAFETY PROFILE: Mildly toxic by ingestion and inhalation. A suspected carcinogen. A dangerous fire hazard. Reaction with solid caustic alkalies or their concentrated solutions produces chloracetylene gas which ignites spontaneously in air. To fight fire, use water spray, foam, $CO_2$, dry chemical. When heated to decomposition it emits toxic fumes of $Cl^-$.

PROP: Colorless liquid, pleasant odor. Mp: −80.5°, bp: 59°, lel: 9.7%, uel: 12.8%, flash p: 39°F, d: 1.2743 @ 25°/4°, vap press: 400 mm @ 41.0°, vap d: 3.34.

**DFJ050** CAS: 111-44-4 $C_4H_8Cl_2O$ HR: 3
**DICHLOROETHYL ETHER**
DOT: 1916
SYNS: BIS(2-CHLOROETHYL) ETHER ◇ DCEE ◇ β,β′-DICHLOROETHYL ETHER (MAK)

OSHA PEL: (Transitional: CL 15 ppm (skin)) TWA 5 ppm; STEL 10 ppm (skin)
ACGIH TLV: TWA 5 ppm; STEL 10 ppm (skin)
DFG MAK: 10 ppm (60 $mg/m^3$)
DOT Class: IMO: Poison B

SAFETY PROFILE: A poison by ingestion, skin contact and inhalation. A skin, eye, and mucous membrane irritant. An experimental carcinogen and tumorigen. Flammable. Dangerous explosion hazard. To fight fire, use water, foam, mist, fog, spray, dry chemical. When heated to decomposition it emits toxic fumes of $Cl^-$.

PROP: Colorless, stable liquid. Bp: 178.5°, fp: −51.9°, flash p: 131°F (CC), d: 1.2220 @ 20°/20°, autoign temp: 696°F, vap press: 0.7 mm @ 20°, vap d: 4.93.

**DFL000** CAS: 75-43-4 $CHCl_2F$ HR: 1
**DICHLOROFLUOROMETHANE**
DOT: 1029
SYNS: DICHLOROMONOFLUOROMETHANE (OSHA, DOT) ◇ FLUORODICHLOROMETHANE ◇ FREON 21

OSHA PEL: (Transitional: TWA 1000 ppm) TWA 10 ppm
ACGIH TLV: TWA 10 ppm
DFG MAK: 10 ppm (45 $mg/m^3$)
DOT Class: Nonflammable Gas

SAFETY PROFILE: Mildly toxic by inhalation. Experimental reproductive effects. When heated to decomposition it emits very toxic fumes of $Cl^-$ and $F^-$.

PROP: Heavy, colorless gas. Mp: −135°, bp: 8.9°, d: 1.48, vap press: 2 atm @ 28.4°, vap d: 3.82.

**DFU000** CAS: 594-72-9 $C_2H_3Cl_2NO_2$ HR: 3
**1,1-DICHLORO-1-NITROETHANE**
DOT: 2650
SYNS: DICHLORONITROETHANE ◇ ETHIDE

OSHA PEL: (Transitional: CL 10 ppm) TWA 2 ppm
ACGIH TLV: TWA 2 ppm
DFG MAK: 10 ppm (60 $mg/m^3$)
DOT Class: Poison B

SAFETY PROFILE: Poison by ingestion. Moderately toxic by inhalation. A strong irritant. Flammable. To fight fire, use water, $CO_2$, dry chemical. When heated to decomposition it emits highly toxic fumes of $Cl^-$ and $NO_x$.

PROP: Liquid. Bp: 124°, flash p: 168°F(OC), d: 1.4153 @ 20°/20°, vap d: 4.97.

**DGH000** CAS: 10061-02-6 $C_3H_4Cl_2$ HR: 3
**trans-1,3-DICHLOROPROPENE**
SYN: trans-1,3-DICHLOROPROPYLENE

OSHA PEL: 1 ppm (skin)

DFG MAK: Animal Carcinogen, Suspected Human Carcinogen.

SAFETY PROFILE: A suspected carcinogen. A dangerous fire hazard. When heated to decomposition it emits toxic fumes of $Cl^-$.

PROP: Flash point 21°C.

**DGH200** CAS: 10061-01-5 $C_3H_4Cl_2$ HR: 3
**cis-1,3-DICHLOROPROPENE**
SYN: cis-1,3-DICHLOROPROPYLENE

OSHA PEL: 1 ppm (skin)

DFG MAK: Animal Carcinogen, Suspected Human Carcinogen.

SAFETY PROFILE: A suspected carcinogen. A dangerous fire hazard. When heated to decomposition it emits toxic fumes of $Cl^-$.

PROP: Flash point 21°C.

**DGI400** CAS: 75-99-0 $C_3H_4Cl_2O_2$ HR: 2
**2,2-DICHLOROPROPIONIC ACID**
DOT: 1760
SYNS: DALAPON (USDA) ◇ α-DICHLOROPROPIONIC ACID ◇ DOWPON

OSHA PEL: TWA 1 ppm
ACGIH TLV: TWA 1 ppm
DFG MAK: 1 ppm (6 mg/$m^3$)
DOT Class: Corrosive Material

SAFETY PROFILE: Moderately toxic by ingestion. Corrosive. A skin irritant. When heated to decomposition it emits toxic fumes of $Cl^-$.

PROP: White to tan powder.

**DGL600** CAS: 1320-37-2 $C_2Cl_2F_4$ HR: 1
**DICHLOROTETRAFLUOROETHANE**
DOT: 1958
SYN: TETRAFLUORODICHLOROETHANE

OSHA PEL: TWA 1000 ppm
ACGIH TLV: TWA 1000 ppm
DOT Class: Nonflammable Gas

SAFETY PROFILE: A mildly toxic irritant; narcotic in high concentrations. An asphyxiant. When heated to decomposition it emits toxic fumes of $F^-$ and $Cl^-$.

PROP: Colorless gas. Bp: 3.5°.

**DGP900** CAS: 62-73-7 $C_4H_7Cl_2O_4P$ HR: 3
**DICHLORVOS**
DOT: 2783
SYNS: CHLORVINPHOS ◇ CYANOPHOS ◇ DDVP ◇ 2,2 DICHLOROVINYL DIMETHYL PHOSPHATE ◇ VAPONA

OSHA PEL: TWA 1 mg/m$^3$ (skin)
ACGIH TLV: TWA 0.1 ppm (skin)
DFG MAK: 0.1 ppm (1 mg/m$^3$)
DOT Class: Poison B

SAFETY PROFILE: Poison by ingestion, inhalation, and skin contact. An experimental teratogen. Experimental reproductive effects. When heated to decomposition it emits very toxic fumes of $Cl^-$ and $PO_x$.

PROP: Liquid. Bp: 120° @ 14 mm, bp: 77° @ 1 mm. Sltly sol in water and glycerin; miscible with aromatic and chlorinated hydrocarbon solvents and alcohol.

**DGQ875** CAS: 141-66-2 $C_8H_{16}NO_5P$ HR: 3
**DICROTOPHOS**
SYNS: DIAPADRIN ◇ 3-HYDROXYDIMETHYL CROTONAMIDE DIMETHYL PHOSPHATE

OSHA PEL: TWA 0.25 mg/m$^3$ (skin)
ACGIH TLV: TWA 0.25 mg/m$^3$ (skin)

SAFETY PROFILE: Poison by ingestion, inhalation and skin contact. When heated to decomposition it emits very toxic fumes of $NO_x$ and $PO_x$.

**DGW000** CAS: 77-73-6 $C_{10}H_{12}$ HR: 3
**DICYCLOPENTADIENE**
DOT: 2048
SYNS: BICYCLOPENTADIENE ◇ 1,3-CYCLOPENTADIENE, DIMER

OSHA PEL: TWA 5 ppm
ACGIH TLV: TWA 5 ppm
DOT Class: IMO: Flammable or Combustible Liquid

SAFETY PROFILE: Poison by ingestion. Moderately toxic by inhalation. A severe skin and moderate eye irritant. Dangerous fire hazard. To fight fire, use alcohol foam.

PROP: Colorless crystals. Mp: 32.9°, bp: 166.6°, d: 0.976 @ 35°, vap press: 10 mm @ 47.6°, vap d: 4.55, flash p: 90°F (OC).

**DHB400** CAS: 60-57-1 $C_{12}H_8Cl_6O$ HR: 3
**DIELDRIN**
DOT: 2761
SYNS: HEXACHLOROEPOXYOCTAHYDRO-ENDO,EXO-DIMETHANO-NAPHTHALENE ◇ ILLOXOL ◇ OCTALOX

OSHA PEL: TWA 0.25 $mg/m^3$ (skin)
ACGIH TLV: TWA 0.25 $mg/m^3$ (skin)
DFG MAK: 0.25 $mg/m^3$
NIOSH REL: Lowest reliable detectable level.
DOT Class: ORM-A

SAFETY PROFILE: A human poison by ingestion. Poison experimentally by inhalation and skin contact. An experimental carcinogen and teratogen. Experimental reproductive effects. Absorbed readily through the skin and by other routes. When heated to decomposition it emits toxic fumes of $Cl^-$.

PROP: White crystals; odorless. Mp: 150°, vap d: 13.2. Insol in water; sol in common organic solvents.

**DHF000** CAS: 111-42-2 $C_4H_{11}NO_2$ HR: 2
**DIETHANOLAMINE**
SYNS: BIS(2-HYDROXY ETHYL)AMINE ◇ DEA ◇ 2,2′-IMINOBISETHANOL

OSHA PEL: TWA 3 ppm
ACGIH TLV: TWA 3 ppm

SAFETY PROFILE: Moderately toxic by ingestion. Mildly toxic by skin contact. A severe eye and mild skin irritant. Combustible. To fight fire, use alcohol foam, water, $CO_2$, dry chemical. When heated to decomposition it emits toxic fumes such as $NO_x$.

PROP: A faintly colored, viscous liquid. Mp: 28°, bp: 269.1° (decomp), flash p: 305°F (OC), d: 1.0919 @ 30°/20°, autoign temp: 1224°F, vap press: 5 mm @ 138°, vap d: 3.65.

**DHJ200** CAS: 109-89-7 $C_4H_{11}N$ HR: 2
**DIETHYLAMINE**
DOT: 1154
SYN: 2-AMINOPENTANE

OSHA PEL: (Transitional: TWA 25 ppm) TWA 10 ppm; STEL 25 ppm
ACGIH TLV: TWA 10 ppm; STEL 25 ppm
DFG MAK: 10 ppm (30 $mg/m^3$)
DOT Class: Flammable Liquid

SAFETY PROFILE: Moderately toxic by ingestion, inhalation, and skin contact. A skin and severe eye irritant. A very dangerous fire hazard. To fight fire, use alcohol foam, $CO_2$, dry chemical. When heated to decomposition it emits toxic fumes of $NO_x$.

PROP: Colorless liquid, ammoniacal odor. Mp: −38.9°, bp: 55.5°, flash p: −0.4°F, d: 0.7108 @ 20°/20°, autoign temp: 594°F, vap press: 400 mm @ 38.0°, vap d: 2.53, lel: 1.8%, uel: 10.1%.

**DHO500** CAS: 100-37-8 $C_6H_{15}NO$ HR: 3
**2-DIETHYLAMINOETHANOL**
DOT: 2686
SYNS: DEAE ◇ β-DIETHYLAMINOETHYL ALCOHOL ◇ 2-HYDROXYTRIETHYLAMINE

OSHA PEL: TWA 10 ppm (skin)
ACGIH TLV: TWA 10 ppm (skin)
DFG MAK: 10 ppm (50 $mg/m^3$)
DOT Class: Flammable or Combustible Liquid

SAFETY PROFILE: Moderately toxic by ingestion and skin contact. A skin and severe eye irritant. Combustible liquid. Flammable. To fight fire, use alcohol foam, $CO_2$, dry chemical. When heated to decomposition it emits toxic fumes of $NO_x$.

PROP: Colorless, hygroscopic liquid. Bp: 162°, flash p: 140°F (OC), d: 0.8851 @ 20°/20°, vap press: 1.4 mm @ 20°, vap d: 4.03.

**DIW400** CAS: 88-10-8 $C_5H_{10}ClNO$ HR: 2
**DIETHYLCARBAMOYL CHLORIDE**
SYN: DIETHYLCARBAMYL CHLORIDE

DFG MAK: Suspected Carcinogen.

SAFETY PROFILE: Reacts with water or steam to produce toxic and corrosive fumes. When heated to decomposition it emits highly toxic fumes of $Cl^-$ and $NO_x$.

PROP: Liquid. Mp: −44°, bp: 190°-195°, vap d: 4.1.

**DJG600** CAS: 111-40-0 $C_4H_{13}N_3$ HR: 3
**DIETHYLENETRIAMINE**
DOT: 2079
SYNS: AMINOETHYLETHANDIAMINE ◇ DETA ◇ 2,2′-IMINOBISETHYLAMINE

OSHA PEL: TWA 1 ppm
ACGIH TLV: TWA 1 ppm (skin)
DOT Class: Corrosive Material

SAFETY PROFILE: Poison by skin contact. Moderately toxic by ingestion. Corrosive. A severe skin and eye irritant. Repeated exposures can cause asthma and sensitization of skin. Combustible. To fight fire, use alcohol foam. When heated to decomposition it emits toxic fumes of $NO_x$.

PROP: Yellow, viscous liquid; mild ammoniacal odor. Mp: −39°, bp: 207°, flash p: 215°F (OC), d: 0.9586 @ 20°/20°, autoign temp: 750°F, vap press: 0.22 mm @ 20°, vap d: 3.48.

**DJN750** CAS: 96-22-0 $C_5H_{10}O$ HR: 3
**DIETHYL KETONE**
DOT: 1156
SYNS: DEK ◇ DIMETHYLACETONE ◇ 3-PENTANONE ◇ PROPIONE

OSHA PEL: TWA 200 ppm
ACGIH TLV: TWA 200 ppm
DOT Class: Label: Flammable Liquid.

SAFETY PROFILE: Moderately toxic by ingestion. A skin and eye irritant. Dangerous fire hazard. To fight fire, use alcohol foam, foam, $CO_2$, dry chemical.

PROP: Colorless, mobile liquid; acetone-like odor. Mp: −42°, bp: 101°, flash p: 55°F, d: 0.8159 @ 19°/4°, vap d: 2.96, autoign temp: 842°F, lel: 1.6%. Sol in water; misc in alc and ether.

**DJX000** CAS: 84-66-2 $C_{12}H_{14}O_4$ HR: 3
**DIETHYL PHTHALATE**
SYNS: ETHYL PHTHALATE ◇ PHTHALOL

OSHA PEL: TWA 5 $mg/m^3$
ACGIH TLV: TWA 5 $mg/m^3$

SAFETY PROFILE: Moderately toxic by ingestion. An eye irritant and systemic irritant by inhalation. An experimental teratogen. Combustible. To fight fire, use water spray, mist, foam.

PROP: Clear, colorless liquid. Mp: −40.5°, bp: 302°, flash p: 325°F (OC), d: 1.110, vap d: 7.66.

**DKB110** CAS: 64-67-5 $C_4H_{10}O_4S$ HR: 3
**DIETHYL SULFATE**
DOT: 1594
SYNS: DIETHYL ESTER SULFURIC ACID ◇ ETHYL SULFATE

DFG TRK: 0.03 ppm; Animal Carcinogen, Human Suspected Carcinogen.

SAFETY PROFILE: Poison by inhalation. Moderately toxic by ingestion and skin contact. A severe skin irritant. An experimental carcinogen and tumorigen. To fight fire, use alcohol foam, $H_2O$ foam, $CO_2$, and dry chemicals. When heated to decomposition it emits toxic fumes of $SO_x$.

PROP: Colorless, oily liquid; faint ethereal odor. Mp: −25°, bp: decomp to ethyl ether, flash p: 220°F (CC), d: 1.172 @ 25°/4°, autoign temp: 817°F, vap press: 1 mm @ 47.0°, vap d: 5.31.

**DKG850** CAS: 75-61-6 $CBr_2F_2$ HR: 1
**DIFLUORODIBROMOMETHANE**
DOT: 1941
SYNS: DIBROMODIFLUOROMETHANE ◇ FREON 12-B2

OSHA PEL: TWA 100 ppm
ACGIH TLV: TWA 100 ppm

DFG MAK: 100 ppm (860 mg/$m^3$)
DOT Class: ORM-A

SAFETY PROFILE: Mildly toxic by inhalation. When heated to decomposition it emits very toxic fumes of $Br^-$ and $F^-$.

PROP: Colorless, heavy liquid. Bp: 23.2°, fp: −141°, d: 2.288 @ 15°/4°.

**DKM200** CAS: 2238-07-5 $C_6H_{10}O_3$ HR: 3
**DIGLYCIDYL ETHER**
SYNS: BIS(2,3-EPOXYPROPYL)ETHER ◇ DGE

OSHA PEL: (Transitional: CL 0.5 ppm) TWA 0.1 ppm
ACGIH TLV: TWA 0.1 ppm
DFG MAK: 0.1 ppm (0.6 mg/$m^3$); Suspected Carcinogen.
NIOSH REL: (Glycidyl Ethers) CL 1 mg/$m^3$/15M

SAFETY PROFILE: Poison by ingestion and inhalation. Moderately toxic by skin contact. An experimental tumorigen. A suspected carcinogen. A severe eye and skin irritant.

PROP: Liquid.

**DNI800** CAS: 108-83-8 $C_9H_{18}O$ HR: 2
**DIISOBUTYL KETONE**
DOT: 1157
SYNS: 2,6-DIMETHYL-4-HEPTANONE ◇ ISOBUTYL KETONE ◇ ISOVALERONE

OSHA PEL: (Transitional: TWA 50 ppm) TWA 25 ppm
ACGIH TLV: TWA 25 ppm
DFG MAK: 50 ppm (290 mg/$m^3$)
NIOSH REL: (Ketones) TWA 140 mg/$m^3$
DOT Class: Flammable or Combustible Liquid

SAFETY PROFILE: Moderately toxic by ingestion and inhalation. Mildly toxic by skin contact. An eye and skin irritant. Flammable. To fight fire, use $CO_2$, dry chemical, water spray, mist or fog.

PROP: Liquid. Bp: 166°, flash p: 140°F, d: 0.81, vap d: 4.9, lel: 0.8% @ 212°F, uel: 6.2% @ 212°F.

**DNJ800** CAS: 822-06-0 $C_8H_{12}N_2O_2$ HR: 3
**1,6-DIISOCYANATOHEXANE**
DOT: 2281
SYNS: 1,6-HEXAMETHYLENE DIISOCYANATE (MAK) ◇ HEXAMETHYLENEDIISOCYANATE (DOT) ◇ HMDI

DFG MAK: 0.01 ppm (0.07 mg/$m^3$)
NIOSH REL: (Diisocyanates) TWA 0.005 ppm; CL 0.02 ppm/10M
DOT Class: Poison B

SAFETY PROFILE: Poison by inhalation. Moderately toxic by ingestion and skin contact. When heated to decomposition it emits toxic fumes of $NO_x$.

**DNM200** CAS: 108-18-9 $C_6H_{15}N$ HR: 2
**DIISOPROPYLAMINE**
DOT: 1158
SYNS: DIPA ◇ N-(1-METHYLETHYL)-2-PROPANAMINE

OSHA PEL: TWA 5 ppm (skin)
ACGIH TLV: TWA 5 ppm (skin)
DOT Class: Flammable Liquid

SAFETY PROFILE: Moderately toxic by ingestion. Mildly toxic by inhalation. A severe eye irritant. A very dangerous fire hazard. To fight fire, use alcohol foam, foam, $CO_2$, dry chemical. When heated to decomposition it emits toxic fumes of $NO_x$.

PROP: Colorless liquid. Bp: 83-84°, flash p: 19.4°F. D: 0.722 @ 220.0°, vap d: 3.5.

**DOA800** CAS: 20325-40-0 $C_{14}H_{16}N_2O_2 \cdot 2ClH$ HR: 3
**3,3′-DIMETHOXYBENZIDINE DIHYDROCHLORIDE**
SYNS: C.I. DISPERSE BLACK-6-DIHYDROCHLORIDE ◇ o-DIANISIDINE DIHYDROCHLORIDE

NIOSH REL: (Benzidine-Based Dye) Reduce to lowest feasible level

SAFETY PROFILE: An experimental carcinogen and teratogen. When heated to decomposition it emits very toxic fumes of $NO_x$ and HCl.

**DOO800** CAS: 127-19-5 $C_4H_9NO$ HR: 3
**N,N-DIMETHYLACETAMIDE**
SYNS: ACETDIMETHYLAMIDE ◇ ACETIC ACID DIMETHYLAMIDE ◇ DMA

OSHA PEL: TWA 10 ppm (skin)
ACGIH TLV: TWA 10 ppm (skin)
DFG MAK: 10 ppm (35 $mg/m^3$)

SAFETY PROFILE: Moderately toxic by skin contact and inhalation. Mildly toxic by ingestion. An experimental teratogen. Experimental reproductive effects. A skin irritant. Flammable. When heated to decomposition it emits toxic fumes of $NO_x$.

PROP: Liquid. Mp: −20°, bp: 165°, d: 0.9448 @ 15.5°, vap d: 3.01, vap press: 1.3 mm @ 25°, flash p: 171°F (TOC), lel: 2.0%, uel: 11.5% @ 740 mm and 160°.

**DOQ800** CAS: 124-40-3 $C_2H_7N$ HR: 3
**DIMETHYLAMINE**
DOT: 1032/1160
SYNS: DMA ◇ N-METHYLMETHANAMINE

OSHA PEL: TWA 10 ppm
ACGIH TLV: TWA 10 ppm
DFG MAK: 10 ppm (18 $mg/m^3$)
DOT Class: Flammable Liquid

SAFETY PROFILE: Poison by ingestion. Moderately toxic by inhalation. Corrosive to the eyes, skin, and mucous membranes. A flammable gas. When heated to decomposition it emits toxic fumes of $NO_x$.

**DOT300** CAS: 60-11-7 $C_{14}H_{15}N_3$ HR: 3
**4-DIMETHYLAMINOAZOBENZENE**
SYNS: C.I. 11020 ◇ DAB ◇ DIMETHYLAMINOAZOBENZENE ◇ DIMETHYL YELLOW

OSHA PEL: Carcinogen

SAFETY PROFILE: Poison by ingestion. An experimental carcinogen and teratogen. When heated to decomposition it emits toxic fumes of $NO_x$.

PROP: Yellow, crystalline tablets. Insol in water; sol in strong mineral acids and oils.

**DQF800** CAS: 121-69-7 $C_8H_{11}N$ HR: 3
**N,N-DIMETHYLANILINE**
DOT: 2253
SYNS: (DIMETHYLAMINO)BENZENE ◇ N,N-DIMETHYLBENZENEAMINE

OSHA PEL: (Transitional: TWA 5 ppm (skin)) TWA 5 ppm; STEL 10 ppm (skin)
ACGIH TLV: TWA 5 ppm; STEL 10 ppm (skin)
DFG MAK: 5 ppm (25 $mg/m^3$)
DOT Class: Poison B

SAFETY PROFILE: Human poison by ingestion. Moderately toxic by inhalation and skin contact. A skin irritant. A suspected carcinogen. Flammable. To fight fire, use foam, $CO_2$, dry chemical. When heated to decomposition it emits highly toxic fumes of aniline and $NO_x$.

PROP: Liquid. Mp: 2.5°, bp: 193.1°, flash p: 145°F (CC), d: 0.9557 @ 20°/4°, ULC: 20-25, autoign temp: 700°F, vap press: 1 mm @ 29.5°, vap d: 4.17.

**DQY950** CAS: 79-44-7 $C_3H_6ClNO$ HR: 3
**DIMETHYL CARBAMOYL CHLORIDE**
DOT: 2262
SYNS: CHLOROFORMIC ACID DIMETHYLAMIDE ◇ DDC ◇ N,N-DIMETHYLCARBAMOYL CHLORIDE (DOT) ◇ DMCC

ACGIH TLV: Suspected Human Carcinogen
DFG MAK: Animal Carcinogen, Suspected Human Carcinogen.
DOT Class: Corrosive Material

SAFETY PROFILE: Moderately toxic by inhalation and ingestion. An experimental carcinogen. A powerful lachrymator. Will react with water or steam to produce toxic and corrosive fumes. When heated to decomposition it emits very toxic fumes of $Cl^-$ and $NO_x$.

PROP: Liquid. Mp: −33°, bp: 165-167°, d: 1.678 @ 20°/4°, vap d: 3.73.

**DSB000** CAS: 68-12-2 $C_3H_7NO$ HR: 3
**DIMETHYLFORMAMIDE**
DOT: 2265
SYNS: N,N-DIMETHYLFORMAMIDE (DOT) ◇ DMF ◇ N-FORMYLDIMETHYLAMINE

OSHA PEL: TWA 10 ppm (skin)
ACGIH TLV: TWA 10 ppm (skin); BEI: 40 mg(N-methylformamide)/g creatinine at end of shift.
DFG MAK: 20 ppm (60 $mg/m^3$)
DOT Class: Flammable or Combustible Liquid

SAFETY PROFILE: Moderately toxic by ingestion. Mildly toxic by skin contact and inhalation. An experimental teratogen. Experimental reproductive effects. A skin and eye irritant. Flammable. To fight fire, use foam, $CO_2$, dry chemical. When heated to decomposition it emits toxic fumes of $NO_x$.

PROP: Colorless, mobile liquid. Bp: 152.8°, lel: 2.2% @ 100°, uel: 15.2% @ 100°, flash p: 136°, fp: −61°, d: 0.9445 @ 25°/4°, autoign temp: 833°F, vap press: 3.7 mm @ 25°, vap d: 2.51.

**DSF400** CAS: 57-14-7 $C_2H_8N_2$ HR: 3
**1,1-DIMETHYLHYDRAZINE**
DOT: 1163
SYNS: DIMAZINE ◇ DIMETHYLHYDRAZINE UNSYMMETRICAL (DOT) ◇ UDMH (DOT)

OSHA PEL: TWA 0.5 ppm (skin)
ACGIH TLV: TWA 0.5 ppm (skin); Suspected Human Carcinogen; (Proposed: TWA 0.01 ppm (skin); Suspected Human Carcinogen)
DFG MAK: Animal Carcinogen, Suspected Human Carcinogen.
NIOSH REL: (Hydrazines) CL 0.15 $mg/m^3$/2H
DOT Class: Flammable Liquid

SAFETY PROFILE: Poison by ingestion. Moderately toxic by inhalation and skin contact. An experimental carcinogen and teratogen. Corrosive. A dangerous fire hazard. It is hypergolic with many oxidants. To fight fire, use alcohol foam, $CO_2$, dry chemical. When heated to decomposition it emits highly toxic fumes of $NO_x$.

PROP: Colorless liquid; ammonia-like odor. Bp: 63.3°, fp: −58°, flash p: 5°F, d: 0.782 @ 25°/4°, vap press: 157 mm @ 25°, vap d: 1.94, autoign temp: 480°F, lel: 2%, uel: 95%. Hygroscopic, water-miscible.

**DSF600** CAS: 540-73-8 $C_2H_8N_2$ HR: 3
**1,2-DIMETHYLHYDRAZINE**
DOT: 2382
SYNS: sym-DIMETHYLHYDRAZINE ◇ DMH ◇ HYDRAZOMETHANE

DFG MAK: Animal Carcinogen, Suspected Human Carcinogen.
DOT Class: Flammable Liquid

SAFETY PROFILE: Poison by ingestion. Moderately toxic by inhalation. An experimental carcinogen and teratogen. Experimental reproductive effects. A very dangerous fire hazard. When heated to decomposition it emits toxic fumes of $NO_x$.

PROP: Clear, colorless, flammable, hygroscopic liquid; fishy ammonia odor. Flash p: < 73.4°F, bp: 81°, mp: −9°, d: 0.8274 @ 20°/4°.

**DTR200** CAS: 131-11-3 $C_{10}H_{10}O_4$ HR: 2
**DIMETHYLPHTHALATE**
SYNS: DIMETHYL-1,2-BENZENEDICARBOXYLATE ◇ METHYL PHTHALATE

OSHA PEL: TWA 5 $mg/m^3$
ACGIH TLV: TWA 5 $mg/m^3$

SAFETY PROFILE: Moderately toxic by ingestion. Mildly toxic by inhalation. An experimental teratogen. Experimental reproductive effects. An eye irritant. Combustible. To fight fire, use $CO_2$, dry chemical.

PROP: Colorless, odorless liquid. Bp: 283.7°, flash p: 295°F (CC), d: 1.189 @ 25°/25°, autoign temp: 1032°F, vap d: 6.69, vap press: 1 mm @ 100.3°.

**DUD100** CAS: 77-78-1 $C_2H_6O_4S$ HR: 3
**DIMETHYL SULFATE**
DOT: 1595
SYNS: DMS ◇ METHYL SULFATE (DOT)

OSHA PEL: (Transitional: TWA 1 ppm (skin)) TWA 0.1 ppm (skin)
ACGIH TLV: TWA 0.1 ppm (skin); Suspected Human Carcinogen
DFG TRK: Production: 0.02 ppm; Use: 0.04 ppm; Animal Carcinogen, Suspected Human Carcinogen.
DOT Classification; Corrosive Material

SAFETY PROFILE: Human poison by inhalation. Experimental poison by ingestion. An experimental carcinogen and teratogen. A corrosive irritant to skin, eyes and mucous membranes. Flammable. To fight fire, use water, foam, $CO_2$, dry chemical. When heated to decomposition it emits toxic fumes of $SO_x$.

PROP: Colorless, odorless liquid. Mp: −31.8°, bp: 188°, flash p: 182°F (OC), d: 1.3322 @ 20°/4°, vap d: 4.35, autoign temp: 370°F.

**DUP300** CAS: 148-01-6 $C_8H_7N_3O_5$ HR: 3
**DINITOLMIDE**
SYNS: 3,5-DINITRO-o-TOLUAMIDE ◇ D.O.T. ◇ ZOALENE

OSHA PEL: TWA 5 $mg/m^3$
ACGIH TLV: TWA 5 $mg/m^3$

SAFETY PROFILE: Moderately toxic by ingestion. When heated to decomposition it emits toxic fumes of $NO_x$.

PROP: Yellowish solid. Mp: 177°. Very sltly sol in water; sol in acetone, acetonitrile, and dimethyl formamide.

**DUQ180** CAS: 25154-54-5 $C_6H_4N_2O_4$ HR: 3
**DINITROBENZENE**
DOT: 1597
SYN: DINITROBENZOL SOLID (DOT)

OSHA PEL: TWA 1 mg/m$^3$ (skin)
ACGIH TLV: TWA 0.15 ppm (skin)
DFG MAK: 0.15 ppm (1 mg/m$^3$); Suspected Carcinogen
DOT Class: Poison B

SAFETY PROFILE: A poison. When heated to decomposition it emits toxic fumes of $NO_x$.

**DUQ200** CAS: 99-65-0 $C_6H_4N_2O_4$ HR: 3
**m-DINITROBENZENE**
DOT: 1597
SYNS: BINITROBENZENE ◇ 1,3-DINITROBENZENE

OSHA PEL: TWA 1 mg/m$^3$ (skin)
ACGIH TLV: TWA 0.15 ppm (skin)
DOT Class: Poison B

SAFETY PROFILE: Human poison by ingestion. Experimental reproductive effects. Mixture with nitric acid is a high explosive. Mixture with tetranitromethane is a high explosive very sensitive to sparks. When heated to decomposition it emits toxic fumes of $NO_x$.

PROP: Yellowish crystals. Mp: 89°, bp: 301°.

**DUQ400** CAS: 528-29-0 $C_6H_4N_2O_4$ HR: 3
**o-DINITROBENZENE**
DOT: 1597
SYN: 1,2-DINITROBENZENE

OSHA PEL: TWA 1 mg/m$^3$ (skin)
ACGIH TLV: TWA 0.15 ppm (skin)
DOT Class: Poison B

SAFETY PROFILE: Poison by inhalation and ingestion. Moderately toxic by skin contact. Combustible. A severe explosion hazard. To fight fire, use water, $CO_2$, dry chemical. Dangerous; when heated to decomposition it emits highly toxic fumes of $NO_x$ and explodes.

PROP: Colorless needles or plates. Mp: 118°, bp: 319°, flash p: 302°F (CC), d: 1.571 @ 0°/4°, vap d: 5.79.

**DUQ600** CAS: 100-25-4 $C_6H_4N_2O_4$ HR: 3
**p-DINITROBENZENE**
DOT: 1597
SYN: DITHANE A-4

OSHA PEL: TWA 1 mg/m$^3$ (skin)
ACGIH TLV: TWA 0.15 ppm (skin)
DOT Class: Poison B

SAFETY PROFILE: Poison by ingestion. Mixture with nitric acid is a high explosive. When heated to decomposition it emits toxic fumes of $NO_x$.

PROP: White crystals. Mp: 173°, bp: 299°. Volatile with steam.

**DUT115** CAS: 534-52-1 $C_7H_6N_2O_5$ HR: 3
**2,4-DINITRO-o-CRESOL**
SYN: DINOC

OSHA PEL: TWA 0.2 mg/m$^3$ (skin)
ACGIH TLV: TWA 0.2 mg/m$^3$ (skin)
DFG MAK: 0.2 mg/m$^3$)
NIOSH REL: (Dinitro-Ortho-Cresol) TWA 0.2 mg/m$^3$

SAFETY PROFILE: Human poison. Experimental poison by ingestion, inhalation, skin contact. An eye and skin irritant.

PROP: Yellow, prismatic crystals. Mp: 85.8°, vap d: 6.82.

**DUX700** CAS: 605-71-0 $C_{10}H_6N_2O_4$ HR: 3
**1,5-DINITRONAPHTHALENE**

DFG MAK: Suspected Carcinogen. (all isomers)

SAFETY PROFILE: Mixtures with sulfur or sulfuric acid (used in commercial reactions) may explode if heated to 120°C. When heated to decomposition it emits toxic fumes of $NO_x$.

**DVD400** CAS: 75321-20-9 $C_{16}H_8N_2O_4$ HR: 3
**1,3-DINITROPYRENE**

DFG MAK: Suspected Carcinogen.

SAFETY PROFILE: An experimental carcinogen. When heated to decomposition it emits toxic fumes of $NO_x$.

**DVD600** CAS: 42397-64-8 $C_{16}H_8N_2O_4$ HR: 3
**1,6-DINITROPYRENE**

DFG MAK: Suspected Carcinogen.

SAFETY PROFILE: An experimental carcinogen. When heated to decomposition it emits toxic fumes of $NO_x$.

**DVD800** CAS: 42397-65-9 $C_{16}H_8N_2O_4$ HR: 3
**1,8-DINITROPYRENE**

DFG MAK: Suspected Carcinogen.

SAFETY PROFILE: An experimental carcinogen. When heated to decomposition it emits toxic fumes of $NO_x$.

**DVG600** CAS: 25321-14-6 $C_7H_6N_2O_4$ HR: 3
**DINITROTOLUENE**
DOT: 1600/2038
SYNS: DINITROPHENYLMETHANE ◇ METHYLDINITROBENZENE

OSHA PEL: TWA 1.5 mg/m$^3$ (skin)
DFG MAK: Animal Carcinogen, Suspected Human Carcinogen.
NIOSH REL: (Dinitrotoluene): Reduce to lowest level
DOT Class: ORM-E; Liquid and Solid; Poison B;

SAFETY PROFILE: A poison. An experimental tumorigen and teratogen. Experimental reproductive effects. Flammable. When heated to decomposition it emits toxic fumes of $NO_x$.

**DVG800** CAS: 602-01-7 $C_7H_6N_2O_4$ HR: 2
**2,3-DINITROTOLUENE**
SYNS: 1-METHYL-2,3-DINITRO-BENZENE (9CI) ◇ 2,3-DNT

OSHA PEL: TWA 1.5 mg/m$^3$ (skin)
NIOSH REL: (Dinitrotoluene): Reduce to lowest level

SAFETY PROFILE: Moderately toxic by ingestion. A skin irritant. When heated to decomposition it emits toxic fumes of $NO_x$.

**DVH000** CAS: 121-14-2 $C_7H_6N_2O_4$ HR: 3
**2,4-DINITROTOLUENE**
SYNS: DNT ◇ 1-METHYL-2,4-DINITROBENZENE

OSHA PEL: TWA 1.5 mg/m$^3$ (skin)
ACGIH TLV: TWA 1.5 mg/m$^3$ (skin)
NIOSH REL: (Dinitrotoluene): Reduce to lowest level

SAFETY PROFILE: Poison by ingestion. An experimental carcinogen. Experimental reproductive effects. An irritant and an allergen. Combustible. To fight fire, use water spray or mist, dry chemical. When heated to decomposition it emits toxic fumes of $NO_x$.

PROP: Yellow needles. Mp: 69.5°, bp: 300°, d: 1.521 @ 15°, vap d: 6.27, flash p: 404°F.

**DVH200** CAS: 619-15-8 $C_7H_6N_2O_4$ HR: 2
**2,5-DINITROTOLUENE**
SYNS: 2,5-DNT ◇ 2-METHYL-1,4-DINITROBENZENE

OSHA PEL: TWA 1.5 mg/m$^3$ (skin)
NIOSH REL: (Dinitrotoluene): Reduce to lowest level

SAFETY PROFILE: Moderately toxic by ingestion. A skin irritant. When heated to decomposition it emits toxic fumes of $NO_x$.

**DVH400** CAS: 606-20-2 $C_7H_6N_2O_4$ HR: 3
**2,6-DINITROTOLUENE**
SYNS: 2,6-DNT ◇ 2-METHYL-1,3-DINITROBENZENE

OSHA PEL: TWA 1.5 μg/m$^3$ (skin)
NIOSH REL: (Dinitrotoluene): Reduce to lowest level

SAFETY PROFILE: Poison by ingestion. An experimental tumorigen. A skin irritant. When heated to decomposition it emits toxic fumes of $NO_x$.

**DVH600** CAS: 610-39-9 $C_7H_6N_2O_4$ HR: 3
**3,4-DINITROTOLUENE**
SYNS: 3,4-DNT ◇ 4-METHYL-1,2-DINITROBENZENE

OSHA PEL: TWA 1.5 mg/m$^3$ (skin)

SAFETY PROFILE: Poison by ingestion. A skin irritant. When heated to decomposition it emits toxic fumes of $NO_x$.

**DVL700** CAS: 117-81-7 $C_{24}H_{38}O_4$ HR: 3
**DI-sec-OCTYL PHTHALATE**
SYNS: DEHP ◇ DIOCTYL PHTHALATE ◇ DOP

OSHA PEL: (Transitional: TWA 5 mg/m$^3$) TWA 5 mg/m$^3$; STEL 10 mg/m$^3$
ACGIH TLV: TWA 5 mg/m$^3$; STEL 10 mg/m$^3$
DFG MAK: 10 mg/m$^3$
NIOSH REL: (DEHP) Reduce to lowest feasible level

SAFETY PROFILE: Suspected human carcinogen and an experimental teratogen. A mild skin and eye irritant.

**DVQ000** CAS: 123-91-1 $C_4H_8O_2$ HR: 3
**DIOXANE**
DOT: 1165
SYNS: DIETHYLENE DIOXIDE ◇ 1,4-DIOXANE (MAK) ◇ GLYCOL ETHYLENE ETHER ◇ TETRAHYDRO-p-DIOXIN

OSHA PEL: (Transitional: TWA 100 ppm (skin)) TWA 25 ppm (skin)
ACGIH TLV: TWA 25 ppm (skin)
DFG MAK: 50 ppm (180 mg/m$^3$); Suspected Carcinogen.
NIOSH REL: CL (Dioxane) 1 ppm/30M
DOT Class: Flammable Liquid

SAFETY PROFILE: Moderately toxic by ingestion and inhalation. Mildly toxic by skin contact. An experimental carcinogen and teratogen. Experimental reproductive effects. An eye and skin irritant. A very dangerous fire and explosion hazard. To fight fire, use alcohol foam, $CO_2$, dry chemical.

PROP: Colorless liquid, pleasant odor. Mp: 12°, bp: 101.1°, lel: 2.0%, uel: 22.2%, flash p: 54°F (CC), d: 1.0353 @ 20°/4°, autoign temp: 356°F, vap press: 40 mm @ 25.2°, vap d: 3.03.

**DVQ709** CAS: 78-34-2 $C_{12}H_{26}O_6P_2S_4$ HR: 3
**DIOXATHION**
SYN: 2,3-p-DIOXANE-S,S-BIS(O,O-DIETHYLPHOSPHOROITHIOATE)

OSHA PEL: TWA 0.2 mg/m$^3$ (skin)
ACGIH TLV: TWA 0.2 mg/m$^3$ (skin)

SAFETY PROFILE: Poison by ingestion, inhalation and skin contact. When heated to decomposition it emits very toxic fumes of $PO_x$ and $SO_x$.

PROP: Nonvolatile, stable solid. Nonflammable. Insol in water.

**DVX800** CAS: 122-39-4 $C_{12}H_{11}N$ HR: 3
**DIPHENYLAMINE**
SYNS: C.I. 10355 ◇ DFA ◇ N-PHENYLANILINE

OSHA PEL: TWA 10 mg/m$^3$
ACGIH TLV: TWA 10 mg/m$^3$

SAFETY PROFILE: Poison by ingestion. An experimental teratogen. Experimental reproductive effects. Combustible. To fight fire, use $CO_2$, dry chemical. When heated to decomposition it emits highly toxic fumes of $NO_x$.

PROP: Crystals; floral odor. Mp: 52.9°, bp: 302.0°, flash p: 307°F (CC), d: 1.16, autoign temp: 1173°F, vap press: 1 mm @ 108.3°, vap d: 5.82. Sol in benzene, ether, and carbon disulfide.

**DWT200** CAS: 34590-94-8 $C_7H_{16}O_3$ HR: 1
**DIPROPYLENE GLYCOL METHYL ETHER**
SYNS: DIPROPYLENE GLYCOL MONOMETHYL ETHER ◇ DOWANOL DPM

OSHA PEL: (Transitional: TWA 100 ppm (skin)) TWA 100 ppm; STEL 150 ppm (skin)
ACGIH TLV: TWA 100 ppm; STEL 150 ppm (skin)
DFG MAK: 50 ppm (300 mg/m$^3$)

SAFETY PROFILE: Mildly toxic by ingestion and skin contact. A skin and eye irritant. A mild allergen. Flammable. To fight fire, use dry chemical, $CO_2$, mist, foam.

PROP: Liquid. Bp: 190°, d: 0.951, vap d: 5.11, flash p: 185°F.

**DWT600** CAS: 123-19-3 $C_7H_{14}O$ HR: 2
**DIPROPYL KETONE**
DOT: 2710
SYNS: BUTYRONE (DOT) ◇ 4-HEPTANONE

OSHA PEL: TWA 50 ppm
ACGIH TLV: TWA 50 ppm
DOT Class: Flammable or Combustible Liquid

SAFETY PROFILE: Moderately toxic by ingestion, inhalation, and skin contact. Flammable. To fight fire, use $CO_2$, dry chemical, alcohol foam, fog and mist.

PROP: Colorless, refractive liquid. Bp: 144°, mp: −32.6°, vap press: 5.2 mm @ 20°, flash p: 120°F (CC), d: 0.815, vap d: 3.93.

**DWX800** CAS: 85-00-7 $C_{12}H_{12}N_2 \cdot 2Br$ HR: 3
**DIQUAT**
DOT: 2781
SYNS: 9,10-DIHYDRO-8a,10-DIAZONIAPHENANTHRENE DIBROMIDE ◇ ETHYLENE DIPYRIDYLIUM DIBROMIDE

OSHA PEL: TWA 0.5 mg/$m^3$
ACGIH TLV: TWA 0.5 mg/$m^3$
DOT Class: ORM-E

SAFETY PROFILE: Poison by ingestion. An experimental teratogen. Experimental reproductive effects. A skin and eye irritant. When heated to decomposition it emits very toxic fumes of $NO_x$ and $Br^-$.

PROP: Yellow crystals. Mp: 355°. Sol in water.

**DXH250** CAS: 97-77-8 $C_{10}H_{20}N_2S_4$ HR: 3
**DISULFIRAM**
SYNS: ANTABUSE ◇ BIS(N,N-DIETHYLTHIOCARBAMOYL)DISULPHIDE ◇ DISULFURAM

OSHA PEL: TWA 2 mg/$m^3$
ACGIH TLV: TWA 2 mg/$m^3$
DFG MAK: 2 mg/$m^3$

SAFETY PROFILE: A human poison by ingestion. An experimental carcinogen. Toxic symptoms when accompanied by ingestion of alcohol.

PROP: Yellow-white crystals; mp: 72°.

**DXH325** CAS: 298-04-4 $C_8H_{19}O_2PS_3$ HR: 3
**DISULFOTON**
DOT: 2783
SYNS: O,O-DIETHYL-S-(2-ETHTHIOETHYL) PHOSPHORODITHIOATE ◇ DISYSTOX ◇ DITHIODEMETON

OSHA PEL: TWA 0.1 mg/$m^3$ (skin)
ACGIH TLV: TWA 0.1 mg/$m^3$
DOT Class: Poison B

SAFETY PROFILE: Poison by ingestion, inhalation, and skin contact. When heated to decomposition it emits very toxic $SO_x$ and $PO_x$.

**DXQ500** CAS: 330-54-1 $C_9H_{10}Cl_2N_2O$ HR: 3
**DIURON**
DOT: 2767
SYNS: DCMU ◇ 3-(3,4-DICHLOROPHENOL)-1,1-DIMETHYLUREA ◇ TELVAR

OSHA PEL: TWA 10 mg/$m^3$
ACGIH TLV: TWA 10 mg/$m^3$
DOT Class: ORM-E

SAFETY PROFILE: An experimental tumorigen and teratogen. Moderately toxic by ingestion. When heated to decomposition emits highly toxic fumes of $Cl^-$ and $NO_x$.

PROP: Crystals. Mp: 159°. Sltly sol in water and hydrocarbon solvents.

**DXQ745** CAS: 1321-74-0 $C_{10}H_{10}$ HR: 2
**DIVINYLBENZENE**
SYN: VINYLSTYRENE

OSHA PEL: 10 ppm
ACGIH TLV: 10 ppm

SAFETY PROFILE: Moderately toxic by ingestion. A skin and eye irritant. Combustible.

PROP: Pale straw-colored liquid. Bp: 195-200°, mp: −87°, d: 0.918, flash p: 165F°. Not misc in water; sol in eth and methanol.

**EAL100** CAS: 1302-74-5 $Al_2O_3$ HR: 3
**EMERY**
SYNS: ALUMINUM OXIDE ◇ CORUNDUM

OSHA PEL: (Transitional: TWA Total Dust: 15 mg/m$^3$; Respirable Fraction: 5 mg/m$^3$) TWA Total Dust: 10 mg/m$^3$; Respirable Fraction: 5 mg/m$^3$
ACGIH TLV: TWA (nuisance particulate) 10 mg/m$^3$ of total dust (when toxic impurities are not present, e.g., quartz $< 1\%$).

SAFETY PROFILE: An experimental carcinogen. May cause a pneumoconiosis. It is mainly a nuisance dust.

PROP: A varicolored mineral. D: 3.95-4.10.

**EAQ750** CAS: 115-29-7 $C_9H_6Cl_6O_3S$ HR: 3
**ENDOSULFAN**
DOT: 2761
SYNS: BENZOEPIN ◇ THIOSULFAN

OSHA PEL: TWA 0.1 mg/m$^3$ (skin)
ACGIH TLV: TWA 0.1 mg/m$^3$ (skin)
DOT Class: Poison B

SAFETY PROFILE: Poison by ingestion, inhalation, and skin contact. An experimental tumorigen and teratogen. When heated to decomposition it emits toxic fumes of $Cl^-$ and $SO_x$.

PROP: A mixture of 2 isomers. Brown crystals. Mp (α): 106°, mp (β): 212°, d: 1.745 @ 20°/20°. Nearly insol in water; sol in most organic solvents.

**EAT500** CAS: 72-20-8 $C_{12}H_8Cl_6O$ HR: 3
**ENDRIN**
DOT: 2761
SYNS: HEXADRIN ◇ MENDRIN

OSHA PEL: TWA 0.1 mg/m$^3$ (skin)
ACGIH TLV: TWA 0.1 mg/m$^3$ (skin)
DFG MAK: 0.1 mg/m$^3$
DOT Class: Poison B

SAFETY PROFILE: Poison by ingestion and skin contact. A suspected human carcinogen. An experimental teratogen. A dangerous fire hazard.

PROP: White crystals. Mp: decomp @ 200°.

**EAT900** CAS: 13838-16-9 $C_3H_2ClF_5O$ HR: 2
**ENFLURANE**
SYNS: 2-CHLORO-1,1,2-TRIFLUOROETHYL DIFLUOROMETHYL ETHER ◇ ETHRANE

ACGIH TLV: TWA 75 ppm
NIOSH REL: (Waste Anesthetic Gases and Vapors) CL 2 ppm/1H

SAFETY PROFILE: Mildly toxic by inhalation and ingestion. An experimental carcinogen and teratogen by inhalation. An eye irritant. When heated to decomposition it emits very toxic fumes of $F^-$ and $Cl^-$.

**EAZ500** CAS: 106-89-8 $C_3H_5ClO$ HR: 3
**EPICHLOROHYDRIN**
DOT: 2023
SYNS: 1-CHLORO-2,3-EPOXYPROPANE ◇ ECH

OSHA PEL: (Transitional: TWA 5 ppm (skin)) TWA 2 ppm (skin)
ACGIH TLV: TWA 2 ppm (skin)
DFG TRK: 3 ppm; Animal Carcinogen, Suspected Human Carcinogen.
NIOSH REL: Minimize exposure
DOT Class: Flammable Liquid

SAFETY PROFILE: Poison by ingestion. Moderately toxic by skin contact and inhalation. An experimental carcinogen and teratogen. A skin and eye irritant. A sensitizer. Flammable. When heated to decomposition it emits toxic fumes of $Cl^-$.

PROP: Colorless, mobile liquid; irritating chloroform-like odor. Bp: 117.9°, fp: − 57.1°, flash p: 105.1°F (OC) (40°C), mp -25.6°C, d: 1.1761 @ 20°/20°, vap press: 10 mm @ 16.6°, vap d: 3.29.

**EBD700** CAS: 2104-64-5 $C_{14}H_{14}NO_4PS$ HR: 3
**EPN**
SYNS: ETHOXY-4-NITROPHENOXYPHENYLPHOSPHINE SULFIDE ◇ ETHYL-p-NITROPHENYL BENZENETHIOPHOSPHONATE

OSHA PEL: TWA 0.5 $mg/m^3$ (skin)
ACGIH TLV: TWA 0.5 $mg/m^3$ (skin)
DFG MAK: 0.5 $mg/m^3$

SAFETY PROFILE: Poison by ingestion and skin contact. This material is extremely hazardous on contact with skin, inhalation or ingestion. When heated to decomposition it emits highly toxic fumes of $SO_x$, $PO_x$, $NO_x$, and phosphine.

PROP: Liquid or pale yellow crystals with an aromatic odor, nearly insol in water, sol in organic solvents. D: 1.268 @ 25°, mp: 36°.

**EEA500** CAS: 107-15-3 $C_2H_8N_2$ HR: 3

**1,2-ETHANEDIAMINE**

DOT: 1604

SYN: 1,2-DIAMINOETHANE ◇ ETHYLENEDIAMINE

OSHA PEL: TWA 10 ppm
ACGIH TLV: TWA 10 ppm
DOT Class: Corrosive Material.

SAFETY PROFILE: An irritant poison in humans by inhalation. Moderately toxic by ingestion and skin contact. Corrosive. A severe skin and eye irritant. An allergen and sensitizer. Flammable. To fight fire, use $CO_2$, dry chemical, alcohol foam. When heated to decomposition it emits toxic fumes of $NO_x$ and $NH_3$.

PROP: Volatile, colorless, hygroscopic liquid; ammonia-like odor. Mp: 8.5°, bp: 117.2°, flash p: 110°F (CC), d: 0.8994 @ 20°/4°, vap press: 10.7 mm @ 20°, vap d: 2.07, autoign temp: 725°F.

**EEC600** CAS: 141-43-5 $C_2H_7NO$ HR: 3

**ETHANOLAMINE**

DOT: 2491

SYNS: 2-AMINOETHANOL (MAK) ◇ GLYCINOL

OSHA PEL: (Transitional: TWA 3 ppm) TWA 3 ppm; STEL 6 ppm
ACGIH TLV: TWA 3 ppm; STEL 6 ppm
DFG MAK: 3 ppm (8 $mg/m^3$)
DOT Class: Corrosive Material

SAFETY PROFILE: Moderately toxic by ingestion and skin contact. A corrosive irritant to skin, eyes, and mucous membranes. Flammable. To fight fire, use foam, alcohol foam, dry chemical. When heated to decomposition it emits toxic fumes of $NO_x$.

PROP: Colorless liquid; ammoniacal odor. Hygroscopic, bp: 170.5°, fp: 10.5°, flash p: 200°F (OC), d: 1.0180 @ 20°/4°, vap press: 6 mm @ 60°, vap d: 2.11. Misc in water and alc; sltly sol in benzene; sol in chloroform.

**EEH600** CAS: 563-12-2 $C_9H_{22}O_4P_2S_4$ HR: 3

**ETHION**

DOT: 2783

SYNS: BLADAN ◇ ETHYL METHYLENE PHOSPHORODITHIOATE

OSHA PEL: TWA 0.4 $mg/m^3$ (skin)
ACGIH TLV: TWA 0.4 $mg/m^3$ (skin)
DOT Class: Poison B

SAFETY PROFILE: Poison by ingestion and skin contact. When heated to decomposition it emits highly toxic fumes of $SO_x$ and $PO_x$.

PROP: Liquid. Mp: −13°, d: 1.220 @ 20°/4°. Sltly sol in water; sol in xylene, chloroform, and acetone.

**EES350** CAS: 110-80-5 $C_4H_{10}O_2$ HR: 2
**2-ETHOXYETHANOL**
DOT: 1171
SYNS: CELLOSOLVE (DOT) ◇ ETHYL CELLOSOLVE ◇ ETHYLENE GLYCOL MONOETHYL ETHER (DOT)

OSHA PEL: TWA 200 ppm (skin)
ACGIH TLV: TWA 5 ppm (skin)
DFG MAK: 20 ppm (75 $mg/m^3$)
NIOSH REL: (Glycol Ethers): Reduce to lowest level
DOT Class: Combustible Liquid

SAFETY PROFILE: Moderately toxic by ingestion and skin contact. Mildly toxic by inhalation. An experimental teratogen. An eye and skin irritant. Combustible. Moderate explosion hazard. To fight fire, use alcohol foam, dry chemical.

PROP: Colorless liquid, practically odorless. Bp: 135.1°, lel: 1.8%, uel: 14%, flash p: 202°F(CC), d: 0.9360 @ 15°/15°, autoign temp: 455°F, vap press: 3.8 mm @ 20°, vap d: 3.10.

**EES400** CAS: 111-15-9 $C_6H_{12}O_3$ HR: 2
**2-ETHOXYETHYL ACETATE**
DOT: 1172
SYNS: CELLOSOLVE ACETATE (DOT) ◇ ETHYLENE GLYCOL MONOETHYL ETHER ACETATE (MAK, DOT) ◇ GLYCOL MONOETHYL ETHER ACETATE

OSHA PEL: TWA 100 ppm (skin)
ACGIH TLV: TWA 5 ppm (skin)
DFG MAK: 20 ppm (110 $mg/m^3$)
DOT Class: Combustible Liquid

SAFETY PROFILE: Moderately toxic by ingestion. Mildly toxic by skin contact, and inhalation. A skin and eye irritant. An experimental teratogen. Flammable. To fight fire, use alcohol foam, $CO_2$, dry chemical.

PROP: Colorless liquid with a mild, pleasant ester-like odor. Bp: 156.4°, flash p: 117°F (COC), lel: 1.7%, fp: −61.7°, d: 0.9748 @ 20°/20°, autoign temp: 715°F, vap press: 1.2 mm @ 20°, vap d: 4.72.

**EFR000** CAS: 141-78-6 $C_4H_8O_2$ HR: 3
**ETHYL ACETATE**
DOT: 1173
SYNS: ACETIC ETHER ◇ ACETOXYETHANE ◇ VINEGAR NAPHTHA

OSHA PEL: TWA 400 ppm
ACGIH TLV: TWA 400 ppm
DFG MAK: 400 ppm (1400 $mg/m^3$)
DOT Class: Flammable Liquid

SAFETY PROFILE: Poison by inhalation. Mildly toxic by ingestion. Human eye irritant. Highly flammable liquid. To fight fire, use $CO_2$, dry chemical or alcohol foam.

PROP: Colorless liquid; fragrant odor. Mp: −83.6°, bp: 77.15°, ULC: 85-90, lel: 2.2%, uel: 11%, flash p: 24°F, d: 0.8946 @ 25°, autoign temp: 800°F, vap press: 100 mm @ 27.0°, vap d: 3.04. Misc with alc, ether, glycerin, volatile oils, and water @ 54°.

**EFT000** CAS: 140-88-5 $C_5H_8O_2$ HR: 3
**ETHYL ACRYLATE**
DOT: 1917
SYNS: ACRYLIC ACID ETHYL ESTER ◇ 2-PROPENOIC ACID, ETHYL ESTER (MAK)

OSHA PEL: (Transitional: TWA 25 mg/m$^3$ (skin)) TWA 5 ppm; STEL 15 ppm (skin)
ACGIH TLV: TWA 5 ppm; STEL 25 ppm (Proposed: 5 ppm; STEL 15 ppm Suspected Human Carcinogen)
DFG MAK: 5 ppm (20 mg/m$^3$)
DOT Class: Flammable Liquid

SAFETY PROFILE: Poison by ingestion and inhalation. Moderately toxic by skin contact. A suspected human carcinogen. An experimental carcinogen. A skin and eye irritant. Flammable liquid. To fight fire, use $CO_2$, dry chemical or alcohol foam.

PROP: Colorless liquid; acrid penetrating odor. Bp: 99.8°, fp: <−72°: lel: 1.8%, flash p: 60°F (OC), 48.2°F d: 0.916-0.919, vap press, 29.3 mm @ 20°, vap d: 3.45. Misc with alc, ether; sltly sol in water.

**EFU000** CAS: 64-17-5 $C_2H_6O$ HR: 3
**ETHYL ALCOHOL**
DOT: 1170
SYNS: ABSOLUTE ETHANOL ◇ ETHANOL (MAK) ◇ GRAIN ALCOHOL

OSHA PEL: TWA 1000 ppm
ACGIH TLV: TWA 1000 ppm
DFG MAK: 1,000 ppm (1,900 mg/m$^3$)
DOT Class: Flammable Liquid

SAFETY PROFILE: Moderately toxic to humans by ingestion. Mildly toxic by inhalation and skin contact. An experimental tumorigen and teratogen. Experimental reproductive effects. An eye and skin irritant. Flammable liquid. To fight fire, use alcohol foam, $CO_2$, dry chemical.

PROP: Clear, colorless liquid; fragrant odor and burning taste. Bp: 78.32°, ULC: 70, lel: 3.3%, uel: 19% @ 60°, fp: <−130°, flash p: 55.6°F, d: 0.7893 @ 20°/4°, autoign temp: 793°F, vap press: 40 mm @ 19°, vap d: 1.59, refr index: 1.364. Misc in water, alc, chloroform, and ether.

**EFU400** CAS: 75-04-7 $C_2H_7N$ HR: 3
**ETHYLAMINE**
DOT: 1036
SYNS: AMINOETHANE ◇ MONOETHYLAMINE (DOT)

OSHA PEL: TWA 10 ppm
ACGIH TLV: TWA 10 ppm

DFG MAK: 10 ppm (18 mg/$m^3$)
DOT Class: Flammable Liquid

SAFETY PROFILE: A poison by ingestion and skin contact. Moderately toxic by inhalation. A severe eye irritant. A very dangerous fire hazard. To fight fire, stop flow of gas, use alcohol foam, dry chemical. When heated to decomposition it emits toxic fumes of $NO_x$.

PROP: Colorless gas or liquid; strong ammoniacal odor. Bp: 16.6°, lel: 4.95%, uel: 20.75%, fp: −80.6°, flash p: −0.4°F, d: 0.662 @ 20°/4°, autoign temp: 725°F, vap d: 1.56. vap press: 400 mm @ 20°. Miscible with water, alc, and ether.

**EGI750** CAS: 541-85-5 $C_8H_{16}O$ HR: 2
**ETHYL AMYL KETONE**
SYNS: AMYL ETHYL KETONE ◇ 3-METHYL-5-HEPTANONE

OSHA PEL: TWA 25 ppm
ACGIH TLV: TWA 25 ppm

SAFETY PROFILE: Moderately irritating to skin, eyes, and mucous membranes by inhalation and ingestion. Combustible. To fight fire use foam, $CO_2$, dry chemical.

PROP: Liquid with mild, fruity odor; sol in many organic solvents. Bp: 157°-162°, d: 0.822 @ 20°/20°, flash p: 138°F.

**EGI755** CAS: 106-68-3 $C_8H_{16}O$ HR: 2
**ETHYL AMYL KETONE**
DOT: 2271
SYNS: AMYL ETHYL KETONE ◇ EAK ◇ 3-OCTANONE

OSHA PEL: TWA 25 ppm
DOT Class: Flammable or Combustible Liquid

SAFETY PROFILE: Moderately irritating to skin, eyes, and mucous membranes by inhalation and ingestion. Dangerously flammable. To fight fire use foam, $CO_2$, dry chemical.

**EGP500** CAS: 100-41-4 $C_8H_{10}$ HR: 2
**ETHYL BENZENE**
DOT: 1175
SYNS: ETHYLBENZOL ◇ PHENYLETHANE

OSHA PEL: (Transitional: TWA 100 ppm (skin)) TWA 100 ppm; STEL 125 ppm
ACGIH TLV: TWA 100 ppm; STEL 125 ppm; BEI: 2 g(mandelic acid)/L in urine at end of shift; 2 ppm ethyl benzene in end-exhaled air prior to next shift.
DFG MAK: 100 ppm (440 mg/$m^3$)
DOT Class: Flammable Liquid

SAFETY PROFILE: Moderately toxic by ingestion. Mildly toxic by inhalation and skin contact. An experimental teratogen. An eye and skin irritant. A very dangerous fire and explosion hazard. To fight fire, use foam, $CO_2$, dry chemical.

PROP: Colorless liquid, aromatic odor. Misc in alc and ether, insol in $NH_3$; sol in $SO_2$. Bp: 136.2°, fp: −94.9°, flash p: 59°F, d: 0.8669 @ 20°/4°, autoign temp: 810°F, vap press: 10 mm @ 25.9°, vap d: 3.66, lel: 1.2%, uel: 6.8%.

**EGV400** CAS: 74-96-4 $C_2H_5Br$ HR: 3
**ETHYL BROMIDE**
DOT: 1891
SYNS: BROMOETHANE ◇ MONOBROMOETHANE

OSHA PEL: (Transitional: TWA 200 ppm) TWA 200 ppm; STEL 250 ppm
ACGIH TLV: TWA 200 ppm; STEL 250 ppm
DFG MAK: 200 ppm (890 $mg/m^3$)
DOT Class: Poison B

SAFETY PROFILE: A poison. Moderately toxic by ingestion. Mildly toxic by inhalation. An eye and skin irritant. Dangerously flammable. Moderately explosive. Reacts with water or steam to producc toxic and corrosive fumes. To fight fire, use $CO_2$, dry chemical. Readily decomposes when heated to emit toxic fumes of $Br^-$.

PROP: Colorless, volatile liquid. Mp: −119°, bp: 38.4°, lel: 6.7%, uel: 11.3%, flash p: < −4°F,d: 1.451 @ 20°/4°, autoign temp: 952°F, vap press: 400 mm @ 21°, vap d 3.76.

**EHA600** CAS: 106-35-4 $C_7H_{14}O$ HR: 2
**ETHYL BUTYL KETONE**
SYNS: n-BUTYL ETHYL KETONE ◇ 3-HEPTANONE

OSHA PEL: TWA 50 ppm
ACGIH TLV: TWA 50 ppm

SAFETY PROFILE: Moderately toxic by ingestion and inhalation. A skin and eye irritant. Combustible. To fight fire, use foam, $CO_2$, dry chemical.

PROP: Clear mobile liquid; fatty odor. Mp: −36.7°, bp: 148°, flash p: 115°F (OC), d: 0.8198 @ 20°/20°, vap d: 3.93. Misc with alc, ether, and water @ 149°.

**EHH000** CAS: 75-00-3 $C_2H_5Cl$ HR: 1
**ETHYL CHLORIDE**
DOT: 1037
SYNS: ETHER MURIATIC ◇ HYDROCHLORIC ETHER

OSHA PEL: TWA 1000 ppm
ACGIH TLV: TWA 1000 ppm
DFG MAK: 1000 ppm (2600 $mg/m^3$)
DOT Class: Flammable Liquid

SAFETY PROFILE: Mildly toxic by inhalation. An irritant to skin, eyes and mucous membranes. A very dangerous fire hazard. Severe explosion hazard. Reacts with water or steam to produce toxic and corrosive fumes. To fight fire, use carbon dioxide. When heated to decomposition it emits toxic fumes of phosgene and $Cl^-$.

PROP: Colorless liquid or gas; ether-like odor, burning taste. Bp: 12.3°, lel: 3.8%, uel: 15.4%, fp: −139°, flash p: −58°F (CC), d: 0.9214 @ 0°/4°, autoign temp: 966°F, vap press: 1000 mm @ 20° vap d: 2.22. Sol in water at 0.45; misc in alcohol and ether.

**EIK000** CAS: 78-78-4 $C_5H_{12}$ HR: 3
**ETHYLDIMETHYLMETHANE**
DOT: 1265
SYNS: ISOAMYLHYDRIDE ◇ 2-METHYLBUTANE

NIOSH REL: (Alkanes) TWA 350 mg/m$^3$
DOT Class: Flammable Liquid

SAFETY PROFILE: Probably mildly toxic and narcotic by inhalation. Flammable Liquid. A very dangerous fire and explosion hazard. To fight fire, use foam, $CO_2$, dry chemical.

PROP: Colorless liquid with pleasant odor. Bp: 27.8°, fp: −160.5°, flash p: < −60°F (CC), vap press: 595 mm @ 21.1°, vap d: 2.48, lel: 1.4%, uel: 7.6%.

**EIU800** CAS: 107-07-3 $C_2H_5ClO$ HR: 3
**ETHYLENE CHLOROHYDRIN**
DOT: 1135
SYNS: Δ-CHLOROETHANOL ◇ 2-CHLOROETHANOL (MAK) ◇ GLYCOL MONOCHLOROHYDRIN

OSHA PEL: (Transitional: TWA 5 ppm (skin)) CL 1 ppm (skin)
ACGIH TLV: CL 1 ppm (skin)
DFG MAK: 1 ppm (3 mg/m$^3$)
DOT Class: Poison B

SAFETY PROFILE: A poison by ingestion, inhalation, and skin contact. Moderately toxic to humans by inhalation. An experimental teratogen. A severe eye and mild skin irritant. Flammable. To fight fire, use alcohol foam, $CO_2$, dry chemical. Reacts with water or steam to produce toxic and corrosive fumes. When heated to decomposition it emits highly toxic fumes of $Cl^-$ and phosgene.

PROP: Colorless liquid; faint, ethereal odor. Mp: −69°, bp: 128.8°, flash p: 140°F (OC), d: 1.197 @ 20°/4°, autoign temp: 797°F, vap press: 10 mm @ 30.3°, vap d: 2.78, lel: 4.9%, uel: 15.9%.

**EIY500** CAS: 106-93-4 $C_2H_4Br_2$ HR: 3
**1,2-ETHYLENE DIBROMIDE**
DOT: 1605
SYNS: 1,2-DIBROMOETHANE (MAK) ◇ GLYCOL BROMIDE

OSHA PEL: TWA 20 ppm; CL 30 ppm; Pk 50 ppm/5M/8H
ACGIH TLV: Suspected Human Carcinogen

DFG TRK: 0.1 ppm; Animal Carcinogen, Suspected Human Carcinogen.
NIOSH REL: (EDB) 0.045 ppm; CL 1 mg/m$^3$/15M
DOT Class: ORM-A

SAFETY PROFILE: Human poison by ingestion. Experimental poison by skin contact. Moderately toxic by inhalation. An experimental carcinogen and teratogen. Experimental reproductive effects. A severe skin and eye irritant. When heated to decomposition it emits toxic fumes of $Br^-$.

PROP: Colorless, heavy liquid; sweet odor. Bp: 131.4°, fp: 9.3°, flash p: none, d: 2.172 @ 25°/25°, 2.1707 @ 25°/4°, vap press: 17.4 mm @ 30°, vap d: 6.48.

**EIY600** CAS: 107-06-2 $C_2H_4Cl_2$ HR: 3
**ETHYLENE DICHLORIDE**
DOT: 1184
SYNS: 1,2-DICHLOROETHANE ◇ GLYCOL DICHLORIDE

OSHA PEL: (Transitional: TWA 50 ppm; CL 100 ppm; PK 200 ppm/5M)) TWA 1 ppm; STEL 2 ppm
ACGIH TLV: TWA 10 ppm
DFG MAK: 20 ppm (80 mg/m$^3$); Suspected Carcinogen.
NIOSH REL: TWA 1 ppm; CL 2 ppm/15M
DOT Class: Flammable Liquid

SAFETY PROFILE: A human poison by ingestion. Moderately toxic by inhalation and skin contact. An experimental carcinogen and teratogen. An experimental transplacental carcinogen. A skin and severe eye irritant. Flammable liquid. When heated to decomposition it emits highly toxic fumes of $Cl^-$ and phosgene.

PROP: Colorless, clear liquid; pleasant odor, sweet taste. Bp: 83.5°, ULC: 60-70, lel: 6.2%, uel: 15.9%, fp: −35.7°, flash p: 56°F, d: 1.257 @ 20°/4°, autoign temp: 775°F, vap press: 100 mm @ 29.4°, vap d: 3.35, refr index: 1.445 @ 20°. Sol in alc, ether, acetone, carbon tetrachloride; sltly sol in water.

**EJC500** CAS: 107-21-1 $C_2H_6O_2$ HR: 3
**ETHYLENE GLYCOL**
SYNS: ETHYLENE ALCOHOL ◇ GLYCOL ALCOHOL

OSHA PEL: CL 50 ppm
ACGIH TLV: CL 50 ppm (vapor)

SAFETY PROFILE: Human poison by ingestion. Mildly toxic by skin contact. A suspected carcinogen. An experimental teratogen. A skin, eye, and mucous membrane irritant. Combustible. To fight fire, use alcohol foam, water, foam, $CO_2$, dry chemical.

PROP: Colorless, sweet-tasting, hygroscopic liquid. Bp: 197.5°, lel: 3.2%, fp: −13°, flash p: 232°F (CC), d: 1.113 @ 25°/25°, autoign temp: 752°F, vap d: 2.14, vap press: 0.05 mm @ 20°.

**EJG000** CAS: 628-96-6 $C_2H_4N_2O_6$ HR: 3
**ETHYLENE GLYCOL DINITRATE**
SYNS: EGDN ◇ ETHANEDIOL DINITRATE ◇ GLYCOL DINITRATE

OSHA PEL: (Transitional: CL 0.2 ppm (skin)) STEL 0.1 $mg/m^3$ (skin)
ACGIH TLV: TWA 0.05 ppm (skin)
DFG MAK: 0.05 ppm (0.3 $mg/m^3$)
NIOSH REL: (Nitroglycerin) CL 0.1 $mg/m^3$/20M
DOT Class: Forbidden.

SAFETY PROFILE: Moderately toxic by ingestion. When heated to decomposition it emits toxic fumes of $NO_x$.

PROP: Yellow liquid. Mp: −20°, bp: explodes @ 114°, d: 1.483 @ 8°, vap d: 5.25.

**EJH500** CAS: 109-86-4 $C_3H_8O_2$ HR: 3
**ETHYLENE GLYCOL METHYL ETHER**
DOT: 1188
SYNS: EGME ◇ ETHYLENE GLYCOL MONOMETHYL ETHER (MAK, DOT) ◇ 2-METHOXYETHANOL (ACGIH) ◇ METHYL CELLOSOLVE (OSHA, DOT)

OSHA PEL: TWA 25 ppm (skin)
ACGIH TLV: TWA 5 ppm (skin)
DFG MAK: 5 ppm (15 $mg/m^3$)
NIOSH REL: TWA (Glycol Ethers): Reduce to lowest level
DOT Class: Combustible Liquid

SAFETY PROFILE: Moderately toxic to humans by ingestion. Moderately toxic experimentally by inhalation, and skin contact. An experimental teratogen. A skin and eye irritant. Flammable or combustible. A moderate explosion hazard. To fight fire, use alcohol foam, $CO_2$, dry chemical.

PROP: Colorless liquid; mild, agreeable odor. Misc in water, alc, ether, benzene. Bp: 124.5°, fp: −86.5°, flash p: 115°F (OC), lel: 2.5%, uel: 14%, d: 0.9660 @ 20°/4°, autoign temp: 545°F, vap press: 6.2 mm @ 20°, vap d: 2.62.

**EJJ500** CAS: 110-49-6 $C_5H_{10}O_3$ HR: 3
**ETHYLENE GLYCOL MONOMETHYL ETHER ACETATE**
DOT: 1189
SYNS: ETHYLENE GLYCOL METHYL ETHER ACETATE ◇ 2-METHOXYETHYL ACETATE (ACGIH) ◇ METHYL CELLOSOLVE ACETATE (OSHA, DOT)

OSHA PEL: TWA 25 ppm (skin)
ACGIH TLV: TWA 5 ppm (skin)
DFG MAK: 5 ppm (25 $mg/m^3$)
DOT Class: Flammable or Combustible Liquid

SAFETY PROFILE: Moderately toxic by ingestion. Mildly toxic by inhalation and skin contact. Experimental reproductive effects. An eye irritant. Flammable. To fight fire, use $CO_2$, dry chemical.

PROP: Colorless liquid. Bp: 143°, fp: −70°, flash p: 111°F (CC), d: 1.005 @ 20°/20°, vap d: 4.07, lel: 1.7%, uel: 8.2%.

**EJN500** CAS: 75-21-8 $C_2H_4O$ HR: 3
**ETHYLENE OXIDE**
DOT: 1040
SYNS: 1,2-EPOXYETHANE ◇ OXIRANE

OSHA PEL: TWA 1 ppm
ACGIH TLV: TWA 1 ppm; Suspected Human Carcinogen.
DFG TRK: 3 ppm; Animal Carcinogen, Suspected Human Carcinogen.
NIOSH REL: (Oxirane) TWA 0.1 ppm; CL 5 ppm/10M/D
DOT Class: Flammable Liquid

SAFETY PROFILE: Poison by ingestion. Moderately toxic by inhalation. A suspected human carcinogen. An experimental carcinogen and teratogen. A skin and eye irritant. Highly flammable liquid or gas. Severe explosion hazard. Incompatible with many substances. To fight fire, use alcohol foam, $CO_2$, dry chemical.

PROP: Colorless gas at room temperature. Mp: −111.3°, bp: 10.7°, ULC: 100, lel: 3.0%, uel: 100%, flash p: −4°F, d: 0.8711 @ 20°/20°, autoign temp: 804°F, vap press: 1095 mm @ 20°, vap d: 1.52. Misc in water and alc; very sol in ether.

**EJR000** CAS: 151-56-4 $C_2H_5N$ HR: 3
**ETHYLENEIMINE**
DOT: 1185
SYN: AZIRIDINE

OSHA PEL: TWA 1 $mg/m^3$ (skin); Carcinogen
ACGIH TLV: TWA 0.5 ppm (skin)
DFG TRK: 0.5 ppm; Animal Carcinogen, Suspected Human Carcinogen.
DOT Class: Flammable Liquid

SAFETY PROFILE: Poison by ingestion, skin contact, and inhalation. An experimental carcinogen and teratogen. A skin, mucous membrane and severe eye irritant. An allergic sensitizer of skin. A very dangerous fire and explosion hazard. To fight fire, use alcohol foam, $CO_2$, dry chemical.

PROP: Oily, water-white liquid; pungent ammoniacal odor. Bp: 55-56°, fp: −71.5°, flash p: 12°F, d: 0.832 @ 20°/4°, autoign temp: 608°F, vap press: 160 mm @ 20°, vap d: 1.48, lel: 3.6%, uel: 46%.

**EJU000** CAS: 60-29-7 $C_4H_{10}O$ HR: 3
**ETHYL ETHER**
DOT: 1155
SYNS: DIETHYL ETHER (DOT) ◇ ETHER

OSHA PEL: (Transitional: TWA 400 ppm) TWA 400 ppm; STEL 500 ppm
ACGIH TLV: TWA 400 ppm; STEL 500 ppm
DFG MAK: 400 ppm (1200 $mg/m^3$)
DOT Class: Flammable Liquid

SAFETY PROFILE: Moderately toxic to humans by ingestion. Mildly toxic by inhalation. A severe eye and moderate skin irritant. A very dangerous fire and explosion hazard. To fight fire, use alcohol foam, $CO_2$, dry chemical.

PROP: A clear, volatile liquid; sweet, pungent odor. Sol in water; misc in alcohol and ether; sol in chloroform. Mp: −116.2°, bp: 34.6°, ULC: 100, lel: 1.85%, uel: 36%, flash p: −49°F, d: 0.7135 @ 20°/4°, autoign temp: 320°F, vap press: 442 mm @ 20°, vap d: 2.56.

**EKL000** CAS: 109-94-4 $C_3H_6O_2$ HR: 3
**ETHYL FORMATE**
DOT: 1190
SYNS: ETHYL FORMIC ESTER ◇ FORMIC ACID, ETHYL ESTER

OSHA PEL: TWA 100 ppm
ACGIH TLV: TWA 100 ppm
DFG MAK: 100 ppm (300 mg/m$^3$)
DOT Class: Flammable Liquid

SAFETY PROFILE: Moderately toxic by ingestion. Mildly toxic by skin contact and inhalation. An experimental tumorigen. A powerful inhalation irritant in humans. A skin and eye irritant. Highly flammable liquid. A very dangerous fire and explosion hazard. To fight fire, use alcohol foam, spray, mist, dry chemical.

PROP: Colorless liquid; sharp, rum-like odor. Mp: −79°, bp: 54.3°, lel: 2.7%, uel: 13.5%, flash p: −4°F (CC), d: 0.9236 @ 20°/20°, refr index: 1.359, autoign temp: 851°F, vap press: 100 mm @ 5.4°, vap d: 2.55. Sol in fixed oils, propylene glycol, water (decomp); sltly sol in mineral oil; insol in glycerin @ 54°.

**ELO500** CAS: 16219-75-3 $C_9H_{12}$ HR: 2
**ETHYLIDENE NORBORNENE**
SYNS: 5-ETHYLIDENEBICYCLO(2.2.1)HEPT-2-ENE ◇ 5-ETHYLIDENE-2-NORBORNENE

OSHA PEL: CL 5 ppm
ACGIH TLV: CL 5 ppm

SAFETY PROFILE: Moderately toxic by ingestion. Mildly toxic by inhalation and skin contact. A skin irritant.

**EMB100** CAS: 75-08-1 $C_2H_6S$ HR: 2
**ETHYL MERCAPTAN**
DOT: 2363
SYNS: ETHANETHIOL ◇ ETHYL SULFHYDRATE ◇ ETHYL THIOALCOHOL

OSHA PEL: (Transitional: CL 10 ppm) TWA 0.5 ppm
ACGIH TLV: TWA 0.5 ppm
DFG MAK: 0.5 ppm (1 mg/m$^3$)
NIOSH REL: (n-Alkane Mono Thiols) CL 0.5 ppm/15M
DOT Class: Flammable Liquid

SAFETY PROFILE: Moderately toxic by ingestion and inhalation. A skin and eye irritant. A very dangerous fire hazard. Will react with water or steam to produce toxic and flammable vapors. To fight fire, use $CO_2$, dry chemical. When heated to decomposition or on contact with acid or acid fumes it emits highly toxic fumes of $SO_x$.

PROP: Colorless liquid, penetrating garlic-like odor. Mp: −147°, bp: 36.2°, lel: 2.8%, uel: 18.2%, d: 0.83907 @ 20°/4°, autoign temp: 570°F, vap d: 2.14. flash p: <−0.4°F.

**ENL000** CAS: 100-74-3 $C_6H_{13}NO$ HR: 3
**N-ETHYLMORPHOLINE**
SYN: 4-ETHYLMORPHOLINE

OSHA PEL: (Transitional: TWA 20 ppm (skin) TWA 5 ppm (skin)
ACGIH TLV: TWA 5 ppm (skin)

SAFETY PROFILE: Moderately toxic by ingestion. Mildly toxic by inhalation. A skin and severe eye irritant. A very dangerous fire hazard. To fight fire, use alcohol foam, foam, $CO_2$, dry chemical. When heated to decomposition it emits toxic fumes of $NO_x$.

PROP: Colorless liquid. Bp: 138°, flash p: 89.6°F (OC), d: 0.916 @ 20°/20°, vap d: 4.00.

**EPF550** CAS: 78-10-4 $C_8H_{20}O_4Si$ HR: 3
**ETHYL SILICATE**
DOT: 1292
SYNS: ETHYL ORTHOSILICATE ◇ TETRAETHYL ORTHOSILICATE (DOT) ◇ TETRAETHYL SILICATE (DOT)

OSHA PEL: (Transitional: TWA 100 ppm) TWA 10 ppm
ACGIH TLV: TWA 10 ppm
DFG MAK: 100 ppm (850 $mg/m^3$)
DOT Class: Flammable or Combustible Liquid

SAFETY PROFILE: Moderately toxic. A skin, mucous membrane, and severe eye irritant. Flammable.

PROP: Colorless, flammable liquid. Mp: −77°, bp: 165-166°. flash p: 125°F (52°C), d: (20/4) 0.933, n (25/D) 1.3818. Viscosity 0.6 cps. Practically insol in water with slow decomp. Miscible with alc.

**FAK000** CAS: 22224-92-6 $C_{13}H_{22}NO_3PS$ HR: 3
**FENAMIPHOS**
SYNS: 1-(METHYLETHYL)-ETHYL 3-METHYL-4-(METHYLTHIO)PHENYL PHOSPHORAMIDATE ◇ NEMACUR

OSHA PEL: TWA 0.1 $mg/m^3$ (skin)
ACGIH TLV: TWA 0.1 $mg/m^3$ (skin)

SAFETY PROFILE: Poison by ingestion, inhalation, and skin contact. When heated to decomposition it emits very toxic fumes of $NO_x$, $PO_x$, and $SO_x$.

**FAQ800** CAS: 115-90-2 $C_{11}H_{17}O_4PS_2$ HR: 3
**FENSULFOTHION**

SYNS: DASANIT ◇ O,O-DIETHYL-O-p-(METHYLSULFINYL)PHENYL THIOPHOSPHATE ◇ DMSP

OSHA PEL: TWA 0.1 mg/m$^3$
ACGIH TLV: TWA 0.1 mg/m$^3$

SAFETY PROFILE: A poison by ingestion and skin contact. When heated to decomposition it emits very toxic fumes of $SO_x$ and $PO_x$.

**FAQ999** CAS: 55-38-9 $C_{10}H_{15}O_3PS_2$ HR: 3
**FENTHION**

SYNS: O,O-DIMETHYL-p-4-(METHYLMERCAPTO)-3-METHYLPHENYL THIOPHOSPHATE ◇ DMTP ◇ MERCAPTOPHOS ◇ MPP

OSHA PEL: TWA 0.2 mg/m$^3$ (skin)
ACGIH TLV: TWA 0.2 mg/m$^3$ (skin)
DFG MAK: 0.2 mg/m$^3$

SAFETY PROFILE: A human poison. Poison experimentally by skin contact. Moderately toxic by inhalation. An experimental tumorigen and teratogen. When heated to decomposition it emits very toxic fumes of $PO_x$ and $SO_x$.

**FAS000** CAS: 14484-64-1 $C_9H_{18}N_3S_6$•Fe HR: 3
**FERBAM**

SYNS: CARBAMATE ◇ DIMETHYLCARBAMODITHIOIC ACID, IRON COMPLEX ◇ FERRIC DIMETHYLDITHIOCARBAMATE

OSHA PEL: (Transitional: TWA Total Dust: 15 mg/m$^3$; Respirable Fraction: 5 mg/m$^3$) TWA Total Dust: 10 mg/m$^3$; Respirable Fraction: 5 mg/m$^3$
ACGIH TLV: TWA 10 mg/m$^3$
DFG MAK: 15 mg/m$^3$

SAFETY PROFILE: Moderately toxic by ingestion. An experimental carcinogen and teratogen. Experimental reproductive effects. When heated to decomposition it emits very toxic fumes of $NO_x$ and $SO_x$.

PROP: Black solid, sltly sol in water. Mp: decomp 180°.

**FBC000** CAS: 102-54-5 $C_{10}H_{10}Fe$ HR: 3
**FERROCENE**

SYNS: BISCYCLOPENTADIENYLIRON ◇ DICYCLOPENTADIENYL IRON (OSHA, ACGIH)

OSHA PEL: (Transitional: TWA Total Dust: 15 mg/m$^3$; Respirable Fraction: 5 mg/m$^3$) TWA Total Dust: 10 mg/m$^3$; Respirable Fraction: 5 mg/m$^3$
ACGIH TLV: TWA 10 mg/m$^3$

SAFETY PROFILE: Moderately toxic by ingestion. An experimental tumorigen. A suspected carcinogen. Flammable.

PROP: Orange crystals; camphor odor. Insol in water; sol in alcohol and ether. Mp: 174°, sublimes @ >100°, volatile in steam.

**FBP000** CAS: 12604-58-9 HR: 3
**FERROVANADIUM DUST**

OSHA PEL: (Transitional: TWA 1 $mg/m^3$) TWA 1 $mg/m^3$; STEL 3 $mg/m^3$
ACGIH TLV: TWA 1 $mg/m^3$; STEL 3 $mg/m^3$
DFG MAK: 1 $mg/m^3$
NIOSH REL: (Vanadium) TWA 1.0 $mg(V)/m^3$

SAFETY PROFILE: Can cause pulmonary damage. Combustible.

PROP: A gray to black dust.

**FBQ000** HR: 2
**FIBROUS GLASS**
SYNS: FIBERGLASS ◇ GLASS FIBERS

OSHA PEL: TWA 15 $mg/m^3$ (total dust); 5 $mg/m^3$ (nuisance dust)
ACGIH TLV: TWA 10 $mg/m^3$ (dust)
NIOSH REL: 3000000 $fb/m^3$

SAFETY PROFILE: An experimental carcinogen. Causes mechanical irritation of the skin and, less frequently, irritation of the eyes, nose, and throat.

**FDR000** CAS: 53-96-3 $C_{15}H_{13}NO$ HR: 3
**N-FLUOREN-2-YL ACETAMIDE**
SYNS: 2-AAF ◇ 2-ACETAMINOFLUORENE ◇ 2-ACETYLAMINOFLUORENE (OSHA) ◇ 2-FLUORENYLACETAMIDE

OSHA PEL: Carcinogen

SAFETY PROFILE: Moderately toxic by ingestion. An experimental carcinogen and teratogen. Experimental reproductive effects. When heated to decomposition it emits toxic fumes of $NO_x$.

**FEX875** CAS: 16984-48-8 F HR: 3
**FLUORIDE**
SYNS: FLUORIDE ION ◇ FLUORIDE ION(1-)

OSHA PEL: TWA 2.5 $mg(F)/m^3$
ACGIH TLV: TWA 2.5 $mg(F)/m^3$
DFG MAK: 2.5 $mg/m^3$
NIOSH REL: TWA 2.5 $mg(F)/m^3$

SAFETY PROFILE: Human poison by ingestion. When heated to decomposition it emits toxic fumes of $F^-$.

**FEY000** HR: D
**FLUORIDES**

OSHA PEL: TWA 2.5 mg(F)/$m^3$
ACGIH TLV: TWA 2.5 mg(F)/$m^3$
DFG MAK: 2.5 mg/$m^3$; BAT: 7 mg/kg creatinine in urine at end of exposure; 4 mg/kg creatinine in urine about 16 hours after end of exposure.
NIOSH REL: TWA 2.5 mg(F)/$m^3$

SAFETY PROFILE: Inorganic fluorides are generally highly irritating and toxic. Irritants to the eyes, skin, and mucous membranes. When heated to decomposition, or on contact with acid or acid fumes, they emit highly toxic fumes of $F^-$. Organic fluorides are generally less toxic than other halogenated hydrocarbons. When heated to decomposition they emit toxic fumes of $F^-$.

**FEZ000** CAS: 7782-41-4 $F_2$ HR: 3
**FLUORINE**

OSHA PEL: TWA 0.1 ppm
ACGIH TLV: TWA 1 ppm; STEL 2 ppm
DFG MAK: 0.1 ppm (0.2 mg/$m^3$)
DOT Class: Nonflammable Gas

SAFETY PROFILE: A poison gas. A skin, eyes, and mucous membrane irritant. A most powerful caustic irritant to tissue. A very dangerous fire and explosion hazard. A powerful oxidizer. Reacts violently with many materials. Reacts with water or steam to produce heat and toxic and corrosive fumes.

PROP: Pale yellow gas. Mp: −218°, bp: −187°, d: 1.14 @ −200°, 1.108 @ −188°, vap d: 1.695.

**FMU045** CAS: 944-22-9 $C_{10}H_{15}OPS_2$ HR: 3
**FONOFOS**
DOT: 1045
SYNS: DIFONATE ◇ DYPHONATE ◇ O-ETHYL-S-PHENYL ETHYLDITHIOPHOSPHONATE

OSHA PEL: TWA 0.1 mg/$m^3$ (skin)
ACGIH TLV: TWA 0.1 mg/$m^3$ (skin)

SAFETY PROFILE: Poison by ingestion and skin contact. When heated to decomposition it emits very toxic fumes of $PO_x$ and $SO_x$.

**FMV000** CAS: 50-00-0 $CH_2O$ HR: 3
**FORMALDEHYDE**
DOT: 1198/2209
SYNS: FORMALIN (DOT) ◇ FORMIC ALDEHYDE ◇ METHANAL ◇ METHYL ALDEHYDE ◇ PARAFORM

OSHA PEL: TWA 1 ppm; STEL 2 ppm (For certain industries: TWA 3 ppm; CL 5 ppm; Pk 10 ppm/30M

ACGIH TLV: TWA 1 ppm; Suspected Human Carcinogen (Proposed: CL 0.3 ppm; Suspected Human Carcinogen)
DFG MAK: 0.5 ppm (0.6 mg/m$^3$); Suspected Carcinogen.
NIOSH REL: (Formaldehyde) Limit to lowest feasible level.
DOT Class: Combustible Liquid

SAFETY PROFILE: Human poison by ingestion. Experimental poison by skin contact and inhalation. A suspected human carcinogen. An experimental carcinogen and teratogen. A human skin and eye irritant. Combustible liquid. A moderate explosion hazard. To fight fire stop flow of gas (for pure form); alcohol foam for 37 % methanol-free form.

PROP: Clear, water-white, very sltly acid gas or liquid; pungent odor. Pure formaldehyde is not available commercially because of its tendency to polymerize. It is sold as aqueous solutions containing from 37% to 50% formaldehyde by weight and varying amounts of methanol. Some alcoholic solns are used industrially and the physical properties and hazards may be greatly influenced by the solvent. Lel: 7.0%, uel: 73.0%, autoign temp: 806°F, d: 1.0, bp: −3°F, flash p: (37% methanol-free): 185°F, flash p: (15% methanol-free): 122°F.

**FMY000** CAS: 75-12-7 $CH_3NO$ HR: 3
**FORMAMIDE**
SYNS: CARBAMALDEHYDE ◇ METHANAMIDE

OSHA PEL: TWA 20 ppm; STEL 30 ppm
ACGIH TLV: TWA 10 ppm (skin)

SAFETY PROFILE: Poison by skin contact. Moderately toxic by ingestion. An irritant to skin, eyes, and mucous membranes. An experimental teratogen. Combustible. When heated to decomposition emits toxic fumes of $NO_x$.

PROP: Colorless, hygroscopic and oily liquid. Mp: 2.5, fp: 2.6°, vap press: 29.7 mm @ 129.4°, flash p: 310°F (COC), bp: 210° decomp, d: 1.134 @ 20°/40°; 1.1292 @ 25°/4°. Misc in water and alc; very sltly sol in ether.

**FNA000** CAS: 64-18-6 $CH_2O_2$ HR: 3
**FORMIC ACID**
DOT: 1779
SYNS: AMINIC ACID ◇ METHANOIC ACID

OSHA PEL: TWA 5 ppm
ACGIH TLV: TWA 5 ppm (Proposed: 5 ppm; STEL 10 ppm)
DFG MAK: 5 ppm (9 mg/m$^3$)
DOT Class: Corrosive Material

SAFETY PROFILE: Moderately toxic by ingestion. Mildly toxic by inhalation. Corrosive. A skin and severe eye irritant. Flammable liquid. To fight fire, use $CO_2$, dry chemical, alcohol foam.

PROP: Colorless, fuming liquid; pungent, penetrating odor. Bp: 100.8°, fp: 8.2°, flash p: 156°F (OC), d: 1.2267 @ 15°/4°, 1.220 @ 20°/4°, autoign temp: 1114°F, vap press: 40 mm @ 24.0°, vap d: 1.59, flash p:(90% soln): 122°F, autoign temp (90% soln): 813°F, lel (90% soln) = 18%, uel (90% soln) = 57%. Misc in water, alc, glycerin, and ether.

**FOO000** CAS: 76-13-1 $C_2Cl_3F_3$ HR: 1
**FREON 113**
SYNS: FLUOROCARBON 113 ◇ FREON 113TR-T ◇ 1,1,2-TRICHLORO-1,2,2-TRIFLUOROETHANE (OSHA, ACGIH, MAK)

OSHA PEL: (Transitional: TWA 1000 ppm) TWA 1000 ppm; STEL 1250 ppm
ACGIH TLV: TWA 1000 ppm; STEL 1250 ppm
DFG MAK: 1000 ppm (7600 mg/m$^3$)

SAFETY PROFILE: Mildly toxic by ingestion and inhalation. A skin irritant. Combustible.

PROP: Colorless gas. Mp: 13.2°, bp: 45.8°, d: 1.5702, autoign temp: 1256°F.

**FOO509** CAS: 76-14-2 $C_2Cl_2F_4$ HR: 1
**FREON 114**
SYNS: 1,2-DICHLORO-1,1,2,2-TETRAFLUOROETHANE (MAK) ◇ DICHLOROTETRAFLUOROETHANE (OSHA, ACGIH) ◇ FLUOROCARBON 114

OSHA PEL: TWA 1000 ppm
ACGIH TLV: TWA 1000 ppm
DFG MAK: 1000 ppm (7000 mg/m$^3$)

SAFETY PROFILE: An asphyxiant.

PROP: Colorless, practically odorless, noncorrosive, nonirritating, nonflammable gas. Faint, ether-like odor in high concentrations. D: 1.5312, mp: −94°, bp: 4.1°, n (0/D) 1.3092. Insol in water; sol in alc, and ether.

**FPQ875** CAS: 98-01-1 $C_5H_4O_2$ HR: 3
**FURFURAL**
DOT: 1199
SYNS: 2-FURALDEHYDE ◇ 2-FURANCARBOXALDEHYDE ◇ PYROMUCIC ALDEHYDE

OSHA PEL: (Transitional: TWA 5 mg/m$^3$ (skin)) TWA 2 ppm (skin)
ACGIH TLV: TWA 2 ppm (skin)
DFG MAK: 5 ppm (20 mg/m$^3$)
DOT Class: Combustible Liquid

SAFETY PROFILE: Poison by ingestion. Moderately toxic by inhalation and skin contact. A skin and eye irritant. Combustible. Moderate explosion hazard. To fight fire, use alcohol foam, $CO_2$, dry chemical.

PROP: Colorless-yellowish liquid; almond-like odor. Bp: 161.7° @ 764 mm, lel: 2.1%, uel: 19.3%, flash p: 140°F (CC), d: 1.154-1.158, refr index: 1.522-1.528, autoign temp: 600°F, vap d: 3.31. Sol in water; misc with alc.

**FPU000** CAS: 98-00-0 $C_5H_6O_2$ HR: 3
**FURFURYL ALCOHOL**
DOT: 2874
SYNS: 2-FURANCARBINOL ◇ 2-FURYLCARBINOL ◇ α-FURYLCARBINOL

OSHA PEL: (Transitional: TWA 50 ppm) TWA 10 ppm; STEL 15 ppm (skin)
ACGIH TLV: TWA 10 ppm; STEL 15 ppm (skin)
DFG MAK: 50 ppm (200 $mg/m^3$)
NIOSH REL: TWA 200 $mg/m^3$
DOT Class: Poison B

SAFETY PROFILE: Poison by ingestion and skin contact. Moderately toxic by inhalation. An eye irritant. Flammable. Moderate explosion hazard. To fight fire, use alcohol foam, $CO_2$, dry chemical.

PROP: Clear, colorless, mobile liquid. Mp: −31°, lel: 1.8%, uel: 16.3%, both between 72-122°, bp: 171° @ 750 mm, flash p: 167°F (OC), d: 1.129 @ 20°/4°, autoign temp: 915°F, vap press: 1 mm @ 31.8°, vap d: 3.37.

**GBY000** CAS: 8006-61-9 HR: 3
**GASOLINE**
DOT: 1203/1257
SYNS: MOTOR FUEL (DOT) ◇ PETROL (DOT)

OSHA PEL: TWA 300 ppm; STEL 500 ppm
ACGIH TLV: TWA 300 ppm; STEL 500 ppm
DOT Class: Flammable Liquid

SAFETY PROFILE: Mildly toxic by inhalation. A suspected carcinogen. Pulmonary aspiration can cause severe pneumonitis. A human eye irritant. A very dangerous fire and explosion hazard. To fight fire use foam, $CO_2$, dry chemical.

PROP: Clear, aromatic, volatile liquid; a mixture of aliphatic HC. Flash p: −50°F, d: <1.0, vap d: 3.0-4.0, ULC: 95-100, lel: 1.3%, uel: 6.0%, autoign temp: 536-853°F, bp: Initially 39°, after 10% distilled = 60°, after 50% = 110°, after 90% = 170°, final bp: 204°. Insol in water; freely sol in absolute alc, ether, chloroform, and benzene.

**GEI100** CAS: 7782-65-2 $GeH_4$ HR: 3
**GERMANIUM TETRAHYDRIDE**
DOT: 2192
SYNS: GERMANE (DOT) ◇ GERMANIUM HYDRIDE

ACGIH TLV: TWA 0.2 ppm
DOT Class: Poison A

SAFETY PROFILE: Poison by inhalation. Ignites spontaneously in air.

PROP: Colorless gas. Mp: −165°, bp: −90°, d: 1.523 @ −142°/4°. Sltly sol in hot HCl; decomp in nitric acid.

**GFQ000** CAS: 111-30-8 $C_5H_8O_2$ HR: 3
**GLUTARALDEHYDE**
SYN: 1,5-PENTANEDIONE

OSHA PEL: CL 0.2 ppm
ACGIH TLV: CL 0.2 ppm
DFG MAK: 0.2 ppm (0.8 $mg/m^3$)

SAFETY PROFILE: Poison by ingestion. Moderately toxic by inhalation and skin contact. An experimental teratogen. A severe eye and human skin irritant.

**GGA000** CAS: 56-81-5 $C_3H_8O_3$ HR: 2
**GLYCERIN**
SYNS: GLYCERINE ◇ GLYCEROL ◇ 1,2,3-PROPANETRIOL

OSHA PEL: (Transitional: TWA Total Mist: 15 $mg/m^3$; Respirable Fraction: 5 $mg/m^3$) TWA Total Mist: 10 $mg/m^3$; Respirable Fraction: 5 $mg/m^3$
ACGIH TLV: TWA 10 $mg/m^3$ (mist)

SAFETY PROFILE: Mildly toxic by ingestion. Experimental reproductive effects. A skin and eye irritant. It is a nuisance particulate and inhalation irritant in the form of mist. Combustible liquid. To fight fire, use alcohol foam, $CO_2$, dry chemical.

PROP: Colorless or pale yellow liquid; odorless, syrupy, sweet and warm taste. Mp: 17.9 (solidifies at a much lower temp), bp: 290°, flash p: 320°F, d: 1.260 @ 20°/4°, autoign temp: 698°F. Misc with water and alc; insol in chloroform, ether, and oils.

**GGW500** CAS: 556-52-5 $C_3H_6O_2$ HR: 3
**GLYCIDOL**
SYNS: 2,3-EPOXYPROPANOL ◇ GLYCIDYL ALCOHOL

OSHA PEL: (Transitional: TWA 50 ppm) TWA 25 ppm
ACGIH TLV: TWA 25 ppm
DFG MAK: 50 ppm (150 $mg/m^3$)

SAFETY PROFILE: Moderately toxic by ingestion, inhalation and skin contact. An experimental teratogen. A suspected carcinogen. A skin irritant.

PROP: Colorless liquid; entirely sol in water, alc, and ether. D: 1.165 @ 0°/4°, bp: 167° (decomp).

**HAC000** CAS: 7440-58-6 Hf HR: 3
**HAFNIUM**

OSHA PEL: TWA 0.5 $mg/m^3$
ACGIH TLV: TWA 0.5 $mg/m^3$
DFG MAK: 0.5 $mg/m^3$
DOT Class: Flammable Solid

SAFETY PROFILE: A poison. Many hafnium compounds are poisons. Dangerous fire hazard. The powder may self-explode.

PROP: A silvery, ductile, lustrous metal. Mp: 2227°, bp: 4602°, d: 13.31 @ 20°.

**HAG500** CAS: 151-67-7 $C_2HBrClF_3$ HR: 3
**HALOTHANE**
DOT: 1326/2545
SYNS: FLUOROTANE ◇ 1,1,1-TRIFLUORO-2-BROMO-2-CHLOROETHANE

ACGIH TLV: TWA 50 ppm
DFG MAK: 5 ppm (40 mg/m$^3$); BAT: 250 μg/dL of trifluoroacetic acid in blood at end of shift.
NIOSH REL: (Waste Anesthetic Gases and Vapors) CL 2 ppm/1H

SAFETY PROFILE: A human poison by ingestion. An experimental teratogen. Other experimental reproductive effects. A severe eye irritant. When heated to decomposition it emits very toxic fumes of $F^-$, $Cl^-$ and $Br^-$.

PROP: Nonflammable, highly volatile liquid; characteristic sweetish, not unpleasant odor. D: 1.871 @ 20°/4°, bp: 50.2°, 20° @ 243 mm. Sensitive to light, misc with petr ether, other fat solvents; sltly sol in water.

**HAR000** CAS: 76-44-8 $C_{10}H_5Cl_7$ HR: 3
**HEPTACHLOR**
DOT: 2761
SYN: 3-CHLOROCHLORDENE

OSHA PEL: TWA 0.5 Mg/m$^3$ (skin)
ACGIH TLV: TWA 0.5 mg/m$^3$ (skin)
DFG MAK: 0.5 mg/m$^3$, Suspected Carcinogen.
DOT Class: ORM-E

SAFETY PROFILE: A poison by ingestion and skin contact. An experimental carcinogen. When heated to decomposition it emits toxic fumes of $Cl^-$. NOTE: The EPA has canceled registration of pesticides containing heptachlor with the exception of its use by subsurface ground insertion external to the dwelling for termite control.

PROP: Crystals. Mp: 96°. Nearly insol in water; sol in organic solvents.

**HBC500** CAS: 142-82-5 $C_7H_{16}$ HR: 3
**HEPTANE**
DOT: 1206
SYNS: DIPROPYL METHANE ◇ HEPTYL HYDRIDE

OSHA PEL: (Transitional: TWA 500 ppm) TWA 400 ppm; STEL 500 ppm
ACGIH TLV: TWA 400 ppm; STEL 500 ppm
DFG MAK: 500 ppm (2000 mg/m$^3$)
NIOSH REL: TWA (Alkanes) 350 mg/m$^3$
DOT Class: Flammable Liquid

SAFETY PROFILE: Narcotic in high concentrations. Flammable. Moderately explosive. To fight fire, use foam, $CO_2$, dry chemical.

PROP: Colorless liquid. Bp: 98.52, lel: 1.05%, uel: 6.7%, fp: −90.5°, flash p: 25°F (CC), d: 0.684 @ 20°/4°, autoign temp: 433.4° F, vap press: 40 mm @ 22.3°, vap d: 3.45. Sltly sol in alc; misc in ether and chloroform; insol in water.

**HBD500** CAS: 1639-09-4 $C_7H_{16}S$ HR: 3
**1-HEPTANETHIOL**
SYN: HEPTYL MERCAPTAN

NIOSH REL: (n-Alkane Mono Thiols) CL 0.5 ppm/15M

SAFETY PROFILE: A poison by intraperitoneal route. When heated to decomposition it emits very toxic fumes of $SO_x$.

**HCB000** CAS: 13007-92-6 $C_6CrO_6$ HR: 3
**HEXACARBONYLCHROMIUM**
SYN: CHROMIUM CARBONYL (MAK)

DFG MAK: Suspected Carcinogen.

SAFETY PROFILE: Explodes at 210°C.

**HCC500** CAS: 118-74-1 $C_6Cl_6$ HR: 3
**HEXACHLOROBENZENE**
DOT: 2729
SYNS: AMATIN ◇ HCB ◇ PENTACHLOROPHENYL CHLORIDE

DFG MAK: BAT: 15 μg/dL in plasma/serum.
DOT Class: IMO: Poison B

SAFETY PROFILE: A human poison. An experimental carcinogen and teratogen. A suspected human carcinogen. To fight fire, use $CO_2$, dry chemical. When heated to decomposition it emits highly toxic fumes of $Cl^-$.

PROP: Monoclinic prisms. Mp: 231°, bp: 323-326°, flash p: 468°F, vap press: 1 mm @ 114.4°, vap d: 9.8. D: 2.44. Insol in water; sol in benzene, hot ether and chloroform; very sltly sol in hot alcohol.

**HCD250** CAS: 87-68-3 $C_4Cl_6$ HR: 3
**HEXACHLORBUTADIENE**
DOT: 2279
SYNS: HCBD ◇ HEXACHLORO-1,3-BUTADIENE (MAK) ◇ PERCHLOROBUTADIENE

OSHA PEL: TWA 0.02 ppm
ACGIH TLV: TWA 0.02 ppm (skin); Suspected Human Carcinogen.
DFG MAK: Suspected Carcinogen.
DOT Class: Poison B

SAFETY PROFILE: Poison by ingestion. Moderately toxic by inhalation and skin contact. A skin and eye irritant. An experimental carcinogen.

Combustible. To fight fire, use dry chemical, $CO_2$, alcohol foam, water spray, fog, mist. When heated to decomposition it emits very toxic fumes of $Cl^-$.

PROP: Autoign temp: 1130°F, vap d: 8.99.

**HCE500** CAS: 77-47-4 $C_5Cl_6$ HR: 3
**HEXACHLOROCYCLOPENTADIENE**
DOT: 2646
SYNS: HCCPD ◇ PCL

OSHA PEL: TWA 0.01 ppm
ACGIH TLV: TWA 0.01 ppm
DOT Class: Corrosive Material

SAFETY PROFILE: A poison by inhalation and ingestion. Moderately toxic by skin contact. Experimental teratogenic effects. Corrosive. A severe skin and eye irritant. When heated to decomposition it emits toxic fumes of $Cl^-$.

PROP: Yellow- to amber-colored liquid, pungent odor. Mp: 9.9°, bp: 239°, fp: −2°, flash p: none (OC), d: 1.715 @ 15.5°/15.5°, vap d: 9.42.

**HCI000** CAS: 67-72-1 $C_2Cl_6$ HR: 3
**HEXACHLOROETHANE**
DOT: 9037
SYNS: CARBON HEXACHLORIDE ◇ ETHYLENE HEXACHLORIDE

OSHA PEL: TWA 1 ppm (skin)
ACGIH TLV: TWA 1 ppm
DFG MAK: 1 ppm (10 $mg/m^3$)
NIOSH REL: (Hexachloroethane) Reduce to lowest level
DOT Class: ORM-A

SAFETY PROFILE: Mildly toxic by ingestion. An experimental carcinogen. Experimental reproductive effects. Dehalogenation of this material by reaction with alkalies, metals, etc., will produce spontaneous explosive chloroacetylenes. When heated to decomposition it emits highly toxic fumes of $Cl^-$ and phosgene.

PROP: Rhombic, triclinic or cubic crystals, colorless, camphor-like odor. Mp: 186.6° (sublimes), d: 2.091, vap press: 1 mm @ 32.7°, bp: 186.8° (triple point). Sol in alc, benzene, chloroform, ether, and oils; insol in water.

**HCK500** CAS: 1335-87-1 $C_{10}H_2Cl_6$ HR: 3
**HEXACHLORONAPHTHALENE**

OSHA PEL: TWA 0.2 $mg/m^3$ (skin)
ACGIH TLV: TWA 0.2 $mg/m^3$ (skin)

SAFETY PROFILE: A poison by ingestion, skin contact, and inhalation. Causes severe acne-form eruptions and toxic narcosis of the liver. Absorbed by skin. When heated to decomposition it emits toxic fumes of $Cl^-$.

PROP: White solid.

**HCZ000** CAS: 684-16-2 $C_3F_6O$ HR: 3
**HEXAFLUOROACETONE**
DOT: 2420

OSHA PEL: TWA 0.1 ppm (skin)
ACGIH TLV: TWA 0.1 ppm (skin)
DOT Class: IMO: Poison A

SAFETY PROFILE: A poison by ingestion and possibly by skin contact. Moderately toxic by inhalation. A poisonous irritant to the skin, eyes and mucous membranes. An experimental teratogen. When heated to decomposition it emits toxic fumes of $F^-$.

PROP: A colorless, nonflammable solvent liquid. D: 1.65 @ 25°.

**HEK000** CAS: 680-31-9 $C_6H_{18}N_3OP$ HR: 3
**HEXAMETHYL PHOSPHORAMIDE**
SYNS: HEXAMETHYLPHOSPHORIC ACID TRIAMIDE (MAK) ◇ HMPA ◇ HMPT

ACGIH TLV: Suspected Human Carcinogen.
DFG MAK: Animal Carcinogen, Suspected Human Carcinogen.

SAFETY PROFILE: Moderately toxic by ingestion and skin contact. An experimental carcinogen. Experimental reproductive effects. When heated to decomposition it emits very toxic fumes of phosphine, $PO_x$ and $NO_x$.

PROP: Clear, colorless, mobile liquid, spicy odor. Bp: 233°, fp: 6°, d: 1.024 @ 25°/25°, vap d: 6.18.

**HEN000** CAS: 110-54-3 $C_6H_{14}$ HR: 3
**n-HEXANE**
DOT: 1208
SYN: HEXANES (FCC)

OSHA PEL: (Transitional: TWA 500 mg/m$^3$) TWA 50 ppm
ACGIH TLV: TWA 50 ppm; BEI: 5 mg(2,5-hexanedione)/L in urine at end of shift; 40 ppm n-hexane in end-exhaled air during shift.
DFG MAK: 50 ppm (180 mg/m$^3$)
NIOSH REL: TWA (Alkanes) 350 mg/m$^3$
DOT Class: Flammable Liquid

SAFETY PROFILE: Slightly toxic by ingestion and inhalation. Experimental teratogenic and reproductive effects. An eye irritant. Can cause a motor neuropathy in exposed workers. Flammable liquid. A very dangerous fire and explosion hazard. To fight fire, use $CO_2$, dry chemical.

PROP: Colorless clear liquid; faint odor. Bp: 69°, ULC: 90-95, lel: 1.2%, uel: 7.5%, fp: −95.6°, flash p: −9.4°F, d: 0.6603 @ 20°/4°, autoign temp: 437°F, vap press: 100 mm @ 15.8°, vap d: 2.97. Insol in water; misc in chloroform, ether, and alc. Very volatile liquid.

**HES000** CAS: 111-31-9 $C_6H_{14}S$ HR: 3
**1-HEXANETHIOL**
SYN: HEXYL MERCAPTAN

NIOSH REL: (n-Alkane Mono Thiols) CL 0.5 ppm/15M

SAFETY PROFILE: Moderately toxic by inhalation and ingestion. When heated to decomposition it emits very toxic fumes of $SO_x$.

**HEV000** CAS: 591-78-6 $C_6H_{12}O$ HR: 3
**2-HEXANONE**
SYNS: BUTYL METHYL KETONE ◇ MBK ◇ METHYL n-BUTYL KETONE (ACGIH)

OSHA PEL: (Transitional: TWA 100 ppm) TWA 5 ppm
ACGIH TLV: TWA 5 ppm
DFG MAK: 5 ppm (21 $mg/m^3$)
NIOSH REL: (Ketones) TWA 4 $mg/m^3$

SAFETY PROFILE: Moderately toxic by ingestion. Mildly toxic by inhalation and skin contact. Experimental teratogenic and reproductive effects. An eye irritant. Dangerous fire and explosion hazard. To fight fire, use alcohol foam, $CO_2$, dry chemical.

PROP: Clear liquid. Mp: −56.9°, bp: 127.2°, lel: 1.22%, uel: 8.0%, flash p: 95°F (OC), d: 0.830 @ 0°/4°, vap press: 10 mm @ 38.8°, vap d: 3.45, autoign temp: 991°F. Slightly sol in water; sol in alc and ether.

**HFG500** CAS: 108-10-1 $C_6H_{12}O$ HR: 3
**HEXONE**
DOT: 1245
SYNS: ISOBUTYL METHYL KETONE ◇ ISOPROPYLACETONE ◇ METHYL ISOBUTYL KETONE (ACGIH, DOT)

OSHA PEL: (Transitional: TWA 100 $mg/m^3$) TWA 50 ppm; STEL 75 ppm
ACGIH TLV: TWA 50 ppm; STEL 75 ppm
NIOSH REL: (Ketones) TWA 200 $mg/m^3$
DOT Class: Flammable Liquid; Label: Flammable Liquid.

SAFETY PROFILE: Moderately toxic by ingestion. Very irritating to the skin, eyes, and mucous membranes. Flammable liquid. To fight fire, use alcohol foam, $CO_2$, dry chemical.

PROP: Colorless mobile liquid; fruity, ethereal odor. Bp: 118°, lel: 1.4%, uel: 7.5%, flash p: 62.6°F, d: 0.796-0.799, fp: −80.2°, autoign temp: 858°F, vap press: 16 mm @ 20°. Misc with alc and ether; sol in alc.

**HFJ000** CAS: 108-84-9 $C_8H_{16}O_2$ HR: 1
**sec-HEXYL ACETATE**
DOT: 1233
SYNS: MAAC ◇ METHYL AMYL ACETATE (DOT) ◇ METHYLISOBUTYLCARBINOL ACETATE ◇ 4-METHYL-2-PENTYL ACETATE

OSHA PEL: TWA 50 ppm
ACGIH TLV: TWA 50 ppm
DFG MAK: 50 ppm (300 mg/$m^3$)
DOT Class: Flammable or Combustible Liquid

SAFETY PROFILE: Mildly toxic by ingestion, skin contact and inhalation. A skin and human eye irritant. Flammable. To fight fire, use alcohol foam, $CO_2$, dry chemical.

PROP: Clear liquid, pleasant odor. Bp: 146.3°, fp: −63.8°, flash p: 113°F (COC), d: 0.8598 @ 20°/20°, vap press: 3.8 mm @ 20°, vap d: 4.97.

**HFP875** CAS: 107-41-5 $C_6H_{14}O_2$ HR: 2

**HEXYLENE GLYCOL**

SYNS: 2,4-DIHYDROXY-2-METHYLPENTANE ◇ 2-METHYL-2,4-PENTANEDIOL

OSHA PEL: CL 25 ppm
ACGIH TLV: CL 25 ppm

SAFETY PROFILE: Moderately toxic by ingestion. Combustible. To fight fire, use foam, $CO_2$, dry chemicals.

PROP: Mild odor; colorless liquid. Bp: 197.1°, fp: −50°, flash p: 205°F (OC), d: 0.9234 @ 20°/20°, vap press: 0.05 mm @ 20°, d: 4. Sol in water.

**HGS000** CAS: 302-01-2 $H_4N_2$ HR: 3

**HYDRAZINE**

DOT: 2029/2030
SYN: DIAMIDE

OSHA PEL: (Transitional: TWA 1 ppm (skin)) TWA 0.1 ppm (skin)
ACGIH TLV: TWA 0.1 ppm (skin); Suspected Human Carcinogen; (Proposed: 0.01 ppm (skin); Suspected Human Carcinogen).
DFG TRK: 0.1 ppm; Animal Carcinogen, Suspected Human Carcinogen.
NIOSH REL: (Hydrazines) CL 0.04 mg/$m^3$/2H
DOT Class: Flammable Liquid

SAFETY PROFILE: A poison by ingestion and skin contact. Moderately toxic by inhalation. An experimental carcinogen and teratogen. Corrosive. May cause skin sensitization. Flammable liquid. Severe explosion hazard. When heated to decomposition it emits highly toxic fumes of $NO_x$ and $NH_3$.

PROP: Colorless, oily, fuming liquid or white crystals. Mp: 1.4°, bp: 113.5°, flash p: 100°F (OC), d: 1.1011 @ 15° (liquid), vap d: 1.1; lel: 4.7%, uel: 100%.

**HGU500** CAS: 7803-57-8 $H_4N_2 \cdot H_2O$ HR: 3

**HYDRAZINE HYDRATE**

SYN: HYDRAZINE MONOHYDRATE

NIOSH REL: (Hydrazines) CL 0.04 mg/$m^3$/2H

SAFETY PROFILE: A poison by ingestion. An experimental carcinogen. Experimental reproductive effects. A corrosive. When heated to decomposition it emits toxic fumes of $NO_x$.

PROP: Colorless fuming, refractive liquid; faint characteristic odor. Mp: −51.7°, bp: 118.5° @ 740 mm. D: 1.03 @ 21°. A strong base, very corrosive; attacks glass, rubber, cork. Very powerful reducing agent. Misc with water and alc.; isol in chloroform and ether.

**HGV000** CAS: 2644-70-4 $H_4N_2 \cdot ClH$ HR: 3
**HYDRAZINE HYDROCHLORIDE**

NIOSH REL: (Hydrazines) CL 0.04 $mg/m^3$/2H

SAFETY PROFILE: A poison by ingestion. When heated to decomposition it emits very toxic fumes of $Cl^-$ and $NO_x$.

**HGW500** CAS: 10034-93-2 $H_4N_2 \cdot H_2O_4S$ HR: 3
**HYDRAZINE SULFATE (1:1)**
SYNS: HYDRAZINE HYDROGEN SULFATE ◇ HYDRAZONIUM SULFATE ◇ HS

NIOSH REL: (Hydrazines) CL 0.04 $mg/m^3$/2H

SAFETY PROFILE: A poison by ingestion. An experimental carcinogen. An eye irritant. When heated to decomposition it emits very toxic fumes of $SO_x$ and $NO_x$.

PROP: Colorless crystals. Very sol in hot water. D: 1.378, mp: 85°. Sol in water; insol in alcohol.

**HHG500** CAS: 7782-79-8 $HN_3$ HR: 3
**HYDRAZOIC ACID**
SYNS: AZOIMIDE ◇ DIAZOIMIDE ◇ HYDROGEN AZIDE ◇ TRIAZOIC ACID

DFG MAK: 0.1 ppm (0.27 $mg/m^3$)

SAFETY PROFILE: Mildly toxic by inhalation. A severe irritant to skin, eyes and mucous membranes. A dangerously sensitive explosive. When heated to decomposition it emits toxic fumes of $NO_x$.

PROP: Colorless liquid; intolerable pungent odor. Mp: −80°, bp: 37°, d: 1.09 @ 25°/4°. Very sol in water.

**HHJ000** CAS: 10035-10-6 BrH HR: 3
**HYDROBROMIC ACID**
DOT: 1048/1788
SYN: HYDROGEN BROMIDE (OSHA ACGIH, MAK, DOT)

OSHA PEL: (Transitional: TWA 3 ppm) CL 3 ppm
ACGIH TLV: CL 3 ppm
DFG MAK: 5 ppm (17 $mg/m^3$)
DOT Class: Corrosive Material

SAFETY PROFILE: Mildly toxic by inhalation. A corrosive. When heated to decomposition or in reaction with water or steam it emits toxic and corrosive fumes of $Br^-$ and HBr.

PROP: Colorless gas or pale yellow liquid. Mp: −87°, bp: −66.5°, d: 3.50 g/L @ 0°. Misc with water and alc. Keep protected from light.

**HHL000** CAS: 7647-01-0 ClH HR: 3
**HYDROCHLORIC ACID**
DOT: 1050/1789/2186
SYNS: HYDROGEN CHLORIDE (OSHA, ACGIH, MAK, DOT) ◇ MURIATIC ACID (DOT)

OSHA PEL: CL 5 ppm
ACGIH TLV: CL 5 ppm
DFG MAK: 5 ppm (7 $mg/m^3$)
DOT Class: Corrosive Material

SAFETY PROFILE: A human poison. A corrosive. Mildly toxic to humans by inhalation. An experimental teratogen. Nonflammable gas. When heated to decomposition it emits toxic fumes of $Cl^-$.

PROP: Colorless, fuming gas or colorless, fuming liquid; strongly corrosive with pungent odor. Mp: −114.3°, bp: −84.8°, d: 1.639 g/L (gas) @ 0°, 1.194 @ −26° (liquid), vap press: 4.0 atm @ 17.8°. Misc with water and alc.

**HHS000** CAS: 74-90-8 CHN HR: 3
**HYDROCYANIC ACID**
DOT: 1614/1051
SYNS: HCN ◇ HYDROGEN CYANIDE (OSHA, ACGIH) ◇ PRUSSIC ACID (DOT)

OSHA PEL: (Transitional: TWA 10 ppm (skin)); STEL 4.7 ppm (skin)
ACGIH TLV: CL 10 ppm (skin)
DFG MAK: 10 ppm (11 $mg/m^3$)
NIOSH REL: (Cyanide) CL 5 mg(CN)/$m^3$/10M
DOT Class: Poison A

SAFETY PROFILE: A deadly human poison. Very dangerous fire hazard. Severe explosion hazard. To fight fire, use $CO_2$, nonalkaline dry chemical, foam. When heated to decomposition or in reaction with water, steam, acid or acid fumes it produces highly toxic fumes of $CN^-$.

PROP: Odor of bitter almonds. Mp: −13.2°, bp: 25.7°, lel: 5.6%, uel: 40%, flash p: 0°F (CC), d: 0.6876 @ 20°/4°, autoign temp: 1000°F, vap press: 400 mm @ 9.8°, vap d: 0.932. Misc in water, alc and ether.

**HHU500** CAS: 7664-39-3 FH HR: 3
**HYDROFLUORIC ACID**
DOT: 1790/1052
SYN: HYDROGEN FLUORIDE (OSHA, ACGIH, MAK, DOT)

OSHA PEL: (Transitional: TWA 3 ppm (F)) TWA 3 ppm; STEL 6 ppm (F)
ACGIH TLV: CL 3 ppm (F)
DFG MAK: 3 ppm (2 $mg/m^3$); BAT 7.0 mg/g creatinine in urine at end of shift.
NIOSH REL: (HF) TWA 2.5 $mg(F)/m^3$; CL 5.0 $mg(F)/m^3$/15M
DOT Class: Corrosive Material

SAFETY PROFILE: A human poison by inhalation. A corrosive. Experimental teratogenic effects. Produces severe skin burns which are slow in healing. Explosive reaction with many substances. Reacts with water or steam to produce toxic and corrosive fumes. When heated to decomposition it emits highly corrosive fumes of $F^-$.

PROP: Clear, colorless, fuming, corrosive liquid or gas. Mp: −83.1°, bp: 19.54°, d: 0.901 g/L(gas); 0.699 @ 22° (liquid), vap press: 400 mm @ 2.5°.

**HHW800** CAS: 61788-32-7 HR: 3
**HYDROGENATED TERPHENYLS**

OSHA PEL: TWA 0.5 ppm
ACGIH TLV: TWA 0.5 ppm

SAFETY PROFILE: Contact with hot coolant can cause severe damage to lungs, skin, and eyes from burns. A suspected carcinogen. Inhalation has caused bronchopneumonia.

PROP: Complex mixtures of o-, m-, and p-terphenyls in various stages of hydrogenation. Five such stages exist for each of the three above isomers.

**HHX000** CAS: 7647-01-0 ClH HR: 3
**HYDROGEN CHLORIDE**

OSHA PEL: CL 5 ppm
ACGIH TLV: CL 5 ppm
DFG MAK: 5 ppm (7 $mg/m^3$)

SAFETY PROFILE: A highly corrosive irritant to the eyes, skin and mucous membranes. Mildly toxic by inhalation. Highly reactive acid gas.

PROP: Colorless, corrosive, nonflammable gas; pungent odor. Fumes in air. D: 1.639 @ −137.77°, bp: −154.37° @ 1.0 mm.

**HIB000** CAS: 7722-84-1 $H_2O_2$ HR: 3
**HYDROGEN PEROXIDE**
DOT: 2015
SYNS: DIHYDROGEN DIOXIDE ◇ HYDROGEN DIOXIDE

OSHA PEL: TWA 1 ppm
ACGIH TLV: TWA 1 ppm

DFG MAK: 1 ppm (1.4 mg/m$^3$)
DOT Class: Oxidizer

SAFETY PROFILE: Moderately toxic by inhalation, ingestion and skin contact. A corrosive irritant. Its pure solutions, vapors, and mists are very irritating to body tissue. An experimental tumorigen and suspected carcinogen. A very powerful oxidizer, particularly in the concentrated state. Severe explosion hazard when highly concentrated.

PROP: Colorless, heavy liquid, or, at low temp, a crystalline solid; bitter taste. D: 1.71 @ −20°, 1.46 @ 0°, vap press: 1 mm @ 15.3°, unstable. Mp: −0.43°, bp: 152°. Misc with water; sol in ether; insol in petroleum ether. Decomposed by many organic solvents.

**HIC000** CAS: 7783-07-5 $H_2Se$ HR: 3
**HYDROGEN SELENIDE**
DOT: 2202
SYN: SELENIUM HYDRIDE

OSHA PEL: TWA 0.05 ppm (Se)
ACGIH TLV: TWA 0.05 ppm (Se)
DFG MAK: 0.05 ppm (0.2 mg/m$^3$)
DOT Classification: Flammable Gas, Poison Gas.

SAFETY PROFILE: A deadly poison by inhalation. A very poisonous irritant to skin, eyes, and mucous membranes. An allergen. Dangerous fire hazard. Dangerous; forms explosive mixtures with air.

PROP: Colorless gas; disagreeable odor. Mp: −64°, bp: −41.4°, d: 3.614 g/L (gas), 2.12 @ −42° (liquid), vap press: 10 atm @ 23.4°. Sol in carbonyl chloride and carbon disulfide.

**HIC500** CAS: 7783-06-4 $H_2S$ HR: 3
**HYDROGEN SULFIDE**
DOT: 1053
SYNS: HYDROGEN SULFURIC ACID ◇ SULFURETED HYDROGEN ◇ SULFUR HYDRIDE

OSHA PEL: (Transitional: CL 20 ppm; Pk 50/10M) TWA 10 ppm; STEL 15 ppm
ACGIH TLV: TWA 10 ppm; STEL 15 ppm
DFG MAK: 10 ppm (15 mg/m$^3$)
NIOSH REL: (Hydrogen Sulfide) CL 15 mg/m$^3$/10M
DOT Class: Flammable Gas

SAFETY PROFILE: A human poison by inhalation. A severe irritant to eyes and mucous membranes. An asphyxiant. Very dangerous fire hazard. Moderately explosive. To fight fire, stop flow of gas.

PROP: Colorless, flammable gas; offensive odor. Mp: −85.5°, bp: −60.4°, lel: 4%, uel: 46%, autoign temp: 500°F, d: 1.539 g/L @ 0°, vap press: 20 atm @ 25.5°, vap d: 1.189.

**HIH000** CAS: 123-31-9 $C_6H_6O_2$ HR: 3
**HYDROQUINONE**
DOT: 2662
SYNS: p-BENZENEDIOL ◇ BENZOHYDROQUINONE ◇ p-DIHYDROXYBENZENE ◇ TECQUINOL

OSHA PEL: TWA 2 mg/m$^3$
ACGIH TLV: TWA 2 mg/m$^3$
DFG MAK: 2 mg/m$^3$
NIOSH REL: (Hydroquinone) CL 2.0 mg/m$^3$/15M
DOT Class: IMO: Poison B

SAFETY PROFILE: A human poison by ingestion. A suspected carcinogen. An active allergen and a strong skin irritant. A severe human skin irritant. Combustible. To fight fire, use water, $CO_2$, dry chemical.

PROP: Colorless, hexagonal prisms. Mp: 170.5°, bp: 286.2°, flash p: 329°F (CC), d: 1.358 @ 20°/4°, autoign temp: 960°F (CC), vap press: 1 mm @ 132.4°, vap d: 3.81. Very sol in alc and ether; sltly sol in benzene. Keep well closed and protected from light.

**HIM500** CAS: 107-16-4 $C_2H_3NO$ HR: 3
**HYDROXYACETONITRILE**
SYNS: CYANOMETHANOL ◇ FORMALDEHYDE CYANOHYDRIN ◇ GLYCOLONITRILE

NIOSH REL: (Nitriles) CL 5 mg/m$^3$/15M

SAFETY PROFILE: A poison by ingestion, skin contact, inhalation. An eye irritant. May undergo spontaneous and violent decomposition. When heated to decomposition it emits toxic fumes of $NO_x$ and $CN^-$.

**HNT600** CAS: 999-61-1 $C_6H_{10}O_3$ HR: 3
**2-HYDROXYPROPYL ACRYLATE**
SYNS: 1,2-PROPANEDIOL-1-ACRYLATE ◇ PROPYLENE GLYCOL MONOACRYLATE

OSHA PEL: TWA 0.5 ppm (skin)
ACGIH TLV: TWA 0.5 ppm (skin)

SAFETY PROFILE: Poison by ingestion and subcutaneous routes.

**IAQ000** CAS: 96-45-7 $C_3H_6N_2S$ HR: 3
**2-IMIDAZOLIDINETHIONE**
SYNS: ETHYLENE THIOUREA ◇ ETU

NIOSH REL: (ETU) Use encapsulated form; minimize exposure.

SAFETY PROFILE: Poison by ingestion. An experimental carcinogen. Experimental teratogenic and reproductive effects. An eye irritant. When heated to decomposition it emits very toxic fumes of $NO_x$ and $SO_x$.

PROP: White crystals. Water solubility: 9 g/100 mL @ 30°.

**IBA000** CAS: 2465-27-2 $C_{17}H_{21}N_3 \cdot ClH \cdot H_2O$ HR: 3
**4,4′-(IMIDOCARBONYL)BIS(N,N-DIMETHYLAMINE) MONOHYDROCHLORIDE**
SYNS: AURAMINE (MAK) ◇ AURAMINE YELLOW ◇ C.I. 41000

DFG MAK: Animal Carcinogen, Suspected Human Carcinogen.

SAFETY PROFILE: Poison by skin contact and ingestion. An experimental tumorigen. When heated to decomposition it emits very toxic fumes of $NO_x$ and HCl.

**IBB000** CAS: 492-80-8 $C_{17}H_{21}N_3$ HR: 3
**4,4′-(IMIDOCARBONYL)BIS(N,N-DIMETHYLANILINE)**
SYNS: AURAMINE (MAK) ◇ C.I. 41000B ◇ YELLOW PYOCTANINE

DFG MAK: Animal Carcinogen, Suspected Human Carcinogen.

SAFETY PROFILE: A human carcinogen and experimental carcinogen. When heated to decomposition it emits toxic fumes of $NO_x$.

PROP: Yellow needles. Mp: 136°. Insol in water.

**IBX000** CAS: 95-13-6 $C_9H_8$ HR: 3
**INDENE**
SYN: INDONAPHTHENE

OSHA PEL: TWA 10 ppm
ACGIH TLV: TWA 10 ppm

SAFETY PROFILE: Moderately toxic by ingestion and inhalation. Irritating to skin, eyes and mucous membranes.

PROP: Liquid from coal tars. Water-insol, but misc in organic solvents. D: 0.9968 @ 20°/4°, mp: −1.8°, bp: 181.6°.

**ICF000** CAS: 7440-74-6 In HR: 3
**INDIUM**

OSHA PEL: TWA 0.1 mg(In)/$m^3$
ACGIH TLV: TWA 0.1 mg(In)/$m^3$

SAFETY PROFILE: Teratogenic effects. Inhalation of indium compounds may cause damage to the respiratory system. Flammable in the form of dust. Incandesces.

PROP: Soft, silvery-white metal. Mp: 156.61°, bp: 2080°, d: 7.31 @ 20°.

**IDM000** CAS: 7553-56-2 $I_2$ HR: 3
**IODINE**

OSHA PEL: CL 0.1 ppm
ACGIH TLV: CL 0.1 ppm
DFG MAK: 0.1 ppm (1 mg/$m^3$)

SAFETY PROFILE: A human poison by ingestion. Moderately toxic by inhalation. Experimental reproductive effects. Reacts vigorously with reducing materials.

PROP: Rhombic, violet-black crystals, metallic luster. Mp: 113.5°, bp: 185.24°, d: 4.93 (Solid @ 25°), vap press: 1 mm @ 38.7°. Characteristic odor, sharp acrid taste, vap press (solid): 0.030 mm @ 0°, very sol in aq solns of HI and iodides.

**IEP000** CAS: 75-47-8 $CHI_3$ HR: 3
**IODOFORM**
SYN: TRIIODOMETHANE

OSHA PEL: TWA 0.6 ppm (skin)
ACGIH TLV: TWA 0.6 ppm

SAFETY PROFILE: A poison by ingestion. Moderately toxic by inhalation and skin contact. When heated to decomposition it emits toxic fumes of $I^-$.

PROP: Yellow powder or crystals, disagreeable odor. D: 4.1, mp: 120° (approx), bp: subl. Decomp at high temp evolving iodine, volatile with steam. Very sol in water, benzene, acetone; sltly sol in petroleum ether.

**IHD000** CAS: 1309-37-1 $Fe_2O_3$ HR: 3
**IRON OXIDE**
SYNS: C.I. 77491 ◇ FERRIC OXIDE ◇ IRON(III) OXIDE ◇ ROUGE

OSHA PEL: Dust and Fume: TWA 10 mg(Fe)/m$^3$; Rouge (Transitional: Total Dust: 15 mg/m$^3$; Respirable Fraction: 5 mg/m$^3$) TWA Total Dust: 10 mg/m$^3$; Respirable Fraction: 5 mg/m$^3$
ACGIH TLV: TWA 5 mg(Fe)/m$^3$ (vapor, dust); Rouge: 10 mg/m$^3$
DFG MAK: 6 mg/m$^3$

SAFETY PROFILE: A suspected human carcinogen. An experimental tumorigen. Catalyzes the potentially explosive polymerization of ethylene oxide.

**IHG500** CAS: 13463-40-6 $C_5FeO_5$ HR: 3
**IRON PENTACARBONYL**
SYNS: IRON CARBONYL ◇ PENTACARBONYLIRON

OSHA PEL: TWA 0.1 ppm (Fe); STEL 0.2 ppm
ACGIH TLV: TWA 0.1 ppm (Fe); STEL 0.2 ppm
DFG MAK: 0.1 ppm (0.8 mg/m$^3$)

SAFETY PROFILE: A poison by inhalation, skin contact, and ingestion. A very dangerous fire and moderate explosion hazard. Pyrophoric in air! Mixtures with nitrogen oxide explode above 50°C. To fight fire, use water, foam, $CO_2$, dry chemical.

PROP: Yellow to dark red viscous liquid. Mp: −25°, bp: 103.0°, flash p: 5°F, d: 1.453 @ 25°/4°, vap press: 40 mm @ 30.3°.

**IHO850** CAS: 123-92-2 $C_7H_{14}O_2$ HR: 3
**ISOAMYL ACETATE**
SYNS: BANANA OIL ◇ ISOAMYL ETHANOATE ◇ 3-METHYLBUTYL ACETATE

OSHA PEL: TWA 100 ppm
ACGIH TLV: TWA 100 ppm

SAFETY PROFILE: Mildly toxic by ingestion and inhalation. Highly flammable liquid. Moderately explosive in the form of vapor. To fight fire, use alcohol foam, $CO_2$, dry chemical.

PROP: Colorless liquid; banana-like odor. Bp: 142.0°, ULC: 55-60, lel: 1% @ 212°F, uel: 7.5%, flash p: 77°F, d: 0.876, refr index: 1.400, autoign temp: 680°F, vap d: 4.49. Misc in alc, ether, ethyl acetate, and fixed oils; sltly sol in water; insol in glycerin and propylene glycol.

**IHP000** CAS: 123-51-3 $C_5H_{12}O$ HR: 3
**ISOAMYL ALCOHOL**
DOT: 1105
SYNS: ISOBUTYLCARBINOL ◇ ISOPENTYL ALCOHOL ◇ 3-METHYL BUTANOL

OSHA PEL: TWA 100 ppm; STEL 125 ppm
ACGIH TLV: TWA 100 ppm; STEL 125 ppm
DFG MAK: 100 ppm (360 $mg/m^3$)
DOT Class: Flammable or Combustible Liquid

SAFETY PROFILE: Moderately toxic by ingestion and skin contact. An experimental carcinogen. A skin and human eye irritant. Flammable. To fight fire, use alcohol foam, $CO_2$, dry chemical.

PROP: Clear liquid; pungent, repulsive taste. Bp: 132°, ULC: 35-40, lel: 1.2%, uel: 9.0% @ 212°F, flash p: 109°F (CC), d: 0.813, autoign temp: 662°F, vap d: 3.04, mp: −117.2°. Sol in water @ 14°; misc in alc and ether.

**IHP010** CAS: 584-02-1 $C_5H_{12}O$ HR: 3
**ISOAMYL ALCOHOL**
DOT: 2706
SYNS: DIETHYLCARBINOL (DOT) ◇ 3-PENTANOL

OSHA PEL: TWA 100 ppm; STEL 125 ppm
ACGIH TLV: TWA 100 ppm; STEL 125 ppm
DFG MAK: 100 ppm (360 $mg/m^3$)
DOT Class: Flammable or Combustible Liquid

SAFETY PROFILE: Moderately toxic by ingestion and skin contact. A severe eye and mild skin irritant. Dangerous fire and explosion hazard.

PROP: Liquid; acetone-like odor. Bp: 115.6°, d: 0.815 @ 25°/4°, flash p: 66°F, lel: 1.2%, uel: 9%. Sol in alc and ether; sltly sol in water.

**IIJ000** CAS: 110-19-0 $C_6H_{12}O_2$ HR: 3
**ISOBUTYL ACETATE**
DOT: 1213/1123
SYNS: 2-METHYLPROPYL ACETATE ◇ β-METHYLPROPYL ETHANOATE

OSHA PEL: TWA 150 ppm
ACGIH TLV: TWA 150 ppm
DFG MAK: 200 ppm (950 mg/m$^3$)
DOT Class: Flammable Liquid

SAFETY PROFILE: Mildly toxic by ingestion and inhalation. A skin and eye irritant. Highly flammable liquid. To fight fire, use alcohol foam, $CO_2$, dry chemical.

PROP: Colorless, neutral liquid; fruit-like odor. Mp: −98.9°, bp: 118°, flash p: 64°F (CC) (18°), d: 0.8685 @ 15°, refr index: 1.389, vap press: 10 mm @ 12.8°, autoign temp: 793°F, vap d: 4.0, lel: 2.4%, uel: 10.5%. Very sol in alc, fixed oils, and propylene glycol; sltly sol in water.

**IIL000** CAS: 78-83-1 $C_4H_{10}O$ HR: 3

**ISOBUTYL ALCOHOL**

DOT: 1212

SYNS: 1-HYDROXYMETHYLPROPANE ◇ ISOBUTANOL (DOT) ◇ 2-METHYL PROPANOL

OSHA PEL: (Transitional: TWA 100 mg/m$^3$) TWA 50 ppm
ACGIH TLV: TWA 50 ppm
DFG MAK: 100 ppm (300 mg/m$^3$)
DOT Class: Flammable or Combustible Liquid

SAFETY PROFILE: Moderately toxic by ingestion and skin contact. Mildly toxic by inhalation. An experimental carcinogen and tumorigen. A severe skin and eye irritant. Dangerous fire hazard. Keep away from heat and open flame. To fight fire, use alcohol foam, $CO_2$, dry chemical.

PROP: Clear mobile liquid; sweet odor. Bp: 107.90°, flash p: 82°F, ULC: 40-45, lel: 1.2%, uel: 10.9% @ 212°F, fp: −108°, d: 0.800, autoign temp: 800°F, vap press: 10 mm @ 21.7°, vap d: 2.55. Sltly sol in water; misc with alc and ether.

**IIM000** CAS: 78-81-9 $C_4H_{11}N$ HR: 3

**ISOBUTYLAMINE**

DOT: 1214

SYNS: 1-AMINO-2-METHYLPROPANE ◇ MONOISOBUTYLAMINE

DFG MAK: 5 ppm (15 mg/m$^3$)
DOT Class: Flammable Liquid

SAFETY PROFILE: A poison by ingestion. A powerful irritant to skin, eyes, and mucous membranes. Skin contact can cause blistering. A very dangerous fire hazard. To fight fire, use dry chemical, foam, $CO_2$, alcohol foam. When heated to decomposition it emits toxic fumes of $NO_x$.

PROP: Colorless liquid. Mp: −85.5°, bp: 68.6°, flash p: 15°F, d: 0.731 @ 20°/20°, vap press: 100 mm @ 18.8°, autoign temp: 712°F, vap d: 2.5. Misc with water, alc, and ether.

**IJX000** CAS: 78-82-0 $C_4H_7N$ HR: 3
**ISOBUTYRONITRILE**
DOT: 2284
SYNS: ISOPROPYL CYANIDE ◇ 2-METHYLPROPIONITRILE

NIOSH REL: (Nitriles) TWA 22 $mg/m^3$
DOT Class: Flammable Liquid

SAFETY PROFILE: A poison by ingestion and skin contact. A skin irritant. A very dangerous fire hazard. When heated to decomposition it emits toxic fumes of $NO_x$ and $CN^-$.

PROP: Colorless liquid, sltly sol in water, very sol in alc and ether. D: 0.773 @ 20°/20°, bp: 107°, mp: −75°, flash p: 46.4°F.

**IKS600** CAS: 107-83-5 $C_6H_{14}$ HR: 3
**ISOHEXANE**
DOT: 2462/1208
SYN: 2-METHYLPENTANE

OSHA PEL: TWA 500 ppm; STEL 1000 ppm
ACGIH TLV: TWA 500 ppm; STEL 1000 ppm (hexane isomer)
NIOSH REL: (Alkanes) TWA 350 $mg/m^3$
DOT Class: Flammable Liquid

SAFETY PROFILE: A human eye irritant. A very dangerous fire hazard. Severe explosion hazard. To fight fire, use foam, $CO_2$, dry chemical.

PROP: Liquid. Bp: 54-60°, lel: 1.0%, uel: 7.0%, flash p: 20°F (CC), d: 0.669, vap d: 3.00, autoign temp: 583°F.

**ILL000** CAS: 26952-21-6 $C_8H_{18}O$ HR: 2
**ISOOCTYL ALCOHOL**
SYN: ISOOCTANOL

OSHA PEL: TWA 50 ppm (skin)
ACGIH TLV: TWA 50 ppm (skin)

SAFETY PROFILE: Moderately toxic by ingestion and skin contact. A skin and severe eye irritant.

**IMF400** CAS: 78-59-1 $C_9H_{14}O$ HR: 3
**ISOPHORONE**
SYNS: ISOACETOPHORONE ◇ 1,1,3-TRIMETHYL-3-CYCLOHEXENE-5-ONE

OSHA PEL: (Transitional: TWA 25 ppm) TWA 4 ppm
ACGIH TLV: CL 5 ppm
DFG MAK: 5 ppm (28 $mg/m^3$)
NIOSH REL: TWA (Ketones) 23 $mg/m^3$

SAFETY PROFILE: Moderately toxic by ingestion and skin contact. An experimental carcinogen and tumorigen. A skin and severe eye irritant. Flammable and explosive. To fight fire, use foam, $CO_2$, dry chemical.

PROP: Practically water-white liquid. Bp: 215.2°, flash p: 184°F (OC), d: 0.9229, autoign temp: 864°F, vap press: 1 mm @ 38.0°, vap d: 4.77, lel: 0.8%, uel: 3.8%.

**IMG000** CAS: 4098-71-9 $C_{12}H_{18}N_2O_2$ HR: 3
**ISOPHORONE DIISOCYANATE**
SYNS: 3-ISOCYANATOMETHYL-3,5,5-TRIMETHYLCYCLOHEXYLISOCYANATE ◇ ISOPHORONE DIAMINE DIISOCYANATE

OSHA PEL: TWA 0.005 ppm (skin)
ACGIH TLV: TWA 0.005 ppm (skin)
DFG MAK: 0.01 ppm (0.09 $mg/m^3$)

SAFETY PROFILE: Poison by inhalation. Moderately toxic by skin contact. When heated to decomposition it emits toxic fumes of $NO_x$.

**INA500** CAS: 109-59-1 $C_5H_{12}O_2$ HR: 2
**2-ISOPROPOXYETHANOL**
SYNS: ETHYLENE GLYCOL ISOPROPYL ETHER ◇ β-HYDROXYETHYL ISOPROPYL ETHER ◇ ISOPROPYL CELLOSOLVE ◇ ISOPROPYL GLYCOL

OSHA PEL: TWA 25 ppm
ACGIH TLV: TWA 25 ppm

SAFETY PROFILE: Moderately toxic by skin contact. Mildly toxic by inhalation and ingestion.

**INE100** CAS: 108-21-4 $C_5H_{10}O_2$ HR: 2
**ISOPROPYL ACETATE**
DOT: 1220
SYN: 2-ACETOXYPROPANE

OSHA PEL: (Transitional: TWA 250 ppm) TWA 250 ppm; STEL 310 ppm
ACGIH TLV: TWA 250 ppm; STEL 310 ppm
DFG MAK: 200 ppm (840 $mg/m^3$)
DOT Class: Flammable Liquid

SAFETY PROFILE: Moderately toxic by ingestion. Mildly toxic by inhalation. Highly flammable liquid. Dangerous fire hazard. Moderately explosive. To fight fire, use foam, $CO_2$, dry chemical.

PROP: Colorless, aromatic liquid. Mp: −73°, bp: 88.4°, lel: 1.8%, uel: 7.8%, fp: −69.3°, flash p: 40°F, d: 0.874 @ 20°/20°, autoign temp: 860°F, vap press: 40 mm @ 17.0°, d: 3.52. Sltly sol in water; misc in alc, ether, and fixed oils.

**INJ000** CAS: 67-63-0 $C_3H_8O$ HR: 3
**ISOPROPYL ALCOHOL**
DOT: 1219
SYNS: DIMETHYLCARBINOL ◇ ISOPROPANOL (DOT) ◇ 2-PROPANOL ◇ sec-PROPYL ALCOHOL (DOT)

OSHA PEL: (Transitional: TWA 400 ppm) TWA 400 ppm; STEL 500 ppm

ACGIH TLV: TWA 400 ppm; STEL 500 ppm
DFG MAK: 400 ppm (980 mg/m$^3$)
NIOSH REL: (Isopropyl Alcohol) TWA 400 ppm; CL 800 ppm/15M
DOT Class: Flammable Liquid

SAFETY PROFILE: Poison by ingestion. Experimental teratogenic and reproductive effects. An eye and skin irritant. Flammable liquid. Moderately explosive. Reacts with air to form dangerous peroxides. To fight fire, use $CO_2$, dry chemical, alcohol foam.

PROP: Clear, colorless liquid; slt odor, sltly bitter taste. Mp: −88.5 to −89.5°, bp: 82.5°, lel: 2.5%, uel: 12%, flash p: 53°F (CC), d: 0.7854 @ 20°/4°, refr index: 1.377 @ 20°, vap d: 2.07, ULC: 70. fp: −89.5°; autoign temp: 852°F. Misc with water, alc, ether, and chloroform; insol in salt solns.

**INK000** CAS: 75-31-0 $C_3H_9N$ HR: 3
**ISOPROPYLAMINE**
DOT: 1221
SYNS: 2-AMINOPROPANE ◇ 1-METHYLETHYLAMINE ◇ 2-PROPYLAMINE

OSHA PEL: (Transitional: TWA 5 ppm) TWA 5 ppm; STEL 10 ppm
ACGIH TLV: TWA 5 ppm; STEL 10 ppm
DFG MAK: 5 ppm (12 mg/m$^3$)
DOT Class: Flammable Liquid

SAFETY PROFILE: Poison by skin contact. Moderately toxic by ingestion. A severe skin and eye irritant. Very dangerous fire hazard and moderate explosion hazard. To fight fire, use alcohol foam, foam, $CO_2$, dry chemical. When heated to decomposition it emits toxic fumes of $NO_x$.

PROP: Colorless liquid, amino odor. Mp: −101.2°, flash p: −35°F (OC), d: 0.694 @ 15°/4°, autoign temp: 756°F, d: 2.03, bp: 33-34°, lel: 2.3%, uel: 10.4%. Misc with water, alc, and ether.

**INX000** CAS: 768-52-5 $C_9H_{13}N$ HR: 2
**N-ISOPROPYLANILINE**
SYNS: o-AMINOISOPROPYLBENZENE ◇ o-CUMIDINE ◇ 2-(1-METHYLETHYL)BENZENAMINE

OSHA PEL: TWA 2 ppm (skin)
ACGIH TLV: TWA 2 ppm (skin)

SAFETY PROFILE: Moderately toxic by ingestion. When heated to decomposition it emits toxic fumes of $NO_x$.

**IOB000** CAS: 80-15-9 $C_9H_{12}O_2$ HR: 3
**ISOPROPYLBENZENE HYDROPEROXIDE**
DOT: 2116
SYNS: CUMENE HYDROPEROXIDE (DOT) ◇ CUMENYL HYDROPEROXIDE ◇ α,α-DIMETHYLBENZYL HYDROPEROXIDE (MAK)

DFG MAK: Moderate Skin Effects.
DOT Class: Organic Peroxide

SAFETY PROFILE: A poison by ingestion. Moderately toxic by skin contact and inhalation. An experimental tumorigen. A skin and eye irritant. A strong oxidizing agent. Flammable. To fight fire, use foam, $CO_2$, dry chemical.

PROP: Bp: 153°, flash p: 175°F, d: 1.05. The hydroperoxide of cumene.

**IOZ750** CAS: 108-20-3 $C_6H_{14}O$ HR: 2
**ISOPROPYL ETHER**
DOT: 1159
SYNS: DIISOPROPYL OXIDE ◇ 2-ISOPROPOXYPROPANE

OSHA PEL: TWA 500 ppm
ACGIH TLV: TWA 250 ppm; STEL 310 ppm
DFG MAK: 500 ppm (2100 mg/m$^3$)
DOT Class: Flammable Liquid

SAFETY PROFILE: Mildly toxic by ingestion, inhalation, and skin contact. A skin irritant. A very dangerous fire hazard and severe explosion hazard. Dangerous: on exposure to air it rapidly forms very sensitive, explosive peroxides. To fight fire, use alcohol foam, $CO_2$, foam, dry chemical.

PROP: Colorless liquid, ethereal odor, misc in water. Mp: −60°, bp: 68.5°, lel: 1.4%, uel: 7.9%, flash p: −18°F (CC), d: 0.719 @ 25°, autoign temp: 830°F, vap press: 150 mm @ 25°, vap d: 3.52.

**IPD000** CAS: 4016-14-2 $C_6H_{12}O_2$ HR: 2
**ISOPROPYL GLYCIDYL ETHER**
SYN: IGE

OSHA PEL: (Transitional: TWA 50 ppm) TWA 50 ppm; STEL 75 ppm
ACGIH TLV: TWA 50 ppm; STEL 75 ppm
DFG MAK: 50 ppm (240 mg/m$^3$)
NIOSH REL: (Glycidyl Ethers) CL 240 mg/m$^3$/15M

SAFETY PROFILE: Moderately toxic by ingestion. Mildly toxic by inhalation and skin contact. A skin and eye irritant.

PROP: A liquid.

**IQU000** HR: 3
**ISOPROPYL OILS**

DFG MAK: Suspected Carcinogen.

SAFETY PROFILE: An experimental neoplastigen. A suspected human carcinogen.

PROP: A by-product of isopropyl alcohol manufacture composed of trimeric and tetrameric polypropylene + small amounts of benzene, toluene, alkyl benzenes, polyaromatic ring compounds, hexane, heptane, acetone, ethanol, isopropyl ether, and isopropyl alcohol.

**KBB600** CAS: 1332-58-7 HR: D
**KAOLIN**
SYN: CHINA CLAY

OSHA PEL: (Transitional: TWA Respirable Fraction: 15 mg/m$^3$; Respirable Fraction: 5 mg/m$^3$) TWA Total Dust: 10 mg/m$^3$; Respirable Fraction: 5 mg/m$^3$
ACGIH TLV: TWA (nuisance particulate) 10 mg/m$^3$ of total dust (when toxic impurities are not present, e.g., quartz $< 1\%$).

SAFETY PROFILE: A nuisance dust.

PROP: Fine white to light yellow powder; earth taste. Insol in ether, alc, dilute acids, and alkali solutions.

**KEA000** CAS: 143-50-0 $C_{10}Cl_{10}O$ HR: 3
**KEPONE**
DOT: 2761
SYNS: CHLORDECONE ◇ DECACHLOROTETRAHYDRO-4,7-METHANOINDENEONE ◇ MEREX

DFG MAK: Suspected Carcinogen.
NIOSH REL: (Kepone) CL 0.001 mg/m$^3$/15M
DOT Class: ORM-E

SAFETY PROFILE: Poison by ingestion and skin contact. An experimental carcinogen. A suspected human carcinogen. Experimental teratogenic and reproductive effects. Registration suspended by the USEPA.

PROP: A chlorinated polycyclic ketone, a crystalline material, sltly water-sol, sol in alc, ketones, and acetic acid. Mp: decomp @ 350°.

**KEK000** CAS: 8008-20-6 HR: 3
**KEROSENE**
DOT: 1223
SYN: COAL OIL

NIOSH REL: (Kerosene) TWA 100 mg/m$^3$
DOT Class: Combustible Liquid

SAFETY PROFILE: Moderately toxic to humans. Combustible. Moderately explosive. To fight fire, use foam, $CO_2$, dry chemical.

PROP: A pale yellow to water-white, oily liquid. Bp: 175-325°, ULC: 40, flash p: 150-185°F, d: 0.80 to <1.0, lel: 0.7%, uel: 5.0%, autoign temp: 410°F, vap d: 4.5. Insol in water; misc with other petroleum solvents.

**KEU000** CAS: 463-51-4 $C_2H_2O$ HR: 3
**KETENE**
SYNS: CARBOMETHENE ◇ ETHENONE

OSHA PEL: (Transitional: TWA 0.5 ppm) TWA 0.5 ppm; STEL 1.5 ppm

ACGIH TLV: TWA 0.5 ppm; STEL 1.5 ppm
DFG MAK: 0.5 ppm (0.9 mg/m$^3$)

SAFETY PROFILE: Poison by inhalation. Moderately toxic by ingestion.

PROP: Colorless gas with disagreeable taste. Mp: −150°, bp: −56°, vap d: 1.45. Decomp in water and alc; sol in ether and acetone.

**LBR000** CAS: 105-74-8 $C_{24}H_{46}O_4$ HR: 3
**LAUROYL PEROXIDE**

DFG MAK: Mild skin effects.
DOT Classification: Organic Peroxide; Label: Organic Peroxide.

SAFETY PROFILE: An experimental tumorigen. A powerful oxidizing agent. A corrosive irritant to the eyes and mucous membranes; can cause burns. A dangerous fire hazard.

PROP: White, coarse powder; tasteless with a faint odor. Mp: 53-55°.

**LBX000** CAS: 112-55-0 $C_{12}H_{26}S$ HR: 3
**LAURYL MERCAPTAN**
DOT: 2124
SYNS: DODECYL MERCAPTAN ◇ 1-MERCAPTODODECANE

NIOSH REL: (n-Alkane Mono Thiols) CL 0.5 ppm/15M

SAFETY PROFILE: Combustible. To fight fire use alcohol foam. When heated to decomposition it emits toxic fumes of $SO_x$.

PROP: Water-white to pale yellow liquid. Mp: −7°, bp: 115-177°, flash p: 262°F (OC), d: 0.849 @ 15.5°/15.5°.

**LCF000** CAS: 7439-92-1 Pb HR: 3
**LEAD**
SYN: C.I. 77575

OSHA PEL: TWA 0.05 mg(Pb)/m$^3$
ACGIH TLV: TWA 0.15 mg(Pb)/m$^3$; BEI: 50 μg(lead)/L in blood; 150 μg(lead)/g creatinine in urine.
DFG MAK: 0.1 mg/m$^3$; BAT: 70 μg(lead)/L in blood, 30 μg(lead)/L in blood of women less than 45 years old.
NIOSH REL: TWA (Inorganic Lead) 0.10 mg(Pb)/m$^3$

SAFETY PROFILE: A cumulative poison by all routes. A suspected carcinogen. An experimental teratogen. Experimental reproductive effects. Organic lead compounds are absorbed through intact skin.

PROP: Bluish-gray, soft metal. Mp: 327.43°, bp: 1740°, d: 11.34 @ 20°/4°. vap press: 1 mm @ 973°.

**LCR000** CAS: 7758-97-6 $CrO_4 \cdot Pb$ HR: 3
**LEAD CHROMATE**
SYNS: CHROME GREEN ◇ CHROME LEMON ◇ CHROME YELLOW ◇ CHROMIC ACID, LEAD(2+) SALT (1:1) ◇ C.I. 77600

OSHA PEL: TWA 0.05 mg(Pb)/$m^3$; CL 0.1 mg($CrO_3$)/$m^3$
ACGIH TLV: 0.05 mg(Cr)/$m^3$; Suspected Carcinogen; (Proposed: TWA 0.05 ppm (Pb), 0.012 ppm (Cr); Suspected Human Carcinogen)
DFG MAK: Suspected Carcinogen.
NIOSH REL: (Chromium(VI)) TWA 0.001 mg(Cr(VI))/$m^3$; (Inorganic Lead) TWA 0.10 mg(Pb)/$m^3$

SAFETY PROFILE: A human carcinogen. An experimental carcinogen. When heated to decomposition it emits toxic fumes of lead.

PROP: Yellow or orange-yellow powder. One of the most insol salts. Insol in acetic acid; sol in solns of fixed alkali hydroxides and dilute nitric acid. Mp: 844°. Bp: decomp, d: 6.3.

**LCS000** CAS: 18454-12-1 $CrO_4Pb \cdot OPb$ HR: 3
**LEAD CHROMATE, BASIC**
SYNS: BASIC LEAD CHROMATE ◇ CHROME ORANGE ◇ CHROMIUM LEAD OXIDE ◇ C.I. 77601 ◇ LEAD CHROMATE OXIDE (MAK)

OSHA PEL: TWA 0.05 mg(Pb)/$m^3$; CL 0.1 mg(CrO3)/$m^3$
ACGIH TLV: TWA 0.05 mg(Cr)/$m^3$; TWA 0.15 mg(Pb)/$m^3$
DFG MAK: Suspected Carcinogen.
NIOSH REL: (Chromium(VI)) TWA 0.001 mg(Cr(VI))/$m^3$; (Inorganic Lead) TWA 0.10 mg(Pb)/$m^3$

SAFETY PROFILE: A human carcinogen. An experimental carcinogen. When heated to decomposition it emits very toxic fumes of Pb.

PROP: Red, amorphous or crystalline solid. Mp: 920°.

**LGM000** CAS: 68476-85-7 HR: 2
**LIQUEFIED PETROLEUM GAS**
DOT: 1075
SYNS: LPG ◇ L.P.G. (OSHA, ACGIH) ◇ PETROLEUM GAS, LIQUEFIED

OSHA PEL: TWA 1000 ppm
NIOSH REL: TWA 350 mg/$m^3$; CL 1800 mg/$m^3$/15M
ACGIH TLV: TWA 1000 ppm
DOT Classification: Flammable Gas; Label: Flammable Gas.

SAFETY PROFILE: A simple asphyxiant. Olefinic impurities may lend a narcotic effect. A very dangerous fire hazard. To fight fire, shut off flow of gas, use $CO_2$, dry chemical, water spray.

**LHH000** CAS: 7580-67-8 HLi HR: 3
**LITHIUM HYDRIDE**
DOT: 2805/1414

OSHA PEL: TWA 0.025 mg/$m^3$
ACGIH TLV: TWA 0.025 mg/$m^3$
DFG MAK: 0.025 mg/$m^3$
DOT Classification: Flammable Solid; Label: Flammable Solid and Dangerous When Wet.

SAFETY PROFILE: Poison by inhalation. A severe eye, skin, and mucous membrane irritant. The powder ignites spontaneously in air. To fight fire, use special mixtures of dry chemical.

PROP: White, translucent, crystals. Mp: 680°, d: 0.76-0.77. Darkens rapidly on exposure to light. Decomp in water liberating LiOH and $H_2$.

**MAC650** CAS: 546-93-0 $CO_3 \cdot Mg$ HR: 1
**MAGNESITE**
SYNS: C.I. 77713 ◇ MAGNESIUM CARBONATE

OSHA PEL: TWA Total Dust: 15 mg/$m^3$; Respirable Fraction: 5 mg/$m^3$
ACGIH TLV: TWA (nuisance particulate) 10 mg/$m^3$ of total dust (when toxic impurities are not present, e.g., quartz < 1%).

SAFETY PROFILE: Incompatible with formaldehyde.

PROP: Very light, white powder; odorless. D: 3.04; decomp @ 350°. Sol in acids; insol in water and alc.

**MAH500** CAS: 1309-48-4 MgO HR: 3
**MAGNESIUM OXIDE**
SYNS: CALCINED MAGNESIA ◇ MAGNESIA

OSHA PEL: Fume: (Transitional: TWA Total Dust: 15 mg/$m^3$; Respirable Fraction: 5 mg/$m^3$) Total Dust: 10 mg/$m^3$; Respirable Fraction: 5 mg/$m^3$
ACGIH TLV: TWA 10 mg/$m^3$ (fume)
DFG MAK: 6 mg/$m^3$ (fume)

SAFETY PROFILE: An experimental tumorigen. Incandescent reaction with phosphorus pentachloride.

PROP: White, bulky, very fine powder; odorless. Mp: 2500-2800°, d: 3.65-3.75. Very sltly sol in water; sol in dil acids; insol in alc.

**MAK700** CAS: 121-75-5 $C_{10}H_{19}O_6PS_2$ HR: 3
**MALATHION**
DOT: 2783
SYNS: CARBETHOXY MALATHION ◇ DIETHYL MERCAPTOSUCCINATE-O,O-DIMETHYL THIOPHOSPHATE

OSHA PEL: (Transitional: TWA Total Dust: 15 mg/$m^3$; Respirable Fraction: 5 mg/$m^3$ (skin)) TWA Total Dust:10 mg/$m^3$' Respirable Fraction: 5 mg/$m^3$ (skin)
ACGIH TLV: TWA 10 mg/$m^3$ (skin)
DFG MAK: 15 mg/$m^3$
NIOSH REL: (Malathion) TWA 15 mg/$m^3$
DOT Class: ORM-A

SAFETY PROFILE: A human poison by ingestion. An experimental poison by inhalation. An allergic sensitizer. When heated to decomposition it emits toxic fumes of $PO_x$ and $SO_x$.

PROP: Brown to yellow liquid; characteristic odor. D: 1.23 @ 25°/4°, mp: 2.9°, bp: 156° @ 0.7 mm. Misc in organic solvents, sltly water-sol.

**MAM000** CAS: 108-31-6 $C_4H_2O_3$ HR: 3
**MALEIC ANHYDRIDE**
DOT: 2215
SYNS: 2,5-DIHYDROFURAN-2,5-DIONE ◇ MALEIC ACID ANHYDRIDE (MAK) ◇ TOXILIC ANHYDRIDE

OSHA PEL: TWA 0.25 ppm
ACGIH TLV: TWA 0.25 ppm
DFG MAK: 0.2 ppm (0.8 mg/m$^3$)
DOT Class: ORM-A

SAFETY PROFILE: Poison by ingestion. Moderately toxic by skin contact. An experimental tumorigen. A corrosive irritant to eyes, skin, and mucous membranes. Combustible. Explosive in the form of vapor. To fight fire, use alcohol foam.

PROP: Fused black or white crystals. Mp: 52.8°, bp: 202°, flash p: 215°F (CC), d: 1.48 @ 20°/4°, autoign temp: 890°F, vap press: 1 mm @ 44.0°, vap d: 3.4, lel: 1.4%, uel: 7.1%. Sol in dioxane, water @ 30° forming maleic acid; very sltly sol in alc.

**MAO250** CAS: 109-77-3 $C_3H_2N_2$ HR: 3
**MALONONITRILE**
DOT: 2647
SYNS: CYANOACETONITRILE ◇ DICYANOMETHANE ◇ MALONIC DINITRILE ◇ METHYLENE CYANIDE

NIOSH REL: (Nitriles) TWA 8 mg/m$^3$
DOT Class: Poison B

SAFETY PROFILE: Poison by ingestion. A severe eye irritant. Combustible. To fight fire, use water, fog, spray, foam. When heated to decomposition it emits toxic fumes of $NO_x$ and $CN^-$.

PROP: White powder. D: 1.049 @ 34°/4°, mp: 30.5°, bp: 220°, flash p: 266°F (TOC). Sol in water, alc, and ether.

**MAP750** CAS: 7439-96-5 Mn HR: 3
**MANGANESE**

OSHA PEL: Fume: (Transitional: CL 5 mg/m$^3$) TWA 1 mg/m$^3$; STEL 3 mg/m$^3$; Compounds: CL 5 mg/m$^3$
ACGIH TLV: Fume: 1 mg/m$^3$; STEL 3 mg/m$^3$; Dust and Compounds: TWA 5 mg/m$^3$
DFG MAK: 5 mg/m$^3$

SAFETY PROFILE: An experimental tumorigen. A skin and eye irritant. Flammable and moderately explosive in the form of dust or powder. The dust may be pyrophoric in air. To fight fire, use special dry chemical.

PROP: Reddish-grey or silvery, brittle, metallic element. Mp: 1260°, bp: 1900°, d: 7.20, vap press: 1 mm @ 1292°.

**MAU800** CAS: 1317-35-7 $Mn_3O_4$ HR: 2
**MANGANESE TETROXIDE**
SYNS: MANGANESE OXIDE ◇ TRIMANGANESE TETROXIDE

OSHA PEL: TWA 1 mg(Mn)/$m^3$
ACGIH TLV: TWA 1 mg(Mn)/$m^3$
DFG MAK: 1 mg/$m^3$

SAFETY PROFILE: Experimental reproductive effects. Reacts violently @ <100°.

PROP: Brownish-black powder. D: 4.7. Insol in water; sol in HCl, liberating chlorine.

**MAV750** CAS: 12108-13-3 $C_9H_7MnO_3$ HR: 3
**MANGANESE TRICARBONYL METHYLCYCLOPENTADIENYL**
SYNS: METHYLCYCLOPENTADIENYL MANGANESE TRICARBONYL (OSHA) ◇ MMT

OSHA PEL: TWA 0.2 mg(Mn)/$m^3$ (skin)
ACGIH TLV: TWA 5 mg(Mn)/$m^3$

SAFETY PROFILE: Poison by ingestion and inhalation, skin contact. A skin irritant. When heated to decomposition it emits toxic fumes of CO.

**MCW250** CAS: 7439-97-6 Hg HR: 3
**MERCURY**
DOT: 2809
SYNS: MERCURY, METALLIC (DOT) ◇ QUICK SILVER

OSHA PEL: Vapor: (Transitional: CL 1 mg/10$m^3$) 0.05 mg/$m^3$ (skin)
ACGIH TLV: TWA 0.05 mg(Hg)/$m^3$ (vapor, skin)
DFG MAK: 0.1 mg/$m^3$; BAT: 5 μg/dL in blood.
NIOSH REL: (Inorganic Mercury) TWA 0.05 mg(Hg)/$m^3$
DOT Class: Corrosive Material

SAFETY PROFILE: Poison by inhalation. An experimental tumorigen. Corrosive to skin, eyes, and mucous membranes. An experimental teratogen. Alkyl mercurials have very high toxicity; aryl compounds, particularly the phenyls, and organomercurials are much less toxic. Fatal poisoning has occurred due to exposure to alkyl mercurials. The alkyls and aryls commonly cause skin burns and other forms of irritation, and both can be absorbed through the skin. Many mercury compounds are explosively unstable or undergo hazardous reactions. When heated to decomposition they emit highly toxic fumes of Hg.

PROP: Silvery, heavy, mobile liquid. A liquid metallic element. Mp: −38.89°, bp: 356.9°, d: 13.534 @ 25°, vap press: $2 \times 10^{-3}$ mm @ 25°. Solid: tin-white, ductile, malleable mass which can be cut with a knife.

**MDJ750** CAS: 141-79-7 $C_6H_{10}O$ HR: 3
**MESITYL OXIDE**
DOT: 1229

SYNS: ISOBUTENYL METHYL KETONE ◇ ISOPROPYLIDENEACETONE ◇ METHYL ISOBUTENYL KETONE ◇ 2-METHYL-2-PENTEN-4-ONE

OSHA PEL: (Transitional: TWA 25 ppm) TWA 15 ppm; STEL 25 ppm
ACGIH TLV: TWA 15 ppm; STEL 25 ppm
DFG MAK: 25 ppm (100 $mg/m^3$)
NIOSH REL: (Ketones) TWA 40 $mg/m^3$
DOT Class: Label: Flammable Liquid.

SAFETY PROFILE: Moderately toxic by ingestion. Mildly toxic by inhalation and skin contact. This compound is highly irritating to all tissues on contact; its vapors also are irritating. Dangerous fire hazard. To fight fire, use alcohol foam, $CO_2$, dry chemical.

PROP: Oily, colorless liquid; strong odor. Mp: −59°, bp: 130.0°, flash p: 87°F (CC), d: 0.8539 @ 20°/4°, autoign temp: 652°F, vap press: 10 mm @ 26.0°, vap d: 3.38. Solidifies @ 41.5°; somewhat sol in water @ 20°. Misc in alc, ether, and with most organic liquids.

**MDN250** CAS: 79-41-4 $C_4H_6O_2$ HR: 3
**METHACRYLIC ACID**
DOT: 2531
SYN: 2-METHYLPROPENOIC ACID

OSHA PEL: TWA 20 ppm (skin)
ACGIH TLV: TWA 20 ppm
DOT Class: Corrosive Material

SAFETY PROFILE: Poison by ingestion. Moderately toxic by skin contact. Corrosive to skin, eyes and mucous membranes. Flammable. To fight fire, use alcohol foam, spray, mist, dry chemical.

PROP: Corrosive liquid or colorless crystals; repulsive odor. Mp: 16°, bp: 163°, flash p: 171°F (COC), d: 1.014 @ 25° (glacial), vap press: 1 mm @ 25.5° Sol in warm water; misc with alc and ether.

**MDU600** CAS: 16752-77-5 $C_5H_{10}N_2O_2S$ HR: 3
**METHOMYL**
SYN: METHYL-N-((METHYLCARBAMOYL)OXY)THIOACETIMIDATE

OSHA PEL: TWA 2.5 $mg/m^3$
ACGIH TLV: TWA 2.5 $mg/m^3$

SAFETY PROFILE: Poison by ingestion and inhalation. When heated to decomposition it emits very toxic fumes of $NO_x$ and $SO_x$.

PROP: White, crystalline solid; slt sulfurous odor. Mp: 79°. Moderately sol in water.

**MEI450** CAS: 72-43-5 $C_{16}H_{15}Cl_3O_2$ HR: 3
**METHOXYCHLOR**
DOT: 2761
SYNS: 1,1-BIS(p-METHOXYPHENYL)-2,2,2-TRICHLOROETHANE ◇ DIMETHOXY-DDT ◇ METHOXY-DDT

OSHA PEL: (Transitional: TWA Total Dust: 15 mg/m$^3$; Respirable Fraction: 5 mg/m$^3$ TWA Total Dust: 10 mg/m$^3$; 5 mg/m$^3$
ACGIH TLV: TWA 10 mg/m$^3$
DFG MAK: 15 mg/m$^3$
DOT Class: ORM-E

SAFETY PROFILE: Moderately toxic by ingestion and skin contact. An experimental carcinogen and teratogen. Experimental reproductive effects. When heated to decomposition it emits highly toxic fumes of $Cl^-$.

PROP: Crystals. Mp: 78°, vap d: 12.

**MFC700** CAS: 150-76-5 $C_7H_8O_2$ HR: 3
**4-METHOXYPHENOL**
SYNS: HYDROQUINONE MONOMETHYL ETHER ◇ MME

OSHA PEL: TWA 5 mg/m$^3$
ACGIH TLV: TWA 5 mg/m$^3$

SAFETY PROFILE: Moderately toxic by ingestion. A skin irritant.

PROP: White, waxy solid. Mp: 52.5°, bp: 246°, d: 1.55 @ 20°/20°.

**MFW100** CAS: 79-20-9 $C_3H_6O_2$ HR: 2
**METHYL ACETATE**
DOT: 1231
SYNS: ACETIC ACID METHYL ESTER ◇ METHYL ETHANOATE

OSHA PEL: (Transitional: TWA 200 ppm) TWA 200 ppm; STEL 250 ppm
ACGIH TLV: TWA 200 ppm; STEL 250 ppm
DFG MAK: 200 ppm (610 mg/m$^3$)
DOT Class: Flammable Liquid

SAFETY PROFILE: Moderately toxic. A human systemic irritant by inhalation. A moderate skin and severe eye irritant. Dangerous fire hazard. Moderate explosion hazard.

PROP: Colorless, volatile liquid. Mp: −98.7°, lel: 3.1%, uel: 16%, bp: 57.8°, ULC: 85-90, flash p: 14°F, d: 0.92438, autoign temp: 935°F, vap press: 100 mm @ 9.4°, vap d: 2.55. Moderately sol in water; misc in alc and ether.

**MFX590** CAS: 74-99-7 $C_3H_4$ HR: 3
**METHYL ACETYLENE**
SYNS: PROPINE ◇ PROPYNE

OSHA PEL: TWA 1000 ppm
ACGIH TLV: TWA 1000 ppm
DFG MAK: 1000 ppm (1650 mg/m$^3$)

SAFETY PROFILE: An asphyxiant. Dangerous fire hazard. Explosive in the form of vapor. To fight fire, stop flow of gas.

PROP: Gas. Mp: −104°, lel: 1.7%, bp: −23.3°, vap press: 3876 mm @ 20°, d: 1.787 g/L @ 0°, vap d: 1.38.

**MFX600** CAS: 59355-75-8 HR: 2
**METHYL ACETYLENE-PROPADIENE MIXTURE**
DOT: 1060
SYNS: MAPP ◇ METHYLACETYLENE-PROPADIENE, STABILIZED (DOT) ◇ PROPYNE mixed with PROPADIENE

OSHA PEL: (Transitional: TWA 1000 ppm) TWA 1000 ppm; STEL 1250 ppm
ACGIH TLV: TWA 1000 ppm; STEL 1250 ppm
DFG MAK: 1000 ppm (1650 mg/$m^3$)
DOT Class: Flammable Gas

SAFETY PROFILE: A flammable gas mixture. To fight fire, stop flow of gas.

**MGA500** CAS: 96-33-3 $C_4H_6O_2$ HR: 3
**METHYL ACRYLATE**
DOT: 1919
SYNS: ACRYLIC ACID METHYL ESTER (MAK) ◇ METHOXYCARBONYLETHYLENE ◇ METHYL PROPENOATE ◇ PROPENOIC ACID METHYL ESTER

OSHA PEL: TWA 10 ppm (skin)
ACGIH TLV: TWA 10 ppm (skin)
DFG MAK: 5 ppm (18 mg/$m^3$)
DOT Class: Flammable Liquid

SAFETY PROFILE: Poison by ingestion. Moderately toxic by skin contact. A suspected carcinogen. A skin and eye irritant. Dangerously flammable. Dangerous explosion hazard in the form of vapor. A storage hazard; forms peroxides which may initiate exothermic polymerization. To fight fire, use foam, $CO_2$, dry chemical.

PROP: Colorless liquid; acrid odor. D: 0.9561 @ 20°/4°, mp: −76.5°, bp: 70° @ 608 mm, lel: 2.8%, uel: 25%, fp: −75°, flash p: 27°F (OC), vap press: 100 mm @ 28°, vap d: 2.97, sol in alc and ether.

**MGA750** CAS: 126-98-7 $C_4H_5N$ HR: 3
**METHYLACRYLONITRILE**
SYNS: 2-CYANOPROPENE-1 ◇ ISOPROPENE CYANIDE ◇ 2-METHYLPROPENENITRILE

OSHA PEL: TWA 1 ppm (skin)
ACGIH TLV: TWA 1 ppm (skin)

SAFETY PROFILE: Poison by ingestion, inhalation, and skin contact. A skin and eye irritant. A dangerous fire hazard. When heated to decomposition it emits toxic fumes of $NO_x$ and $CN^-$.

PROP: Mp: −36°, bp: 90.3°, d: 0.805, vap press: 40 mm @ 12.8°, flash p: 55°F.

**MGA850** CAS: 109-87-5 $C_3H_8O_2$ HR: 2
**METHYLAL**
DOT: 1234
SYNS: DIMETHOXYMETHANE ◇ DIMETHYL FORMAL ◇ FORMAL ◇ METHYLENE DIMETHYL ETHER

OSHA PEL: TWA 1000 ppm
ACGIH TLV: TWA 1000 ppm
DFG MAK: 1000 ppm (3100 mg/m$^3$)
DOT Class: Flammable Liquid

SAFETY PROFILE: Mildly toxic by ingestion and inhalation. A very dangerous fire hazard. Moderately explosive. To fight fire, use foam, $CO_2$, dry chemical.

PROP: Colorless liquid, pungent odor. Mp: −104.8°, bp: 42.3°, d: 0.864 @ 20°/4°, vap press: 330 mm @ 20°, vap d: 2.63, autoign temp: 459°F. flash p.: −0.4°F.

**MGB150** CAS: 67-56-1 $CH_4O$ HR: 3
**METHYL ALCOHOL**
DOT: 1230
SYNS: COLUMBIAN SPIRITS (DOT) ◇ METHANOL ◇ WOOD ALCOHOL (DOT) ◇ WOOD NAPHTHA

OSHA PEL: (Transitional: TWA 200 ppm) TWA 200 ppm; STEL 250 ppm (skin)
ACGIH TLV: TWA 200 ppm; STEL 250 ppm (skin)
DFG MAK: 200 ppm (260 mg/m$^3$); BAT: 30 mg/L in urine at end of shift.
NIOSH REL: TWA 200 ppm; CL 800 ppm/15M
DOT Class: Flammable Liquid

SAFETY PROFILE: A human poison by ingestion. Poison experimentally by skin contact. An experimental teratogen. Experimental reproductive effects. An eye and skin irritant. Dangerous fire hazard. To fight fire, use alcohol foam.

PROP: Clear, colorless, very mobile liquid; slt alcoholic odor when pure; crude material may have a repulsive pungent odor. Mp: 64.8°, lel: 6.0%, uel: 36.5%, ULC: 70, fp: −97.8°, d: 0.7915 @ 20°/4°, autoign temp: 878°F, vap press: 100 mm @ 21.2°, vap d: 1.11. Misc in water, ethanol, ether, benzene, ketones, and most other organic solvents.

**MGC250** CAS: 74-89-5 $CH_5N$ HR: 3
**METHYLAMINE**
DOT: 1061/1235
SYNS: AMINOMETHANE ◇ METHANAMINE (9CI)

OSHA PEL: TWA 10 ppm
ACGIH TLV: TWA 10 ppm
DFG MAK: 10 ppm (12 mg/m$^3$)

DOT Class: Flammable Gas

SAFETY PROFILE: Moderately toxic by inhalation. A severe skin irritant. Flammable and explosive gas. To fight fire, stop flow of gas. When heated to decomposition it emits toxic fumes of $NO_x$.

PROP: Colorless gas or liquid; powerful ammoniacal odor. Bp: 6.3°, lel: 4.95%, uel: 20.75%, mp: −93.5°, flash p: 32°F (CC), d: 0.662 @ 20°/4°, autoign temp: 806°F, vap d: 1.07. Fuming liquid, d: 0.699 @ −10.8°/4°. Sol in alc; misc with ether.

**MGN500** CAS: 110-43-0 $C_7H_{14}O$ HR: 2
**METHYL n-AMYL KETONE**
DOT: 1110
SYNS: n-AMYL METHYL KETONE ◇ AMYL METHYL KETONE (DOT) ◇ 2-HEPTANONE ◇ METHYL PENTYL KETONE

OSHA PEL: TWA 100 ppm
ACGIH TLV: TWA 50 ppm
NIOSH REL: (Ketones) TWA 465 mg/m$^3$
DOT Class: Combustible Liquid

SAFETY PROFILE: Moderately toxic by ingestion. A skin irritant. Flammable or combustible liquid. To fight fire, use foam, $CO_2$, dry chemical.

PROP: Colorless, mobile liquid; penetrating, fruity odor. Bp: 151.5°, flash p: 120°F (OC), autoign temp: 991°F, vap d: 3.94, d: 0.8197 @ 15°/4°. Very sltly sol in water; sol in alc and ether.

**MGN750** CAS: 100-61-8 $C_7H_9N$ HR: 3
**METHYLANILINE**
DOT: 2294
SYNS: ANILINOMETHANE ◇ N-METHYL ANILINE (MAK) ◇ MONOMETHYL ANILINE (OSHA) ◇ N-PHENYLMETHYLAMINE

OSHA PEL: (Transitional: TWA 2 ppm (skin)) TWA 0.5 ppm (skin)
ACGIH TLV: TWA 0.5 ppm (skin)
DFG MAK: 0.5 ppm (2 mg/m$^3$)
DOT Class: Poison B

SAFETY PROFILE: Poison by ingestion. When heated to decomposition it emits toxic fumes of $NO_x$.

PROP: Colorless or sltly yellow liquid which becomes brown on exposure to air. Mp: −57°, d: 0.989 @ 20°/4°, bp: 194-197°. Sol in alc and ether; sltly sol in water.

**MHR200** CAS: 74-83-9 $CH_3Br$ HR: 3
**METHYL BROMIDE**
DOT: 1062
SYN: BROMO METHANE

OSHA PEL: (Transitional: CL 20 ppm (skin)) TWA 5 ppm (skin)
ACGIH TLV: TWA 5 ppm (skin)

DFG MAK: 5 ppm (20 mg/$m^3$), Suspected Carcinogen.
NIOSH REL: Reduce to lowest level
DOT Class: Poison A

SAFETY PROFILE: A human poison by inhalation. An experimental carcinogen. Corrosive to skin; can produce severe burns. Mixtures of 10-15 % with air may be ignited with difficulty. Moderately explosive. To fight fire, use foam, water, $CO_2$, dry chemical. When heated to decomposition it emits toxic fumes of $Br^-$.

PROP: Colorless, transparent, volatile liquid or gas; burning taste, chloroform-like odor. Bp: 3.56°, lel: 13.5%, uel: 14.5%, fp: −93°, flash p: none, d: 1.732 @ 0°/0°, autoign temp: 998°F, vap d: 3.27, vap press: 1824 mm @ 25°.

**MIF765** CAS: 74-87-3 $CH_3Cl$ HR: 3
**METHYL CHLORIDE**
DOT: 1063
SYNS: CHLOROMETHANE ◇ MONOCHLOROMETHANE

OSHA PEL: (Transitional: TWA 100; CL 200 ppm; Pk 300 ppm/5M)TWA 50 ppm; STEL 100 ppm
ACGIH TLV: TWA 50 ppm; STEL 100 ppm
DFG MAK: 50 ppm (105 mg/$m^3$); Suspected Carcinogen.
NIOSH REL: (Monohalomethanes) TWA Reduce to lowest level.
DOT Class: Flammable Gas

SAFETY PROFILE: Very mildly toxic by inhalation. An experimental teratogen. Flammable gas. Moderate explosion hazard. To fight fire, stop flow of gas and use $CO_2$, dry chemical or water spray. When heated to decomposition it emits highly toxic fumes of $Cl^-$.

PROP: Colorless gas; ethereal odor and sweet taste. D: 0.918 @ 20°/4°, mp: −97°, bp: −23.7°, flash p: <32°F, lel: 8.1%, uel: 17%, autoign temp: 1170°F, vap d: 1.78. Sltly sol in water; misc with chloroform, ether, and glacial acetic acid; sol in alcohol.

**MIH275** CAS: 71-55-6 $C_2H_3Cl_3$ HR: 3
**METHYL CHLOROFORM**
DOT: 2831
SYNS: CHLOROTHANE NU ◇ CHLOROTHENE NU ◇ METHYLTRICHLOROMETHANE ◇ 1,1,1-TRICHLOROETHANE

OSHA PEL: (Transitional: TWA 350 ppm) TWA 350 ppm; STEL 450 ppm
ACGIH TLV: TWA 350 ppm; STEL 450 ppm (Proposed: BEI: 10 mg/L trichloroacetic acid in urine at end of workweek.)
DFG MAK: 200 ppm (1080 mg/$m^3$); BAT: 55 μg/dL in blood after several shifts.
NIOSH REL: (1,1,1-Trichloroethane) CL 350 ppm/15M
DOT Class: ORM-A

SAFETY PROFILE: Moderately toxic by ingestion, inhalation, and skin contact. An experimental teratogen. A skin and severe eye irritant. When heated to decomposition it emits toxic fumes of $Cl^-$.

PROP: Colorless liquid. Bp: 74.1°, fp: −32.5°, flash p: none, d: 1.3376 @ 20°/4°, vap press: 100 mm @ 20.0°. Insol in water; sol in acetone, benzene, carbon tetrachloride, methanol, and ether.

**MIQ075** CAS: 137-05-3 $C_9H_{13}NO_2$ HR: 3
**METHYL 2-CYANOACRYLATE**
SYNS: α-CYANOACRYLATE ACID METHYL ESTER ◇ 2-CYANOACRYLATE ACID METHYL ESTER

OSHA PEL: TWA 2 ppm; STEL 4 ppm
ACGIH TLV: TWA 2 ppm; STEL 4 ppm
DFG MAK: 2 ppm (8 $mg/m^3$)

SAFETY PROFILE: Experimental reproductive effects. A human eye irritant. When heated to decomposition it emits toxic fumes of $NO_x$.

**MIQ740** CAS: 108-87-2 $C_7H_{14}$ HR: 2
**METHYLCYCLOHEXANE**
DOT: 2296
SYNS: CYCLOHEXYLMETHANE ◇ HEXAHYDROTOLUENE ◇ TOLUENE HEXAHYDRIDE

OSHA PEL: (Transitional: TWA 500 ppm) TWA 400 ppm
ACGIH TLV: TWA 400 ppm
DFG MAK: 500 ppm (2000 $mg/m^3$)
DOT Class: Flammable Liquid

SAFETY PROFILE: Moderately toxic by ingestion. Dangerous fire hazard and moderate explosion hazard. To fight fire, use foam, $CO_2$, dry chemical.

PROP: Colorless liquid. Mp: −126.4°, lel: 1.2%, uel: 6.7%, bp: 100.3°, flash p: 25°F (CC), d: 0.7864 @ 0°/4°, 0.769 @ 20°/4°, vap press: 40 mm @ 22.0°, vap d: 3.39, autoign temp: 482°F.

**MIQ745** CAS: 25639-42-3 $C_7H_{14}O$ HR: 2
**METHYLCYCLOHEXANOL**
DOT: 2617
SYNS: HEXAHYDROCRESOL ◇ HEXAHYDROMETHYLPHENOL

OSHA PEL: (Transitional: TWA 100 ppm) TWA 50 ppm
ACGIH TLV: TWA 50 ppm
DFG MAK: 50 ppm (235 $mg/m^3$)
DOT Class: Flammable or Combustible Liquid

SAFETY PROFILE: Moderately toxic by ingestion. Flammable. To fight fire, use alcohol foam, $CO_2$, dry chemical.

PROP: Colorless, viscous liquid; aromatic, menthol-like odor. Bp: 155-180°, flash p: 154°F (CC), autoign temp: 565°F, d: 0.924 @ 15.5°/15.5°, vap d: 3.93.

**MIR500** CAS: 583-60-8 $C_7H_{12}O$ HR: 3
**2-METHYLCYCLOHEXANONE**
SYN: 1-METHYLCYCLOHEXAN-2-ONE

OSHA PEL: (Transitional: TWA 100 ppm (skin)) TWA 50 ppm; STEL 75 ppm (skin)
ACGIH TLV: TWA 50 ppm (skin)
DFG MAK: 50 ppm (230 mg/m$^3$)

SAFETY PROFILE: Moderately toxic by ingestion and skin contact.

PROP: Liquid. D: 0.925 @ 20/4°, mp: −14°, bp: 165.1°. Insol in water, sol in alc and ether.

**MIW100** CAS: 8022-00-2 HR: 3
**METHYL DEMETON**
SYNS: DEMETON METHYL ◇ S(and O)-2-(ETHYLTHIO)ETHYL-O,O-DIMETHYL PHOSPHOROTHIOATE ◇ METHYL SYSTOX

OSHA PEL: TWA 0.5 mg/m$^3$ (skin)
ACGIH TLV: TWA 0.5 mg/m$^3$ (skin)
DFG MAK: 0.5 ppm (5 mg/m$^3$)

SAFETY PROFILE: Poison by ingestion, skin contact, and inhalation.

PROP: An oily liquid. D: 1.20. Sltly sol in water.

**MJM200** CAS: 101-14-4 $C_{13}H_{12}Cl_2N_2$ HR: 3
**4,4′-METHYLENE BIS(2-CHLOROANILINE)**
SYNS: 4,4′-DIAMINO-3,3′-DICHLORODIPHENYLMETHANE ◇ MBOCA ◇ MOCA ◇ 4,4′-METHYLENEBIS(o-CHLOROANILINE)

OSHA PEL: TWA 0.02 ppm (skin)
ACGIH TLV: TWA 0.02 ppm (skin); Suspected Human Carcinogen.
DFG MAK: Animal Carcinogen, Suspected Human Carcinogen.
NIOSH REL: (MOCA) Lowest detectable limit.

SAFETY PROFILE: Moderately toxic by ingestion. An experimental carcinogen and tumorigen. When heated to decomposition it emits very toxic fumes of $Cl^-$ and $NO_x$.

**MJM600** CAS: 5124-30-1 $C_{15}H_{22}NO_2$ HR: 3
**METHYLENE BIS(4-CYCLOHEXYLISOCYANATE**
SYN: BIS(4-ISOCYANATOCYCLOHEXYL)METHANE

OSHA PEL: CL 0.01
ACGIH TLV: TWA 0.005 ppm

SAFETY PROFILE: Poison by inhalation. When heated to decomposition it emits very toxic fumes of $NO_x$.

**MJN000** CAS: 101-61-1 $C_{17}H_{22}N_2$ HR: 3
**4,4′-METHYLENE BIS(N,N′-DIMETHYLANILINE)**
SYNS: 4,4′-METHYLENEBIS(N,N-DIMETHYL)BENZENAMINE ◇ MICHLER'S BASE ◇ p,p-TETRAMETHYLDIAMINODIPHENYLMETHANE

DFG MAK: Suspected Carcinogen.

SAFETY PROFILE: Moderately toxic by ingestion. An experimental carcinogen and tumorigen. When heated to decomposition it emits toxic fumes of $NO_x$.

**MJN750** CAS: 139-25-3 $C_{17}H_{14}N_2O_2$ HR: 3
**5,5′-METHYLENEBIS(2-ISOCYANATO)TOLUENE**
SYNS: 3,3′-DIMETHYLDIPHENYLMETHANE-4,4′-DIISOCYANATE ◇ ISOCYANIC ACID, ESTER WITH DI-o-TOLUENEMETHANE

NIOSH REL: (Diisocyanates) TWA 0.005 ppm; CL 0.02 ppm/10M

SAFETY PROFILE: When heated to decomposition it emits toxic fumes of $NO_x$.

**MJO250** CAS: 838-88-0 $C_{15}H_{18}N_2$ HR: 3
**4,4′-METHYLENEBIS(2-METHYLANILINE)**
SYNS: MBOT ◇ 4,4′-METHYLENEBIS(2-METHYLBENZENAMINE) ◇ 4,4′-METHYLENE DI-o-TOLUIDINE

DFG MAK: Animal Carcinogen; Suspected Human Carcinogen.

SAFETY PROFILE: Moderately toxic by ingestion. An experimental carcinogen. An eye irritant. When heated to decomposition it emits toxic fumes of $NO_x$.

PROP: Mp: 149°.

**MJP400** CAS: 101-68-8 $C_{15}H_{10}N_2O_2$ HR: 3
**METHYLENE BISPHENYL ISOCYANATE**
DOT: 2489
SYNS: BIS(p-ISOCYANATOPHENYL)METHANE ◇ DIPHENYL METHANE DIISOCYANATE ◇ MDI ◇ METHYLENEBIS(4-ISOCYANATOBENZENE)

OSHA PEL: CL 0.02 ppm
ACGIH TLV: 0.005 ppm
DFG MAK: 0.01 ppm (0.1 $mg/m^3$)
NIOSH REL: (Diisocyanates) TWA 0.005 ppm; CL 0.02 ppm/10M
DOT Class: Poison B

SAFETY PROFILE: Poison by inhalation. A skin and eye irritant. An allergic sensitizer. When heated to decomposition it emits toxic fumes of $NO_x$ and $SO_x$.

PROP: Crystals or yellow fused solid. Mp: 37.2°, bp: 194-199° @ 5 mm, d: 1.19 @ 50°, vap press: 0.001 mm @ 40°.

**MJP450** CAS: 75-09-2 $CH_2Cl_2$ HR: 3
**METHYLENE CHLORIDE**
DOT: 1593
SYNS: DCM ◇ DICHLOROMETHANE (MAK, DOT) ◇ FREON 30

OSHA PEL: (Transitional: TWA 500 ppm; CL 1000 ppm; Pk 2000/5M/2H)

ACGIH TLV: TWA 50 ppm, Suspected Human Carcinogen
DFG MAK: 100 ppm (360 mg/m$^3$); BAT: 5% CO-Hb in blood at end of shift; Suspected Carcinogen.
NIOSH REL: (Methylene Chloride) Reduce to lowest feasible level.
DOT Class: Poison B

SAFETY PROFILE: Moderately toxic by ingestion. An experimental carcinogen and tumorigen. An experimental teratogen. An eye and severe skin irritant. It is flammable in the range of 12-19 percent in air but ignition is difficult. When heated to decomposition it emits highly toxic fumes of phosgene and $Cl^-$.

PROP: Colorless, volatile liquid; odor of chloroform. Bp: 39.8°, lel: 15.5% in $O_2$, uel: 66.4% in $O_2$, fp: −96.7°, d: 1.326 @ 20°/4°, autoign temp: 1139°F, vap press: 380 mm @ 22°, vap d: 2.93, refr index: 1.424 @ 20L. Sol in water; misc with alcohol, acetone, chloroform, ether, and carbon tetrachloride.

**MJP750** CAS: 1208-52-2 $C_{13}H_{14}N_2$ HR: 2
**2,4′-METHYLENEDIANILINE**
SYNS: 2,4′-DIAMINODIPHENYLMETHANE ◇ 2,4′-METHYLENEBIS(ANILINE)

DFG MAK: Animal Carcinogen, Suspected Human Carcinogen.

SAFETY PROFILE: Suspected human carcinogen. When heated to decomposition it emits toxic fumes of $NO_x$.

**MJQ000** CAS: 101-77-9 $C_{13}H_{14}N_2$ HR: 3
**4,4′-METHYLENEDIANILINE**
DOT: 2651
SYNS: 4-(4-AMINOBENZYL)ANILINE ◇ BIS(p-AMINOPHENYL)METHANE ◇ DDM ◇ MDA ◇ METHYLENEBIS(ANILINE)

ACGIH TLV: TWA 0.1 ppm (skin); Suspected Human Carcinogen.
DFG MAK: Animal Carcinogen, Suspected Human Carcinogen.
DOT Class: Poison B

SAFETY PROFILE: Poison by ingestion. An experimental carcinogen and tumorigen. An eye irritant. Combustible. When heated to decomposition it emits highly toxic fumes of aniline and $NO_x$.

PROP: Tan flakes or lumps; faint amine-like odor. Mp: 90°, flash p: 440°F.

**MKA400** CAS: 78-93-3 $C_4H_8O$ HR: 3
**METHYL ETHYL KETONE**
DOT: 1193/1232
SYNS: 2-BUTANONE ◇ ETHYL METHYL KETONE (DOT) ◇ MEK ◇ METHYL ACETONE (DOT)

OSHA PEL: (Transitional: TWA 200 ppm) TWA 200 ppm; STEL 300 ppm
ACGIH TLV: TWA 200 ppm; STEL 300 ppm; BEI: 2 mg(MEK)/L in urine at end if shift.

DFG MAK: 200 ppm (590 mg/m$^3$)
NIOSH REL: (Ketones) TWA 590 mg/m$^3$
DOT Class: Flammable Liquid

SAFETY PROFILE: Moderately toxic by ingestion and skin contact. An experimental teratogen. Experimental reproductive effects. A strong irritant. Highly flammable liquid. To fight fire, use alcohol foam, $CO_2$, dry chemical.

PROP: Colorless liquid; acetone-like odor. Bp: 79.57°, fp: −85.9°, lel: 1.8%, uel: 11.5%, flash p: 22°F (TOC), d: 0.80615 @ 20°/20°, vap press: 71.2 mm @ 20°, autoign temp: 960°F, vap d: 2.42, ULC: 85-90. Misc with alc, ether, fixed oils, and water.

**MKA500** CAS: 1338-23-4 $C_8H_{16}O_4$ HR: 3
**METHYL ETHYL KETONE PEROXIDE**
SYNS: MEKP ◇ MEK PEROXIDE

OSHA PEL: CL 0.7 ppm
ACGIH TLV: CL 0.2 ppm
DFG MAK: Organic Peroxide, moderate skin irritant.
DOT Class: Forbidden.

SAFETY PROFILE: Moderately toxic by ingestion and inhalation. An experimental tumorigen. A moderate skin and eye irritant. A shock-sensitive explosive.

**MKB000** CAS: 10595-95-6 $C_3H_8N_2O$ HR: 3
**N,N-METHYLETHYLNITROSAMINE**
SYNS: ETHYLMETHYLNITROSAMINE ◇ NEMA ◇ N-NITROSOMETHYL-ETHYLAMINE (MAK) ◇ NMEA

DFG MAK: Animal Carcinogen, Suspected Human Carcinogen.

SAFETY PROFILE: Poison by ingestion. An experimental carcinogen. When heated to decomposition it emits toxic fumes of $NO_x$.

**MKG750** CAS: 107-31-3 $C_2H_4O_2$ HR: 3
**METHYL FORMATE**
DOT: 1243
SYNS: METHYL FORMATE (DOT) ◇ METHYL METHANOATE

OSHA PEL: (Transitional: TWA 100 ppm) TWA 100 ppm; STEL 150 ppm
ACGIH TLV: TWA 100 ppm; STEL 150 ppm
DFG MAK: 100 ppm (250 mg/m$^3$)
DOT Class: Flammable Liquid

SAFETY PROFILE: Moderately toxic by ingestion. Flammable liquid. Explosive in the form of vapor. To fight fire, use alcohol foam, $CO_2$, dry chemical.

PROP: Colorless liquid; agreeable odor. Mp: −99.8°, bp: 31.5°, lel: 5.9%, uel: 20%, flash p: −2.2°F, d: 0.98149 @ 15°/4°, 0.975 @ 20°/

4°, autoign temp: 869°F, vap press: 400 mm @ 16°/0°, vap d: 2.07. Solidifies at about 100°. Moderately sol in water and methyl alcohol; misc in alc.

**MKN000** CAS: 60-34-4 $CH_6N_2$ HR: 3
**METHYL HYDRAZINE**
DOT: 1244
SYNS: HYDRAZOMETHANE ◇ 1-METHYL HYDRAZINE ◇ MMH

OSHA PEL: CL 0.2 ppm (skin))
ACGIH TLV: CL 0.2 ppm; Suspected Human Carcinogen
NIOSH REL: CL 0.08 $mg/m^3$/2H
DOT Class: Flammable Liquid

SAFETY PROFILE: Poison by inhalation, ingestion, and skin contact. An experimental carcinogen and teratogen. Corrosive to skin, eyes, and mucous membranes. Very dangerous fire hazard. May self-ignite in air. To fight fire, use alcohol foam, $CO_2$, dry chemical. When heated to decomposition it emits toxic fumes of $NO_x$.

PROP: Colorless, hydroscopic liquid; ammonia-like odor. D: 0.874 @ 25°, mp: −20.9°, bp: 87.8°, vap d: 1.6, flash p: 73.4°F, fp: −52.4°, autoign temp: 196°, lel: 2.5%, uel: 97 ±2%. Sltly sol in water, sol in alc, hydrocarbons, and ether. Misc with water and hydrazine. Strong reducing agent.

**MKN250** CAS: 7339-53-9 $CH_6N_2 \cdot ClH$ HR: 3
**METHYLHYDRAZINE HYDROCHLORIDE**

NIOSH REL: CL 0.08 $mg/m^3$/2H

SAFETY PROFILE: Poison by ingestion. When heated to decomposition it emits very toxic fumes of $Cl^-$ and $NO_x$.

**MKW200** CAS: 74-88-4 $CH_3I$ HR: 3
**METHYL IODIDE**
DOT: 2644
SYN: IODOMETHANE

OSHA PEL: (Transitional: TWA 5 ppm (skin)) TWA 2 ppm (skin)
ACGIH TLV: TWA 2 ppm (skin); Suspected Human Carcinogen.
DFG MAK: Animal Carcinogen, Suspected Human Carcinogen.
NIOSH REL: (Methylbromide) Reduce to lowest level.
DOT Class: Poison B

SAFETY PROFILE: A poison by ingestion. Moderately toxic by inhalation and skin contact. A suspected carcinogen. A human skin irritant. When heated to decomposition it emits toxic fumes of $I^-$.

PROP: Colorless liquid, turns brown on exposure to light. Mp: −66.4°, bp: 42.5°, d: 2.279 @ 20°/4°, vap press: 400 mm @ 25.3°, vap d: 4.89. Sol in water @ 15°, misc in alc and ether.

**MKW450** CAS: 110-12-3 $C_7H_{14}O$ HR: 2
**METHYL ISOAMYL KETONE**
DOT: 2302

SYNS: ISOAMYL METHYL KETONE ◇ ISOPENTYL METHYL KETONE ◇ 2-METHYL-5-HEXANONE ◇ MIAK

OSHA PEL: TWA 50 ppm
ACGIH TLV: TWA 50 ppm
NIOSH REL: TWA 50 ppm
DOT Class: Flammable or Combustible Liquid

SAFETY PROFILE: Moderately toxic by ingestion. Flammable. To fight fire, use dry chemical, $CO_2$, foam, fog.

PROP: Colorless, stable liquid; pleasant odor. Bp: 144°, d: 0.8132 @ 20°/20°, fp: −73.9°, flash p: 110°F (OC). Sltly sol in water; misc with most organic solvents.

**MKW600** CAS: 108-11-2 $C_6H_{14}O$ HR: 3
**METHYL ISOBUTYL CARBINOL**
DOT: 2053
SYNS: ISOBUTYL METHYL CARBINOL ◇ MAOH ◇ METHYL AMYL ALCOHOL ◇ 4-METHYL-2-PENTANOL (MAK) ◇ MIBC

OSHA PEL: (Transitional: TWA 25 ppm (skin)) TWA 25 ppm; STEL 40 ppm (skin)
ACGIH TLV: TWA 25 ppm; STEL 40 ppm (skin)
DFG MAK: 25 ppm (100 mg/m$^3$)
DOT Class: Flammable or Combustible Liquid

SAFETY PROFILE: Moderately toxic by ingestion, skin contact. A skin and severe eye irritant. Flammable. A moderate explosion hazard. To fight fire, use alcohol foam.

PROP: Clear liquid. Bp: 131.8°, fp: $<-90°$ (sets to a glass), flash p: 106°F, d: 0.8079 @ 20°/20°, vap press: 2.8 mm @ 20°, vap d: 3.53, lel: 1.0%, uel: 5.5%.

**MKX250** CAS: 624-83-9 $C_2H_3NO$ HR: 3
**METHYL ISOCYANATE**
DOT: 2480
SYN: ISOCYANIC ACID, METHYL ESTER

OSHA PEL: TWA 0.02 ppm (skin)
ACGIH TLV: TWA 0.02 ppm (skin)
DFG MAK: 0.01 ppm (0.025 mg/m$^3$)
DOT Class: Flammable Liquid

SAFETY PROFILE: Poison by inhalation, ingestion, and skin contact. A severe eye, skin, and mucous membrane irritant and a sensitizer. A very dangerous fire hazard. To fight fire, use spray, foam, $CO_2$, dry chemical. When heated to decomposition it emits toxic fumes of $NO_x$.

PROP: Liquid. D: 0.9599 @ 20°/20°, bp: 39.1°, flash p: $<5°F$.

**MLA750** CAS: 563-80-4 $C_5H_{10}O$ HR: 3
**METHYL ISOPROPYL KETONE**
DOT: 2397

SYNS: ISOPROPYL METHYL KETONE ◇ 3-METHYL BUTAN 2-ONE (DOT) ◇ MIPK

OSHA PEL: TWA 200 ppm
ACGIH TLV: TWA 200 ppm
DOT Class: Flammable Liquid

SAFETY PROFILE: Poison by ingestion. Mildly toxic by inhalation and skin contact. A skin and eye irritant. Flammable.

**MLC750** CAS: 75-86-5 $C_4H_7NO$ HR: 3
**2-METHYLLACTONITRILE**
DOT: 1541
SYNS: ACETONE CYANOHYDRIN (DOT) ◇ 2-HYDROXY-2-METHYLPROPIONITRILE

NIOSH REL: (Nitriles) CL 4 $mg/m^3$/15M
DOT Class: Poison B

SAFETY PROFILE: Poison by ingestion and skin contact. Readily decomposes to HCN and acetone. Combustible. To fight fire, use $CO_2$, dry chemical, alcohol foam. When heated to decomposition it emits toxic fumes of $CN^-$.

PROP: Mp: −20°, bp: 82° @ 23 mm, d: 0.932 @ 19°, autoign temp: 1270°F, flash p: 165°F, vap d: 2.93.

**MLE650** CAS: 74-93-1 $CH_4S$ HR: 3
**METHYL MERCAPTAN**
DOT: 1064
SYN: METHANETHIOL

OSHA PEL: (Transitional: CL 10 ppm) TWA 0.5 ppm
ACGIH TLV: TWA 0.5 ppm
DFG MAK: 0.5 ppm (1 $mg/m^3$)
NIOSH REL: (n-Alkane Monothiols) CL 0.5 ppm/15M
DOT Class: Flammable Gas

SAFETY PROFILE: Poison by inhalation. Very dangerous fire hazard. Explosive in the form of vapor. Reacts with water, steam or acids to produce toxic and flammable vapors. To fight fire, use alcohol foam, $CO_2$, dry chemical. Upon decomposition it emits highly toxic fumes of $SO_x$.

PROP: Flammable gas; odor of rotten cabbage. Mp: −123.1°, vap d: 1.66, lel: 3.9%, uel: 21.8%, bp: 5.95°, d: 0.8665 @ 20°/4°, solidifies @ −123°, flash p: −0.4°F.

**MLF550** CAS: 22967-92-6 $CH_3Hg$ HR: 3
**METHYLMERCURY**
SYNS: METHYLMERCURY(II) CATION ◇ METHYLMERCURY ION

OSHA PEL: (Transitional: CL 1 mg/10$m^3$) TWA 0.01 mg(Hg)/$m^3$; STEL 0.03 $mg/m^3$ (skin)

ACGIH TLV: TWA 0.01 mg(Hg)/m$^3$; STEL 0.03 mg(Hg)/m$^3$
DFG MAK: 0.01 mg/m$^3$
NIOSH REL: TWA 0.05 mg(Hg)/m$^3$

SAFETY PROFILE: A poison. Experimental reproductive effects. When heated to decomposition it emits toxic fumes of Hg.

**MLH750** CAS: 80-62-6 $C_5H_8O_2$ HR: 3
**METHYL METHACRYLATE**
DOT: 1247
SYNS: METHACRYLIC ACID, METHYL ESTER (MAK) ◇ METHYL-α-METHYL-ACRYLATE ◇ METHYL-2-METHYL-2-PROPENOATE ◇ MME

OSHA PEL: TWA 100 ppm
ACGIH TLV: TWA 100 ppm
DFG MAK: 50 ppm (210 mg/m$^3$)
DOT Class: Flammable Liquid

SAFETY PROFILE: Moderately toxic by inhalation. An experimental tumorigen and teratogen. A skin and eye irritant. A very dangerous fire hazard. Explosive in the form of vapor. The monomer may undergo spontaneous, explosive polymerization. To fight fire, use foam, $CO_2$, dry chemical.

PROP: Colorless liquid, very sltly sol in water. Mp: −50°, bp: 101.0°, flash p: 50°F (OC), d: 0.936 @ 20°/4°, vap press: 40 mm @ 25.5°, vap d: 3.45, lel: 2.1%, uel: 12.5%.

**MMU250** CAS: 614-00-6 $C_7H_8N_2O$ HR: 3
**N-METHYL-N-NITROSOANILINE**
SYNS: N-METHYL-N-NITROSOBENZENAMINE ◇ MNA ◇ N-NITROSOMETHYL-PHENYLAMINE (MAK) ◇ PHENYLMETHYLNITROSAMINE

DFG MAK: Animal Carcinogen, Suspected Human Carcinogen.

SAFETY PROFILE: Poison by ingestion. An experimental carcinogen and teratogen. When heated to decomposition it emits toxic fumes of $NO_x$.

**MNH000** CAS: 298-00-0 $C_8H_{10}NO_5PS$ HR: 3
**METHYL PARATHION**
DOT: 2783
SYNS: O,O-DIMETHYL-O-(p-NITROPHENYL) PHOSPHOROTHIOATE ◇ PARA-THION METHYL

OSHA PEL: TWA 0.2 mg/m$^3$ (skin)
ACGIH TLV: TWA 0.2 mg/m$^3$ (skin)
NIOSH REL: (Methyl Parathion) TWA 0.2 mg/m$^3$
DOT Class: Poison B

SAFETY PROFILE: Poison by inhalation, ingestion, and skin contact. An experimental teratogen. When heated to decomposition it emits very toxic fumes of $NO_x$, $PO_x$, and $SO_x$.

PROP: Crystals. Vap d: 9.1, mp: 37-38°, d: 1.358 @ 20°/4°. Sol in most organic solvents.

**MPF200** CAS: 872-50-4 $C_5H_9NO$ HR: 2
**N-METHYLPYRROLIDONE**
SYNS: 1-METHYL-2-PYRROLIDONE ◇ NMP

DFG MAK: 100 ppm (400 mg/$m^3$)

SAFETY PROFILE: Mildly toxic by ingestion and skin contact. An experimental teratogen. Combustible. To fight fire, use foam, $CO_2$, dry chemical. When heated to decomposition it emits toxic fumes of $NO_x$.

PROP: Colorless liquid; mild odor. Bp: 202°, fp: −24°, flash p: 204°F (OC), d: 1.027 @ 25°/4°, vap d: 3.4.

**MPI750** CAS: 681-84-5 $C_4H_{12}O_4Si$ HR: 3
**METHYL SILICATE**
DOT: 2606
SYNS: METHYL ESTER of o-SILICIC ACID ◇ METHYL ORTHOSILICATE ◇ TETRAMETHYLSILICATE

OSHA PEL: TWA 1 ppm
ACGIH TLV: TWA 1 ppm
DOT Class: Flammable Liquid

SAFETY PROFILE: Moderately toxic by ingestion and inhalation. Mildly toxic by skin contact. A severe eye irritant. Flammable.

PROP: Clear liquid. Vap d: 5.25.

**MPK250** CAS: 98-83-9 $C_9H_{10}$ HR: 1
**α-METHYL STYRENE**
DOT: 2303
SYNS: ISOPROPENYLBENZENE ◇ 2-PHENYLPROPENE ◇ β-PHENYLPROPYLENE

OSHA PEL: (Transitional: TWA CL 100 ppm) TWA 50 ppm; STEL 100 ppm
ACGIH TLV: TWA 50 ppm; STEL 100 ppm
DFG MAK: 100 ppm (480 mg/$m^3$)
DOT Class: Flammable Liquid.

SAFETY PROFILE: Mildly toxic by inhalation. A skin and eye irritant. Flammable.

PROP: Colorless liquid. D: 0.862 @ 20°/4°, mp: −96.0°, bp: 152.4°. Insol in water; misc in alc and ether.

**MPK500** CAS: 25013-15-4 $C_9H_{10}$ HR: 2
**METHYL STYRENE (mixed isomers)**
SYN: VINYLTOLUENE (MIXED ISOMERS)

OSHA PEL: TWA 100 ppm
ACGIH TLV: TWA 50 ppm; STEL 100 ppm
DFG MAK: 100 ppm (480 mg/$m^3$)

SAFETY PROFILE: Moderately toxic by ingestion.

PROP: A mixture containing 55-70% m-vinyltoluene and 30-45% p-vinyltoluene (AMIHAB 14,387,56).

**MQR275** CAS: 21087-64-9 $C_8H_{14}N_4OS$ HR: 3
**METRIBUZIN**
SYN: 4-AMINO-6-tert-BUTYL-3-METHYLTHIO-as-TRIAZIN-5-ONE

OSHA PEL: TWA 5 $mg/m^3$
ACGIH TLV: TWA 5 $mg/m^3$

SAFETY PROFILE: Poison by ingestion. When heated to decomposition it emits very toxic fumes of $NO_x$ and $SO_x$.

**MQR750** CAS: 7786-34-7 $C_7H_{13}O_6P$ HR: 3
**MEVINPHOS**
SYNS: CMDP ◇ DIMETHYL-1-CARBOMETHOXY-1-PROPEN-2-YL PHOSPHATE ◇ PHOSDRIN (OSHA)

OSHA PEL: (Transitional: TWA 0.1 $mg/m^3$ (skin)) TWA 0.01 ppm; STEL 0.03 ppm (skin)
ACGIH TLV: TWA 0.01 ppm; STEL 0.03 ppm (skin)
DFG MAK: 0.01 ppm (0.1 $mg/m^3$)
DOT Class: Poison B

SAFETY PROFILE: Poison by ingestion, inhalation, and skin contact. When heated to decomposition it emits toxic fumes of $PO_x$.

**MQS250** CAS: 12001-26-2 HR: 2
**MICA**
SYN: MICA SILICATE

OSHA PEL: (Transitional: TWA 20 mppcf) TWA Respirable Fraction: 3 $mg/m^3$
ACGIH TLV: TWA Respirable Fraction: 3 $mg/m^3$

SAFETY PROFILE: The dust is injurious to lungs.

PROP: Contains less than 1% crystalline silica.

**MQS500** CAS: 90-94-8 $C_{17}H_{20}N_2O$ HR: 3
**MICHLER'S KETONE**
SYNS: 4,4′-BIS(DIMETHYLAMINO)BENZOPHENONE ◇ TETRAMETHYLDIAMINOBENZOPHENONE

DFG MAK: Suspected Carcinogen.

SAFETY PROFILE: Poison by ingestion. An experimental carcinogen. When heated to decomposition it emits toxic fumes of $NO_x$.

PROP: Leaves from ethanol. Mp: 172°, bp: >360° decomp. Insol in water; very sol in benzene; sol in alc; very sltly sol in ether.

**MQV750** CAS: 8012-95-1 HR: 3
**MINERAL OIL**
SYNS: MINERAL OIL, WHITE (FCC) ◇ OIL MIST, MINERAL (OSHA, ACGIH) ◇ PARRAFIN OIL ◇ PETROLATUM, LIQUID ◇ WHITE MINERAL OIL

OSHA PEL: Oil Mist: TWA 5 mg/m$^3$
ACGIH TLV: Oil Mist: TWA 5 mg/m$^3$; STEL 10 mg/m$^3$

SAFETY PROFILE: A human carcinogen by inhalation. Inhalation of vapor or particulates can cause aspiration pneumonia. An eye irritant. Combustible liquid. To fight fire, use dry chemical, $CO_2$, foam.

PROP: A mixture of liquid hydrocarbons from petroleum. Colorless, oily liquid; practically tasteless and odorless. D: 0.83-0.86 (light), 0.875-0.905 (heavy), flash p: 444°F (OC), ULC: 10-20. Insol in water and alc; sol in benzene, chloroform, and ether.

**MRC250** CAS: 7439-98-7 Mo HR: 3
**MOLYBDENUM**
SYN: MOLYBDATE

OSHA PEL: Soluble Compounds: TWA 5 mg(Mo)/m$^3$; Insoluble Compounds: (Transitional: TWA Total Dust: 15 mg/m$^3$; Respirable Fraction: 5 mg/m$^3$) TWA Total Dust: 10 mg/m$^3$; Respirable Fraction: 5 mg/m$^3$
ACGIH TLV: Soluble Compounds: TWA 5 mg(Mo)/m$^3$; Insoluble Compounds: TWA 10 mg(Mo)/m$^3$
DFG MAK: (insoluble compounds) 15 mg/m$^3$; (soluble compounds) 5 mg/m$^3$

SAFETY PROFILE: An experimental teratogen. Experimental reproductive effects. Flammable or explosive in the form of dust. When heated to decomposition it emits toxic fumes of Mo.

PROP: Cubic, silver-white metallic crystals or gray-black powder. Mp: 2622°, bp: approx 4825°, d: 10.2, vap press: 1 mm @ 3102°.

**MRG000** CAS: 55398-86-2 $C_{12}H_9ClO$ HR: 2
**MONOCHLORO DIPHENYL OXIDE**
SYNS: MONOCHLOROPHENYLETHER ◇ PHENYL ETHER MONO-CHLORO

OSHA PEL: TWA 500 µg/m$^3$

SAFETY PROFILE: Moderately toxic by ingestion. When heated to decomposition it emits toxic fumes of $Cl^-$.

**MRH209** CAS: 6923-22-4 $C_7H_{14}NO_5P$ HR: 3
**MONOCROTOPHOS**
SYNS: AZODRIN ◇ 3-(DIMETHOXYPHOSPHINYLOXY)N-METHYL-cis-CROTONAMIDE

OSHA PEL: TWA 0.25 mg/m$^3$
ACGIH TLV: TWA 0.25 mg/m$^3$

SAFETY PROFILE: Poison by ingestion, inhalation, and skin contact. When heated to decomposition it emits very toxic $NO_x$ and $PO_x$.

PROP: A reddish-brown solid; mild ester odor. Bp: 125°.

**MRP750** CAS: 110-91-8 $C_4H_9NO$ HR: 3
**MORPHOLINE**
DOT: 2054/1760

SYNS: DIETHYLENE IMIDOXIDE ◇ 1-OXA-4-AZACYCLOHEXANE ◇ TETRAHYDRO-p-ISOXAZINE

OSHA PEL: (Transitional: TWA 20 ppm (skin) TWA 20 ppm (skin); STEL 30 ppm (skin)
ACGIH TLV: TWA 20 ppm; STEL 30 ppm (skin); (Proposed: TWA 20 ppm (skin))
DFG MAK: 20 ppm (70 $mg/m^3$)
DOT Class: Flammable Liquid

SAFETY PROFILE: Moderately toxic by ingestion, inhalation, skin contact. A corrosive irritant to skin, eyes, and mucous membranes. Flammable liquid. To fight fire, use alcohol foam, $CO_2$, dry chemical. When heated to decomposition it emits highly toxic fumes of $NO_x$.

PROP: Colorless, hygroscopic oil; amine odor. Bp: 128.9°, fp: −7.5°, flash p: 100°F (OC), autoign temp: 590°F, vap press: 10 mm @ 23°, vap d: 3.00, mp: −4.9°, d: 1.007 @ 20°/4°. Volatile with steam; misc with water evolving some heat, acetone, benzene, and ether.

**NAG400** CAS: 300-76-5 $C_4H_7Br_2Cl_2O_4P$ HR: 3
**NALED**
DOT: 2783
SYN: DIMETHYL-1,2-DIBROMO-2,2-DICHLOROETHYL PHOSPHATE

OSHA PEL: (Transitional: TWA 3 $mg/m^3$) TWA 3 $mg/m^3$ (skin)
ACGIH TLV: TWA 3 $mg/m^3$ (skin)

SAFETY PROFILE: Poison by ingestion and inhalation. Moderately toxic by skin contact. A skin irritant. When heated to decomposition it emits very toxic fumes of $Br^-$, $Cl^-$ and $PO_x$.

PROP: Slightly sol in aliphatic hydrocarbons, very sol in aromatic hydrocarbons. Mp: 27.0°.

**NAI500** CAS: 8030-30-6 HR: 3
**NAPHTHA**
DOT: 1255/1256/1271/2553
SYNS: BENZIN ◇ COAL TAR NAPHTHA ◇ PETROLEUM DISTILLATES (NAPHTHA) ◇ PETROLEUM SPIRIT (DOT) ◇ VM & P NAPHTHA (ACGIH)

OSHA PEL: TWA 100 ppm
ACGIH TLV: TWA 300 ppm
NIOSH REL: TWA 350 $mg/m^3$; CL 1800 $mg/m^3$/15M
DOT Class: Flammable Liquid

SAFETY PROFILE: Mildly toxic by inhalation. Flammable. To fight fire, use foam, $CO_2$, dry chemical.

PROP: Dark straw-colored to colorless liquid. Bp: 149-216°, flash p: 107°F (CC), d: 0.862-0.892, autoign temp: 531°F. Sol in benzene, toluene, xylene, etc. Made from American coal oil and consists chiefly of pentane, hexane, and heptane.

**NAJ500** CAS: 91-20-3 $C_{10}H_8$ HR: 3

**NAPHTHALENE**

DOT: 1334/2304

SYNS: NAPHTHALINE ◇ TAR CAMPHOR ◇ WHITE TAR

OSHA PEL: (Transitional: TWA 10 ppm) TWA 10 ppm; STEL 15 ppm

ACGIH TLV: TWA 10 ppm; STEL 15 ppm

DFG MAK: 10 ppm (50 mg/$m^3$)

DOT Class: ORM-A

SAFETY PROFILE: Human poison by ingestion. An experimental tumorigen. Experimental reproductive effects. An eye and skin irritant. Flammable. Explosive in the form of vapor or dust. To fight fire, use water, $CO_2$, dry chemical.

PROP: White, crystalline, volatile flakes; aromatic odor. Mp: 80.1°, bp: 217.9°, flash p: 174°F (OC), d: 1.162, lel: 0.9%, uel: 5.9%, vap press: 1 mm @ 52.6°, vap d: 4.42, autoign temp: 1053°F (567°C). Sol in alc and benzene; insol in water; very sol in ether, $CCl_4$, $CS_2$, hydronaphthalenes, and in fixed and volatile oils.

**NAM500** CAS: 3173-72-6 $C_{12}H_6N_2O_2$ HR: 2

**1,5-NAPHTHALENE DIISOCYANATE**

SYNS: 1,5-DIISOCYANATONAPHTHALENE ◇ ISOCYANIC ACID-1,5-NAPHTHYLENE ESTER

DFG MAK: 0.01 ppm (0.09 mg/$m^3$)

SAFETY PROFILE: A powerful allergen. An irritant. When heated to decomposition it emits toxic fumes of $NO_x$.

PROP: White to light yellow crystals.

**NBE000** CAS: 134-32-7 $C_{10}H_9N$ HR: 3

**1-NAPHTHYLAMINE**

DOT: 2077

SYNS: 1-AMINONAPHTHALENE ◇ α-NAPHTHYLAMINE

OSHA PEL: Carcinogen

DOT Class: Poison B

SAFETY PROFILE: Moderately toxic by ingestion. A human carcinogen. Combustible. To fight fire, use dry chemical, $CO_2$, mist, spray. When heated to decomposition it emits toxic fumes of $NO_x$.

PROP: White crystals, reddening on exposure to air; unpleasant odor. Mp: 50°, bp: 300.8°, flash p: 315°F, d: 1.131, vap press: 1 mm @ 104.3°, vap d: 4.93. Sublimes, volatile with steam. Sol in 590 parts $H_2O$; very sol in alc and ether. Keep well closed and away from light.

**NBE500** CAS: 91-59-8 $C_{10}H_9N$ HR: 3

**β-NAPHTHYLAMINE**

DOT: 1650

SYNS: 2-AMINONAPHTHALENE ◇ C.I. 37270

OSHA PEL: Carcinogen
ACGIH TLV: Confirmed Human Carcinogen.
DFG MAK: Human Carcinogen.
DOT Class: Poison B

SAFETY PROFILE: Moderately toxic by ingestion. A human carcinogen. A teratogen. Combustible. When heated to decomposition it emits toxic fumes of $NO_x$.

PROP: White to faint pink, lustrous leaflets; faint aromatic odor. Mp: 111.5°, d: 1.061 @ 98°/4°, vap press: 1 mm @ 108.0°. Bp: 306° also listed as 294°. Sol in hot water, alc, and ether.

**NCG500** CAS: 7440-01-9 Ne HR: 1
**NEON**
DOT: 1913/1065

ACGIH TLV: Simple Asphyxiant
DOT Class: Nonflammable Gas

SAFETY PROFILE: An inert asphyxiant gas.

PROP: Colorless inert gas; odorless. Mp: −248.67°, bp: −245.9°. D: (liquid): 1.204 @ −245.9°; (gas) 0.89994 g/L @ 0°.

**NCH000** CAS: 463-82-1 $C_5H_{12}$ HR: 3
**NEOPENTANE**
DOT: 1265/2044
SYNS: 2,2-DIMETHYLPROPANE (DOT) ◇ tert-PENTANE (DOT)

NIOSH REL: TWA 120 ppm; CL 610 ppm/15M
DOT Class: Flammable Liquid

SAFETY PROFILE: Gas and the liquid are flammable.

PROP: Liquid or gas. Solidifies @ −19.8°, bp: 9.5°, d: 0.613° @ 0°/0° (liquid), flash p: <19.4°F, lel: 1.4%, uel: 7.5%. Insol in water.

**NCI500** CAS: 126-99-8 $C_4H_5Cl$ HR: 3
**NEOPRENE**
DOT: 1991
SYNS: CHLOROBUTADIENE ◇ β-CHLOROPRENE (OSHA, MAK)

OSHA PEL: (Transitional: TWA 25 ppm (skin)) TWA 10 ppm (skin)
ACGIH TLV: TWA 10 ppm (skin)
DFG MAK: 10 ppm (36 $mg/m^3$)
NIOSH REL: CL (Chloroprene) 1 ppm/15M
DOT Class: Flammable Liquid

SAFETY PROFILE: Poison by ingestion. Moderately toxic by inhalation. An experimental teratogen. Experimental reproductive effects. A very dangerous fire hazard. Explosive in the form of vapor. To fight fire, use alcohol foam. Autooxidizes in air to form an unstable peroxide which catalyzes exothermic polymerization of the monomer. When heated to decomposition it emits toxic fumes of $Cl^-$.

PROP: Colorless liquid. D: 0.958 @ 20°/20°, bp: 59.4°, flash p: −4°F, lel: 4.0%, uel: 20%, vap d: 3.0, brittle point: −35°, softens @ approx 80°. Sltly sol in water; misc in alc and ether.

**NCS000** CAS: 62765-93-9 HR: 2
**NIAX CATALYST ESN**

NIOSH REL: (Niax ESN): Exposure to be minimized

SAFETY PROFILE: Moderately toxic by ingestion and skin contact. When heated to decomposition it emits toxic fumes of $NO_x$ and $CN^-$.

PROP: Mixture of 95% dimethylaminopropionitrile and 5% bis-dimethylaminoethyl ether.

**NCW500** CAS: 7440-02-0 Ni HR: 3
**NICKEL and COMPOUNDS**
SYNS: C.I. 77775 ◇ RANEY NICKEL

OSHA PEL: TWA Soluble: Compounds: 0.1 mg(Ni)/$m^3$; Insoluble Compounds: 1 mg(Ni)/$m^3$
ACGIH TLV: TWA 1 mg(Ni)/$m^3$; (Proposed: TWA 0.05 mg(Ni)/$m^3$; Human Carcinogen)
DFG TRK: 0.5 mg/$m^3$; Human Carcinogen.
NIOSH REL: (Inorganic Nickel) TWA 0.015 mg(Ni)/$m^3$

SAFETY PROFILE: Poison by ingestion. An experimental carcinogen and teratogen. Experimental reproductive effects. All airborne nickel contaminating dusts are regarded as carcinogenic by inhalation. Nickel carbonyl is probably the most hazardous compound of nickel in the workplace. Hypersensitivity to nickel is common and can cause allergic contact dermatitis and pulmonary asthma. Powders may ignite spontaneously in air.

PROP: A silvery-white, hard, malleable and ductile metal. D: 8.90 @ 25°, vap press: 1 mm @ 1810°. Crystallizes as metallic cubes. Mp: 1455°, bp: 2730°. Stable in air at room temp.

**NCZ000** CAS: 13463-39-3 $C_4NiO_4$ HR: 3
**NICKEL CARBONYL**
DOT: 1259
SYN: NICKEL TETRACARBONYL

OSHA PEL: TWA 0.001 ppm (Ni)
ACGIH TLV: TWA 0.05 ppm (Ni); (Proposed: TWA 0.05 mg(Ni)/$m^3$; Human Carcinogen)
DFG TRK: 0.1 ppm; Animal Carcinogen, Suspected Human Carcinogen.
NIOSH REL: (Nickel Carbonyl) TWA 0.001 ppm
DOT Class: Poison B

SAFETY PROFILE: A human poison by inhalation. A human carcinogen. An experimental teratogen. Sensitization dermatitis is fairly common. A very dangerous fire hazard. Moderate explosion hazard. To fight fire, use water, foam, $CO_2$, dry chemical. When heated to decomposition or on contact with acid or acid fumes it emits highly toxic fumes of CO.

PROP: Colorless, volatile liquid or needles. Bp: 43°, mp: −19.3°, lel: 2% @ 20°, d: 1.3185 @ 17°, vap press: 400 mm @ 25.8°, flash p: <−4°. Oxidizes in air. Sol in alc, benzene, chloroform, acetone, and carbon tetrachloride.

**NDN000** CAS: 54-11-5 $C_{10}H_{14}N_2$ HR: 3
**NICOTINE**
DOT: 1654
SYN: 3-(N-METHYLPYRROLIDINO)PYRIDINE

OSHA PEL: TWA 0.5 mg/m$^3$ (skin)
ACGIH TLV: TWA 0.5 mg/m$^3$ (skin)
DFG MAK: 0.07 ppm; 0.5 mg/m$^3$
DOT Class: Poison B

SAFETY PROFILE: A deadly human poison. Experimental poison by skin contact. Human teratogenic effects. Can be absorbed by intact skin. Combustible. Moderately explosive in the form of vapor. To fight fire, use alcohol foam, dry chemical, $CO_2$.

PROP: An alkaloid from tobacco. In its pure state, a colorless and almost odorless oil; sharp burning taste. Mp: <−80°, bp: 247.3° (partial decomp), lel: 0.75%, uel: 4.0%, d: 1.0092 @ 20°, autoign temp: 471°F, vap press: 1 mm @ 61.8°, vap d: 5.61. Volatile with steam; misc with water below 60°; very sol in alc, chloroform ether, petr ether, and kerosene oils.

**NED500** CAS: 7697-37-2 $HNO_3$ HR: 3
**NITRIC ACID**
DOT: 2031
SYNS: AZOTIC ACID ◇ HYDROGEN NITRATE

OSHA PEL: (Transitional: TWA 2 ppm) TWA 2 ppm; STEL 4 ppm
ACGIH TLV: TWA 2 ppm; STEL 4 ppm
DFG MAK: 10 ppm (25 mg/m$^3$)
NIOSH REL: (Nitric Acid) TWA 2 ppm
DOT Class: Corrosive Material

SAFETY PROFILE: Human poison. An experimental teratogen. Experimental reproductive effects. Corrosive to eyes, skin, mucous membranes and teeth. Flammable by chemical reaction with reducing agents. It is a powerful oxidizing agent. Will react with water or steam to produce heat and toxic and corrosive fumes. To fight fire, use water. When heated to decomposition it emits highly toxic fumes of $NO_x$ and hydrogen nitrate.

PROP: Transparent, colorless or yellowish, fuming, suffocating, caustic and corrosive liquid. Mp: −42°, bp: 86°, d: 1.50269 @ 25°/4°.

**NEG100** CAS: 10102-43-9 NO HR: 3
**NITRIC OXIDE**
DOT: 1660
SYN: NITROGEN MONOXIDE

OSHA PEL: TWA 25 ppm
ACGIH TLV: TWA 25 ppm
NIOSH REL: (Oxides of Nitrogen) TWA 25 ppm
DOT Class: Poison A

SAFETY PROFILE: A poison gas. A severe eye, skin, and mucous membrane irritant. An oxidizer. The liquid is a sensitive explosive.

PROP: Colorless gas, blue liquid and solid. Mp: −161°, bp: −151.18, d: 1.3402 g/L; liquid, 1.269 @ −150°; gas, 1.04.

**NEJ500** CAS: 602-87-9 $C_{12}H_9NO_2$ HR: 3
**5-NITROACENAPHTHENE**
SYNS: 5-NAN ◇ 5-NITROACENAPHTHYLENE

DFG MAK: Animal Carcinogen, Suspected Human Carcinogen.

SAFETY PROFILE: An experimental carcinogen and teratogen. When heated to decomposition it emits toxic fumes of $NO_x$.

**NEM480** CAS: 119-34-6 $C_6H_6N_2O_3$ HR: 3
**2-NITRO-4-AMINOPHENOL**
SYNS: 4-AMINO-2-NITROPHENOL ◇ C.I. 76555

DFG MAK: Suspected Carcinogen.

SAFETY PROFILE: Moderately toxic by ingestion. An experimental carcinogen. A severe eye irritant. When heated to decomposition it emits toxic fumes of $NO_x$.

**NEO500** CAS: 100-01-6 $C_6H_6N_2O_2$ HR: 3
**p-NITROANILINE**
DOT: 1661 ◇ SYNS: p-AMINONITROBENZENE ◇ C.I. 37035 ◇ 4-NITROANILINE (MAK) ◇ PARANITROANILINE, SOLID (DOT) ◇ PNA

OSHA PEL: (Transitional: TWA 6 mg/m$^3$ (skin)) TWA 3 mg/m$^3$ (skin)
ACGIH TLV: TWA 3 mg/m$^3$ (skin)
DFG MAK: 1 ppm (6 mg/m$^3$)
DOT Class: Poison B

SAFETY PROFILE: Poison by ingestion. Combustible. To fight fire, use water spray or mist, foam, dry chemical, $CO_2$. When heated to decomposition it emits toxic fumes of $NO_x$.

PROP: Bright yellow powder. Mp: 148.5°, bp: 332°, flash p: 390°F (CC), d: 1.424, vap press: 1 mm @ 142.4°. Sol in water, alc, ether, benzene, and methanol.

**NEX000** CAS: 98-95-3 $C_6H_5NO_2$ HR: 3
**NITROBENZENE**
DOT: 1662
SYNS: MIRBANE OIL ◇ OIL of MIRBANE (DOT)

OSHA PEL: TWA 1 ppm (skin)
ACGIH TLV: TWA 1 ppm (skin)
DFG MAK: 1 ppm (5 mg/$m^3$); BAT: 100 μg/L of aniline in blood after several shifts.
DOT Class: Poison B

SAFETY PROFILE: Human poison. Moderately toxic by skin contact. Experimental reproductive effects. An eye and skin irritant. It is absorbed rapidly through the skin. Combustible. Moderate explosion hazard. When heated to decomposition it emits toxic fumes of $NO_x$.

PROP: Bright yellow crystals or yellow, oily liquid; odor of volatile almond oil. Mp: 6°, bp: 210-211°, ULC: 20-30%, lel: 1.8% @ 200°F, flash p: 190°F (CC), d: 1.205 @ 15°/4°, autoign temp: 900°F, vap press: 1 mm @ 44.4°, vap d: 4.25. Volatile with steam; sol in about 500 parts water; very sol in alc, benzene, ether, and oils.

**NFQ000** CAS: 92-93-3 $C_{12}H_9NO_2$ HR: 3
**4-NITROBIPHENYL**
SYNS: p-PHENYL-NITROBENZENE ◇ PNB

OSHA-Carcinogen.
ACGIH TLV: Confirmed Human Carcinogen.
DFG MAK: Animal Carcinogen, Suspected Human Carcinogen.

SAFETY PROFILE: Moderately toxic by ingestion. A human carcinogen. When heated to decomposition it emits toxic fumes of $NO_x$.

PROP: Needles from alc. Mp: 113-114°, bp: 340°. Insol in water; sltly sol in cold alc; very sol in ether.

**NFS525** CAS: 100-00-5 $C_6H_4ClNO_2$ HR: 3
**p-NITROCHLOROBENZENE**
DOT: 1578
SYNS: p-CHLORONITROBENZENE ◇ 4-CHLORO-1-NITROBENZENE ◇ PNCB

OSHA PEL: TWA 1 mg/$m^3$ (skin)
ACGIH TLV: TWA 0.1 ppm (skin)
DFG MAK: 1 mg/$m^3$
DOT Class: Poison B

SAFETY PROFILE: A poison by ingestion. An experimental carcinogen. May explode on heating. When heated to decomposition it emits very toxic fumes of $NO_x$ and $Cl^-$.

PROP: D: 1.520, mp: 83°, bp: 242°, flash p: 110°. Insol in water; sltly sol in alc; very sol in $CS_2$ and ether.

**NFY500** CAS: 79-24-3 $C_2H_5NO_2$ HR: 3
**NITROETHANE**
DOT: 2842

OSHA PEL: TWA 100 ppm
ACGIH TLV: TWA 100 ppm
DFG MAK: 100 ppm (310 $mg/m^3$)
DOT Class: Flammable or Combustible Liquid

SAFETY PROFILE: Moderately toxic by ingestion. An eye and mucous membrane irritant. A dangerous fire hazard. Explodes when heated. To fight fire, use alcohol foam, $CO_2$, dry chemical; water can blanket fire. When heated to decomposition it emits toxic fumes of $NO_x$.

PROP: Oily, colorless liquid; agreeable odor. Mp: −90°, bp: 114.0°, fp: −50°, d: 1.052 @ 20°/20°, autoign temp: 778°F, flash p: 106°F, decomp @ 335-382°, lel: 4.0%, vap press: 15.6 mm @ 20°, vap d: 2.58. Sol in water, acid, and alkali; misc in alc, chloroform, and ether.

**NGR500** CAS: 10102-44-0 $NO_2$ HR: 3
**NITROGEN DIOXIDE**
DOT: 1067
SYN: NITROGEN PEROXIDE, LIQUID (DOT)

OSHA PEL: (Transitional: CL 5 ppm) STEL 1 ppm
ACGIH TLV: TWA 3 ppm; STEL 5 ppm
DFG MAK: 5 ppm (9 $mg/m^3$)
NIOSH REL: CL (Oxides of Nitrogen) 1 ppm/15M
DOT Class: Poison A

SAFETY PROFILE: A poison by inhalation. An experimental teratogen. Experimental reproductive effects. When heated to decomposition it emits toxic fumes of $NO_x$.

PROP: Colorless solid to yellow liquid; irritating odor. Mp: −9.3° (yellow liquid), bp: 21° (red-brown gas with decomp), d: 1.491 @ 0°, vap press: 400 mm @ 80°. Liquid below 21.15°. Sol in concentrated sulfuric acid and nitric acid. Corrosive to steel when wet.

**NGU000** CAS: 10024-97-2 $N_2O$ HR: 2
**NITROGEN OXIDE**
DOT: 1070/2201
SYNS: DINITROGEN MONOXIDE ◇ LAUGHING GAS ◇ NITROUS OXIDE (DOT)

ACGIH TLV: 50 ppm
NIOSH REL: (Waste Anesthetic Gases and Vapors) TWA 25 ppm
DOT Class: Nonflammable Gas

SAFETY PROFILE: Moderately toxic by inhalation. An experimental teratogen. Does not burn but is flammable by chemical reaction and supports combustion. Moderate explosion hazard. Also self-explodes at high temperatures.

PROP: Colorless gas, liquid or cubic crystals; slightly sweet odor. Mp: −90.8°, bp: −88.49°, d: 1.977 g/L (liquid 1.226 @ −89°).

**NGW000** CAS: 7783-54-2 $F_3N$ HR: 3
**NITROGEN TRIFLUORIDE**
DOT: 2451
SYN: NITROGEN FLUORIDE

OSHA PEL: TWA 10 ppm
ACGIH TLV: TWA 10 ppm
NIOSH REL: (Inorganic Fluorides) TWA 2.5 mg(F)/$m^3$
DOT Classification: Poison A; Nonflammable Gas.

SAFETY PROFILE: A poison. Mildly toxic by inhalation. Severe explosion hazard by chemical reaction with reducing agents. A very dangerous fire hazard. Particularly hazardous under pressure. When heated to decomposition it emits highly toxic fumes of $F^-$.

PROP: Colorless gas; odor of mold. Mp: −208.5°, bp: −129°, d (liquid): 1.537 @ −129°; d: (liquid @ bp:) 1.885; insol in $H_2O$.

**NGY000** CAS: 55-63-0 $C_3H_5N_3O_9$ HR: 3
**NITROGLYCERIN**
DOT: 0143
SYNS: BLASTING GELATIN (DOT) ◇ GLYCEROL TRINITRATE

OSHA PEL: (Transitional: TWA CL 0.2 ppm (skin)) SREL: 0.1 mg/$m^3$ (skin)
ACGIH TLV: TWA 0.05 ppm (skin)
DFG MAK: 0.05 ppm (0.5 mg/$m^3$) (skin)
NIOSH REL: CL (Nitroglycerin or EGDN) 0.1 mg/$m^3$/20M
DOT Class: Flammable Liquid

SAFETY PROFILE: Human poison. An experimental tumorigen and teratogen. A skin irritant. Toxic effects may occur by ingestion, inhalation of dust, or absorption through intact skin. A very dangerous fire hazard. Severe explosion hazard. When heated to decomposition it emits toxic fumes of $NO_x$.

PROP: Colorless to yellow liquid; sweet taste. Mp: 13°, bp: explodes @ 218°, d: 1.599 @ 15°/15°, vap press: 1 mm @ 127°, vap d: 7.84, autoign temp: 518°F, decomp @ 50-60°, volatile @ 100°. Misc with ether, acetone, glacial acetic acid, sltly sol in petr ether and glycerol.

**NGY500** CAS: 53569-64-5 $C_3H_5N_3O_9 \cdot C_2H_4N_2O_6$ HR: 3
**NITROGLYCERIN mixed with ETHYLENE GLYCOL DINITRATE (1:1)**
SYN: ETHYLENE GLYCOL DINITRATE mixed with NITROGLYCERIN (1:1)

OSHA PEL: (Transitional: TWA CL 0.2 ppm (skin)) STEL: 0.1 mg/$m^3$ (skin)
NIOSH REL: CL 0.1 mg/$m^3$/20M

SAFETY PROFILE: Human poison. A high explosive. When heated to decomposition it emits very toxic fumes of $NO_x$.

**NHI500** CAS: 7046-61-9 $C_6H_8N_4O_4$ HR: 3
**NITROIMINODIETHYLENEDIISOCYANIC ACID**
SYN: 3-NITRO-3-AZAPENTANE-1,5-DIISOCYANATE

NIOSH REL: (Diisocyanates) TWA 0.005 ppm; CL 0.02 ppm/10M

SAFETY PROFILE: When heated to decomposition it emits toxic fumes of $NO_x$.

**NHM500** CAS: 75-52-5 $CH_3NO_2$ HR: 3
**NITROMETHANE**
DOT: 1261
SYN: NITROCARBOL

OSHA PEL: TWA 100 ppm
ACGIH TLV: TWA 100 ppm
DFG MAK: 100 ppm (250 $mg/m^3$)
DOT Class: Flammable Liquid

SAFETY PROFILE: Poison by ingestion. A very dangerous fire hazard. May explode by detonation, heat, or shock. When heated to decomposition it emits toxic fumes of $NO_x$.

PROP: An oily liquid; moderate to strong, disagreeable odor. Bp: 101°, lel: 7.3%, fp: −29°, flash p: 95°F (CC), d: 1.1322 @ 25°/4°, autoign temp: 785°F, vap press: 27.8 mm @ 20°, vap d: 2.11. Sltly sol in water; sol in alc, and ether.

**NHQ000** CAS: 86-57-7 $C_{10}H_7NO_2$ HR: 3
**1-NITRONAPHTHALENE**
SYN: α-NITRONAPHTHALENE

DFG MAK: Suspected Carcinogen.

SAFETY PROFILE: Poison by ingestion. A skin, eye and mucous membrane irritant. Combustible. To fight fire, use $CO_2$, dry chemical or water spray. When heated to decomposition it emits toxic fumes of $NO_x$.

PROP: Yellow crystals. Bp: 304°, flash p: 327°F (CC), d: 1.331 @ 4°/4°, vap d: 5.96, mp: 59-61°. Insol in water; sol in $CS_2$, alc, chloroform and ether.

**NHQ500** CAS: 581-89-5 $C_{10}H_7NO_2$ HR: 3
**2-NITRONAPHTHALENE**
SYN: β-NITRONAPHTHALENE

DFG TRK: 0.035 ppm; Animal Carcinogen, Suspected Human Carcinogen.

SAFETY PROFILE: Moderately toxic by ingestion. An experimental tumorigen. A skin and lung irritant. Combustible. When heated to decomposition it emits toxic fumes of $NO_x$.

PROP: Colorless in ethanol. Mp: 79°, bp: 165° @ 15 mm. Insol in water, very sol in alc and ether.

**NIX500** CAS: 108-03-2 $C_3H_7NO_2$ HR: 3
**1-NITROPROPANE**

OSHA PEL: TWA 25 ppm
ACGIH TLV: TWA 25 ppm
DFG MAK: 25 ppm (90 mg/$m^3$)
DOT Class: Flammable or Combustible Liquid

SAFETY PROFILE: Poison by ingestion. A human eye irritant. Very dangerous fire hazard. May explode on heating. To fight fire, use alcohol foam, $CO_2$, dry chemical, water spray. When heated to decomposition it emits toxic fumes of $NO_x$.

PROP: Colorless liquid. Bp: 132°, fp: −108°, flash p: 93°F (TCC), d: 1.003 @ 20°/20°, autoign temp: 789°F, vap press: 7.5 mm @ 20°, vap d: 3.06, lel: 2.2%. Sltly sol in water; misc with alc, ether, and many organic solvents.

**NIY000** CAS: 79-46-9 $C_3H_7NO_2$ HR: 3
**2-NITROPROPANE**
DOT: 2608
SYNS: DIMETHYLNITROMETHANE ◇ ISONITROPROPANE

OSHA PEL: (Transitional: TWA 25 ppm) TWA 10 ppm
ACGIH TLV: TWA 10 ppm; Suspected Human Carcinogen.
DFG TRK: 5 ppm; Animal Carcinogen, Suspected Human Carcinogen.
DOT Class: Flammable or Combustible Liquid

SAFETY PROFILE: Moderately toxic by ingestion and inhalation. An experimental carcinogen and teratogen. Very dangerous fire hazard. May explode on heating. To fight fire, use alcohol foam, $CO_2$, dry chemical, water spray. When heated to decomposition it emits toxic fumes of $NO_x$.

PROP: Colorless liquid. Bp: 120°, fp: −93°, flash p: 82°F (TCC), d: 0.992 @ 20°/20°, autoign temp: 802°F, vap press: 10 mm @ 15.8°, vap d: 3.06, lel: 2.6%. Misc with organic solvents; sol in water, alc, and ether.

**NJA000** CAS: 5522-43-0 $C_{16}H_9NO_2$ HR: 3
**3-NITROPYRENE**

DFG MAK: Suspected Carcinogen.

SAFETY PROFILE: An experimental carcinogen. When heated to decomposition it emits toxic fumes of $NO_x$.

**NJW500** CAS: 55-18-5 $C_4H_{10}N_2O$ HR: 3
**N-NITROSODIETHYLAMINE**
SYNS: DIETHYLNITROSAMINE ◇ NDEA

DFG MAK: Animal Carcinogen, Suspected Human Carcinogen.

SAFETY PROFILE: Poison by ingestion. An experimental carcinogen and teratogen. When heated to decomposition it emits toxic fumes of $NO_x$.

PROP: Yellow oil. D: 0.9422 @ 20°/4°, bp: 47° @ 5 mm, bp: 176.9°. Sol in water, alc, and ether.

**NKA000** CAS: 601-77-4 $C_6H_{14}N_2O$ HR: 3
**N-NITROSODIISOPROPYLAMINE**
SYN: N-NITROSODI-i-PROPYLAMINE (MAK)

DFG MAK: Animal Carcinogen, Suspected Human Carcinogen.

SAFETY PROFILE: Moderately toxic by ingestion. An experimental carcinogen. When heated to decomposition it emits toxic fumes of $NO_x$.

**NKA600** CAS: 62-75-9 $C_2H_6N_2O$ HR: 3
**N-NITROSODIMETHYLAMINE**
SYNS: DIMETHYLNITROSAMINE ◇ DMNA ◇ NDMA

OSHA PEL: Carcinogen
ACGIH TLV: Suspected Human Carcinogen
DFG MAK: Animal Carcinogen, Suspected Human Carcinogen.

SAFETY PROFILE: Human poison by ingestion. Experimental poison by inhalation. An experimental carcinogen and transplacental teratogen. When heated to decomposition it emits toxic fumes of $NO_x$.

PROP: Yellow liquid; sol in water, alc, and ether. Bp: 152°, d: 1.005 @ 20°/4°.

**NKB700** CAS: 621-64-7 $C_6H_{14}N_2O$ HR: 3
**N-NITROSODI-N-PROPYLAMINE**
SYNS: DI-n-PROPYLNITROSAMINE ◇ DPNA

DFG MAK: Animal Carcinogen, Suspected Human Carcinogen.

SAFETY PROFILE: Moderately toxic by ingestion. An experimental carcinogen and teratogen. When heated to decomposition it emits toxic fumes of $NO_x$.

**NKD000** CAS: 612-64-6 $C_8H_{10}N_2O$ HR: 3
**N-NITROSO-N-ETHYL ANILINE**
SYNS: ETHYLNITROSOANILINE ◇ NEA ◇ N-NITROSOETHYLPHENYLAMINE (MAK)

DFG MAK: Animal Carcinogen, Suspected Human Carcinogen.

SAFETY PROFILE: Poison by ingestion. An experimental teratogen. When heated to decomposition it emits toxic fumes of $NO_x$.

PROP: Yellow oil. D: 1.087 @ 20°/4°, bp: 119-120° @ 15 mm. Insol in water.

**NKM000** CAS: 1116-54-7 $C_4H_{10}N_2O_3$ HR: 3
**NITROSOIMINO DIETHANOL**
SYNS: NDELA ◇ N-NITROSODIETHANOLAMINE (MAK)

DFG MAK: Animal Carcinogen, Suspected Human Carcinogen.

SAFETY PROFILE: Mildly toxic by ingestion. An experimental carcinogen. When heated to decomposition it emits toxic fumes of $NO_x$.

**NKZ000** CAS: 59-89-2 $C_4H_8N_2O_2$ HR: 3
**4-NITROSOMORPHOLINE**
SYNS: N-NITROSOMORPHOLINE (MAK) ◇ NMOR

DFG MAK: Animal Carcinogen, Suspected Human Carcinogen.

SAFETY PROFILE: Poison by ingestion. Moderately toxic by inhalation. An experimental carcinogen. When heated to decomposition it emits toxic fumes of $NO_x$.

**NLJ500** CAS: 100-75-4 $C_5H_{10}N_2O$ HR: 3
**N-NITROSOPIPERIDINE**
SYNS: HEXAHYDRO-N-NITROSOPYRIDINE ◇ NPIP

DFG MAK: Animal Carcinogen, Suspected Human Carcinogen.

SAFETY PROFILE: Poison by ingestion. An experimental carcinogen. Experimental reproductive effects. When heated to decomposition it emits toxic fumes of $NO_x$.

PROP: Light yellow oil. D: 1.063 @ 18.5°/4°, bp: 217-218°. Sol in water, very sol in acid solns.

**NLP500** CAS: 930-55-2 $C_4H_8N_2O$ HR: 3
**N-NITROSOPYRROLIDINE**
SYN: NPYR ◇ TETRAHYDRO-N-NITROSOPYRROLE

DFG MAK: Animal Carcinogen, Suspected Human Carcinogen.

SAFETY PROFILE: Poison by ingestion. An experimental carcinogen. When heated to decomposition it emits toxic fumes of $NO_x$.

**NMO500** CAS: 99-08-1 $C_7H_7NO_2$ HR: 3
**m-NITROTOLUENE**
DOT: 1664
SYNS: m-METHYLNITROBENZENE ◇ MNT ◇ 3-NITROTOLUOL

OSHA PEL: (Transitional: TWA 5 ppm (skin)) TWA 2 ppm (skin)
ACGIH TLV: TWA 2 ppm (skin)
DFG MAK: 5 ppm (30 $mg/m^3$)
DOT Class: Poison B

SAFETY PROFILE: Poison by ingestion. Combustible. To fight fire, use water, $CO_2$, dry chemical. Probably an explosive. When heated to decomposition it emits toxic fumes of $NO_x$.

PROP: Liquid. Mp: 15.1°, flash p: 233°F (CC), d: 1.1630 @ 15°/4°, vap press: 1 mm @ 50.2°, vap d: 4.72, bp: 231.9°. Misc with alc and ether; sol in benzene and water @ 30°.

**NMO525** CAS: 88-72-2 $C_7H_7NO_2$ HR: 3
**o-NITROTOLUENE**
DOT: 1664
SYNS: 2-METHYLNITROBENZENE ◇ ONT

OSHA PEL: (Transitional: TWA 5 ppm (skin)) TWA 2 ppm (skin)
ACGIH TLV: TWA 2 ppm (skin)

DFG MAK: 5 ppm (30 mg/m$^3$)
DOT Class: Poison B

SAFETY PROFILE: A poison. Moderately toxic by ingestion. Combustible. To fight fire, use water spray, fog, foam, $CO_2$. When heated to decomposition it emits toxic fumes of $NO_x$.

PROP: Yellowish liquid. Mp: −10°, bp: 222.3°, flash p: 223°F (CC), d: 1.1622 @ 19°/15°, vap press: 1 mm @ 50°, vap d: 4.72. Insol in water; sol in $SO_2$ and petr ether; misc in alc, benzene, and ether; sltly sol in $NH_3$.

**NMO550** CAS: 99-99-0 $C_7H_7NO_2$ HR: 3
**p-NITROTOLUENE**
DOT: 1664
SYNS: 4-METHYLNITROBENZENE ◇ PNT

OSHA PEL: (Transitional: TWA 5 ppm (skin)) TWA 2 ppm (skin)
ACGIH TLV: TWA 2 ppm (skin)
DFG MAK: 5 ppm (30 mg/m$^3$)
DOT Class: Poison B

SAFETY PROFILE: A poison. Combustible. To fight fire, use $CO_2$, dry chemical, foam. May explode on standing. When heated to decomposition it emits toxic fumes of $NO_x$.

PROP: Yellowish crystals. Bp: 238.3°, flash p: 223°F (CC), d: 1.286, vap press: 1 mm @ 53.7°, vap d: 4.72. Mp: 53-54°. Insol in water; sol in alc, benzene, ether, chloroform, and acetone.

**NMO600** CAS: 1321-12-6 $C_7H_7NO_2$ HR: 3
**mixo-NITROTOLUENE**

DFG MAK: 5 ppm (30 mg/m$^3$)

SAFETY PROFILE: May decompose explosively if heated above 190°C. When heated to decomposition it emits toxic fumes of $NO_x$.

**NMP500** CAS: 99-55-8 $C_7H_8N_2O_2$ HR: 3
**5-NITRO-o-TOLUIDINE**
SYNS: 2-AMINO-4-NITROTOLUENE ◇ C.I. 37105 ◇ 6-METHYL-3-NITROANILINE

DFG MAK: Animal Carcinogen, Suspected Human Carcinogen.

SAFETY PROFILE: A poison. Moderately toxic by ingestion. An experimental carcinogen. When heated to decomposition it emits toxic fumes of $NO_x$.

**NMX000** CAS: 111-84-2 $C_9H_{20}$ HR: 3
**NONANE**
DOT: 1920
SYN: NONANE (DOT)

OSHA PEL: TWA 200 ppm
ACGIH TLV: TWA 200 ppm
DOT Class: Flammable or Combustible Liquid

SAFETY PROFILE: Mildly toxic by inhalation. A very dangerous fire hazard. Explosive in the form of vapor. To fight fire, use $CO_2$, dry chemical.

PROP: Colorless liquid. Mp: −53.7°, bp: 150.7°, lel: 0.8%, uel: 2.9%, flash p: 88°F (CC), d: 0.718 @ 20°/4°, autoign temp: 374°F, vap press: 10 mm @ 38.0°, vap d: 4.41. Insol in water; sol in abs alc and ether.

**OAP000** CAS: 2234-13-1 $C_{10}Cl_8$ HR: 3
**OCTACHLORONAPHTHALENE**

OSHA PEL: (Transitional: TWA 0.1 $mg/m^3$ (skin)) TWA 0.1 $mg/m^3$ (skin); STEL 0.3 $mg/m^3$
ACGIH TLV: TWA 0.1 $mg/m^3$ (skin); STEL 0.3 $mg/m^3$

SAFETY PROFILE: Poison by inhalation, ingestion, and skin contact. When heated to decomposition it emits highly toxic fumes of $Cl^-$.

**OCU000** CAS: 111-65-9 $C_8H_{18}$ HR: 3
**OCTANE**
DOT: 1262

OSHA PEL: (Transitional: TWA 500 ppm) TWA 300 ppm; STEL 375 ppm
ACGIH TLV: TWA 300 ppm; STEL 375 ppm
DFG MAK: 500 ppm (2350 $mg/m^3$)
NIOSH REL: (Alkanes) TWA 350 $mg/m^3$
DOT Class: Flammable Liquid

SAFETY PROFILE: May act as a simple asphyxiant. A narcotic in high concentration. A very dangerous fire hazard and severe explosion hazard.

PROP: Clear liquid. Bp: 125.8°, lel: 1.0%, uel: 4.7%; fp: −56.5°, flash p: 56°F, d: 0.7036 @ 20°/4°, autoign temp: 428°F, vap press: 10 mm @ 19.2°, vap d: 3.86. Insol in water, sltly sol in alc and ether; misc with benzene.

**OKK000** CAS: 20816-12-0 $O_4Os$ HR: 3
**OSMIUM TETROXIDE**
DOT: 2471
SYN: OSMIC ACID

OSHA PEL: (Transitional: TWA 0.002 $mg/m^3$ (Os)) TWA 0.0002 ppm; STEL 0.0006 ppm (Os)
ACGIH TLV: TWA 0.0002 ppm; STEL 0.0006 ppm (Os)
DFG MAK: 0.0002 ppm (0.002 $mg/m^3$)
DOT Class: Poison B

SAFETY PROFILE: Poison by ingestion, inhalation. Experimental reproductive effects.

PROP: (A) Monoclinic, colorless crystals; (B) yellow mass; pungent, chlorine-like odor. Mp (A): 39.5°, mp: (B): 41°, bp: 130° (sublimes), d: 4.906 @ 22°, vap press (A): 10 mm @ 26.0°, vap press (B): 10 mm @ 31.3°. Sol in benzene.

**OLA000** CAS: 144-62-7 $C_2H_2O_4$ HR: 3
**OXALIC ACID**
SYNS: ETHANEDIOIC ACID ◇ ETHANEDIONIC ACID

OSHA PEL: (Transitional: TWA 1 mg/m$^3$) TWA 1 mg/m$^3$; STEL 2 mg/m$^3$
ACGIH TLV: TWA 1 mg/m$^3$; STEL 2 mg/m$^3$

SAFETY PROFILE: Poison by ingestion and skin contact. A skin and severe eye irritant.

PROP: Orthorhombic colorless crystals. Mp: 101°, sublimes @ 150°, d: 1.653. Sol in water, abs alc and ether

**OPM000** CAS: 101-80-4 $C_{12}H_{12}N_2O$ HR: 3
**4,4′-OXYDIANILINE**
SYNS: p-AMINOPHENYL ETHER ◇ 4,4-DIAMINODIPHENYL ETHER

DFG MAK: Animal Carcinogen, Suspected Human Carcinogen.

SAFETY PROFILE: Moderately toxic by ingestion. An experimental carcinogen. Experimental reproductive effects. When heated to decomposition it emits toxic fumes of $NO_x$.

PROP: Colorless crystals. Mp: 187°, bp: >300°.

**ORA000** CAS: 7783-41-7 $F_2O$ HR: 3
**OXYGEN DIFLUORIDE**
DOT: 2190
SYNS: FLUORINE MONOXIDE ◇ FLUORINE OXIDE ◇ OXYGEN FLUORIDE

OSHA PEL: (Transitional: TWA 0.1 mg/m$^3$) CL 0.05 ppm
ACGIH TLV: CL 0.05 ppm
DOT Class: Poison A

SAFETY PROFILE: Poison by inhalation. A corrosive skin, eye and mucous membrane irritant. Attacks lungs with delayed appearance of symptoms. When heated to decomposition it emits highly toxic fumes of $F^-$.

PROP: Colorless gas or yellowish-brown liquid. Reacts slowly with water. D: (liquid) 1.90 @ −224°, mp: −223.8°, bp: −144.8°.

**ORW000** CAS: 10028-15-6 $O_3$ HR: 3
**OZONE**
SYN: TRIATOMIC OXYGEN

OSHA PEL: (Transitional: TWA 0.1 ppm) TWA 0.1 ppm; STEL 0.3 ppm
ACGIH TLV: CL 0.1 ppm
DFG MAK: 0.1 ppm (0.2 mg/m$^3$)

SAFETY PROFILE: A human poison by inhalation. Experimental reproductive effects. A skin, eye, upper respiratory system and mucous membrane irritant. A powerful oxidizing agent. A severe explosion hazard in liquid form.

PROP: Unstable colorless gas or dark bluc liquid; characteristic odor. Mp: −193°, bp: −111.9°, d (gas): 2.144 g/L, 1.71 @ −183°. D: (liquid) 1.614 g/mL @ −195.4°.

**PAH750** CAS: 8002-74-2 HR: 3
**PARAFFIN**
SYN: PARAFFIN WAX FUME (ACGIH)

OSHA PEL: Fume: TWA 2 $mg/m^3$ (fume)
ACGIH TLV: Fume: TWA 2 $mg/m^3$ (fume)

SAFETY PROFILE: The semi-refined, fully refined, and the crude paraffins are experimental tumorigens by the implant route. Many paraffin waxes contain carcinogens.

PROP: Colorless or white, translucent wax; odorless. D: approx 0.90, mp: 50-57°. Insol in water, alc; sol in benzene, chloroform, ether, carbon disulfide, and oils; misc with fats.

**PAI990** CAS: 4685-14-7 $C_{12}H_{14}N_2$ HR: 3
**PARAQUAT**
SYNS: DIMETHYL VIOLOGEN ◇ METHYL VIOLOGEN (2+)

OSHA PEL: Respirable Dust: (Transitional: TWA 0.5 $mg/m^3$ (skin)) TWA 0.1 $mg/m^3$ (skin)
ACGIH TLV: TWA 0.1 $mg/m^3$

SAFETY PROFILE: Poison by ingestion. When heated to decomposition it emits toxic fumes of $NO_x$.

**PAJ000** CAS: 1910-42-5 $C_{12}H_{14}N_2 \cdot 2Cl$ HR: 3
**PARAQUAT DICHLORIDE**
SYN: DIMETHYL VICLOGEN CHLORIDE

OSHA PEL: TWA 500 $\mu g/m^3$ (skin)
ACGIH TLV: TWA 0.1 $mg/m^3$
DFG MAK: 0.1 $mg/m^3$

SAFETY PROFILE: A human poison by ingestion. Poison experimentally by skin contact. Experimental reproductive effects. An eye irritant. When heated to decomposition it emits toxic fumes of $Cl^-$ and $NO_x$.

PROP: Yellow solid. Sol in water.

**PAJ250** CAS: 2074-50-2 $C_{12}H_{14}N_2 \cdot 2CH_3O_4S$ HR: 3
**PARAQUAT DIMETHYL SULPHATE**
SYNS: PARAQUAT BIS(METHYL SULFATE) ◇ PARAQUAT DIMETHYL SULFATE

ACGIH TLV: TWA 0.1 $mg/m^3$

SAFETY PROFILE: Poison by ingestion. When heated to decomposition it emits very toxic fumes of $SO_x$ and $NO_x$.

**PAK000** CAS: 56-38-2 $C_{10}H_{14}NO_5PS$ HR: 3
**PARATHION**
DOT: 2783
SYNS: O,O-DIETHYL-O-4-NITROPHENYLPHOSPHOROTHIOATE ◇ DNTP ◇ ETHYL PARATHION

OSHA PEL: TWA 0.1 mg/m$^3$ (skin)
ACGIH TLV: TWA 0.1 mg/m$^3$ (skin) (Proposed: BEI: 0.5 mg/L total p-nitrophenol in urine at end of shift.)
DFG MAK: 0.1 mg/m$^3$; BAT: 500 μg/L p-nitrophenol in urine after several shifts.
NIOSH REL: TWA 0.05 mg/m$^3$
DOT Class: Poison B

SAFETY PROFILE: A deadly human poison. An experimental carcinogen and teratogen. Experimental reproductive effects. Combustible. When heated to decomposition it emits highly toxic fumes of $NO_x$, $PO_x$, $SO_x$.

PROP: Pale-yellow liquid. Bp: 375°, mp: 6°. Very sol in alcs, esters, ethers, ketones, and aromatic hydrocarbons; insol in water, petroleum ether, and kerosene.

**PAT750** CAS: 19624-22-7 $B_5H_9$ HR: 3
**PENTABORANE**
DOT: 1380

OSHA PEL: (Transitional: 0.005 ppm) TWA 0.005 ppm; STEL 0.015 ppm
ACGIH TLV: TWA 0.005 ppm; STEL 0.013 ppm
DFG MAK: 0.005 ppm (0.01 mg/m$^3$)
DOT Class: Flammable Liquid

SAFETY PROFILE: Poison by inhalation. Dangerous fire hazard by chemical reaction; spontaneously flammable in air. Dangerous explosion hazard. To fight fire, use special fire-fighting materials; water is not effective.

PROP: Colorless gas or liquid; bad odor. Mp: −46.6°, d: 0.61 @ 0°, vap d: 2.2, vap press: 66 mm @ 0°, lel: 0.42%, bp: 60.°

**PAW250** CAS: 42279-29-8 $C_{12}H_5Cl_5O$ HR: 3
**PENTACHLORO DIPHENYL OXIDE**
SYNS: ETHER, PENTACHLOROPHENYL ◇ PHENYL ETHER PENTACHLORO

OSHA PEL: TWA 0.5 mg/m$^3$

SAFETY PROFILE: Poison by ingestion. When heated to decomposition-it emits toxic fumes of $Cl^-$.

**PAW500** CAS: 76-01-7 $C_2HCl_5$ HR: 3
**PENTACHLOROETHANE**
DOT: 1669
SYNS: ETHANE PENTACHLORIDE ◇ PENTALIN

DFG MAK: 5 ppm (40 mg/m$^3$)
DOT Class: Poison B

SAFETY PROFILE: Poison by inhalation. An experimental carcinogen. Moderately toxic by ingestion. An irritant. Flammable. Moderately explosive. To fight fire, use water, $CO_2$, dry chemical. Dangerous; when heated to decomposition it emits highly toxic fumes of $Cl^-$.

PROP: Colorless liquid; chloroform-like odor. Mp: −29°, bp: 161-162°, d: 1.6728 @ 25°/4°. Insol in water; misc in alc and ether.

**PAW750** CAS: 1321-64-8 $C_{10}H_3Cl_5$ HR: 3
**PENTACHLORONAPHTHALENE**

OSHA PEL: TWA 0.5 mg/m$^3$ (skin)
ACGIH TLV: TWA 0.5 mg/m$^3$
DFG MAK: 0.5 mg/m$^3$

SAFETY PROFILE: Poison by ingestion, inhalation, and skin contact. An irritant. When heated to decomposition it emits highly toxic fumes of $Cl^-$.

PROP: White solid.

**PAX250** CAS: 87-86-5 $C_6HCl_5O$ HR: 3
**PENTACHLOROPHENOL**
DOT: 2020
SYNS: PCP ◇ PENTA

OSHA PEL: TWA 0.5 mg/m$^3$ (skin)
ACGIH TLV: TWA 0.5 mg/m$^3$ (skin); BEI: 2 mg(total PCP)/L in urine prior to last shift of workweek; 5 mg(free PCP)/L in plasma at end of shift.
DFG MAK: 0.05 ppm (0.5 mg/m$^3$); BAT: 1000 μg/L in plasma/serum.
DOT Class: ORM-E

SAFETY PROFILE: Human poison by ingestion. Poison experimentally by skin contact. A suspected human carcinogen. An experimental teratogen. A skin irritant. When heated to decomposition it emits highly toxic fumes of $Cl^-$.

PROP: Dark-colored flakes and sublimed needle crystals; characteristic odor. Mp: 191°, bp: 310° (decomp), d: 1.978, vap press: 40 mm @ 211.2°. Sol in ether and benzene; very sol in alc; insol in water; sltly sol in cold petroleum ether.

**PBB750** CAS: 115-77-5 $C_5H_{12}O_4$ HR: 1
**PENTAERYTHRITOL**
SYNS: 2,2-BIS(HYDROXYMETHYL)-1,3-PROPANEDIOL ◇ METHANE TETRAMETHYLOL

OSHA PEL: (Transitional: TWA Total Dust: 15 mg/m$^3$; Respirable Fraction: 5 mg/m$^3$) TWA Total Dust: 10 mg/m$^3$; Respirable Fraction: 5 mg/m$^3$
ACGIH TLV: TWA (nuisance particulate) 10 mg/m$^3$ of total dust (when toxic impurities are not present, e.g., quartz $< 1\%$).

SAFETY PROFILE: Mildly toxic by ingestion. A nuisance dust. Flammable.

PROP: Crystals. Mp: 262°, d: 1.38 @ 25°/4°.

**PBK250** CAS: 109-66-0 $C_5H_{12}$ HR: 3
**PENTANE**
DOT: 1265
SYN: AMYL HYDRIDE (DOT)

OSHA PEL: (Transitional: TWA 1000 ppm; STEL 750 ppm) TWA 600 ppm; STEL 750 ppm
ACGIH TLV: TWA 600 ppm; STEL 750 ppm
DFG MAK: 1000 ppm (2950 $mg/m^3$)
NIOSH REL: TWA 350 $mg/m^3$
DOT Class: Flammable Liquid

SAFETY PROFILE: The liquid can cause blisters on contact. Flammable liquid. Severe explosion hazard. To fight fire, use foam, $CO_2$, dry chemical.

PROP: Colorless liquid. Bp: 36.1°, flash p: $<-40$°F, fp: $-129.8$°, d: 0.626 @ 20°/4°, autoign temp: 588°F, vap press: 400 mm @ 18.5°, vap d: 2.48, lel: 1.5%, uel: 7.8%. Sol in water; misc in alc, ether, and organic solvents.

**PBM000** CAS: 110-66-7 $C_5H_{12}S$ HR: 2
**1-PENTANETHIOL**
DOT: 1111
SYNS: AMYL MERCAPTAN (DOT) ◇ AMYL SULFHYDRATE ◇ PENTYL MERCAPTAN

NIOSH REL: CL 0.5 ppm/15M
DOT Class: Label: Flammable Liquid.

SAFETY PROFILE: Moderately toxic by inhalation. A weak sensitizer and allergen. May cause contact dermatitis. Dangerous fire hazard. To fight fire, use foam, $CO_2$, dry chemical.

PROP: Water-white to yellow liquid. D: 0.857 @ 20°, bp: 123.64°, flash p: 65°F, vap press: 13.8 mm @ 25°, vap d: 3.59. Insol in water; misc in alc and ether.

**PBN250** CAS: 107-87-9 $C_5H_{10}O$ HR: 3
**2-PENTANONE**
DOT: 1249
SYNS: ETHYL ACETONE ◇ METHYL PROPYL KETONE (ACGIH, DOT) ◇ MPK

OSHA PEL: (Transitional: TWA 200 ppm; STEL 250 ppm) TWA 200 ppm; STEL 250 ppm
ACGIH TLV: TWA 200 ppm; STEL 250 ppm
DFG MAK: 200 ppm (700 $mg/m^3$)
NIOSH REL: TWA 530 $mg/m^3$
DOT Class: Label: Flammable Liquid.

SAFETY PROFILE: Moderately toxic by ingestion. A skin irritant. A highly flammable liquid. An explosion hazard in vapor form. To fight fire, use alcohol foam.

PROP: Water-white liquid; fruity, ethereal odor. D: 0.801-0.806, vap d: 3.0, bp: 216°F, flash p: 45°F, autoign temp: 941°F, lel: 1.5%, uel: 8.2%. Sltly sol in water; misc with alc and ether.

**PCF275** CAS: 127-18-4 $C_2Cl_4$ HR: 3
**PERCHLOROETHYLENE**
DOT: 1897
SYNS: ETHYLENE TETRACHLORIDE ◇ PERCHLORETHYLENE ◇ TETRACHLOROETHYLENE (DOT)

OSHA PEL: (Transitional: TWA 100 ppm; CL 200 ppm; Pk 600 ppm/5M)TWA 25 ppm
ACGIH TLV: TWA 50 ppm; STEL 200 ppm (Proposed: BEI: 7 mg/L trichloroacetic acid in urine at end of workweek.)
DFG MAK: 50 ppm (345 mg/m$^3$); BAT: blood 100 μg/dl
NIOSH REL: (Tetrachloroethylene) Minimize workplace exposure.
DOT Class: Poison B

SAFETY PROFILE: Moderately toxic to humans by inhalation. An experimental carcinogen and teratogen. An eye and severe skin irritant. When heated to decomposition it emits highly toxic fumes of $Cl^-$.

PROP: Colorless liquid; chloroform-like odor. Mp: −23.35°, bp: 121.20°, d: 1.6311 @ 15°/4°, vap press: 15.8 mm @ 22°, vap d: 5.83.

**PCF300** CAS: 594-42-3 $CCl_4S$ HR: 3
**PERCHLOROMETHYL MERCAPTAN**
DOT: 1670
SYNS: PCM ◇ TRICHLOROMETHANE SULFENYL CHLORIDE

OSHA PEL: TWA 0.1 ppm
ACGIH TLV: TWA 0.1 ppm
DOT Class: Poison B

SAFETY PROFILE: Poison by ingestion and inhalation. A severe skin, eye, and mucous membrane irritant. When heated to decomposition it emits very toxic fumes of $Cl^-$ and $SO_x$.

PROP: Yellow, oily liquid. Bp: slt decomp @ 149°, d: 1.700 @ 20°, vap d: 6.414.

**PCF750** CAS: 7616-94-6 $ClFO_3$ HR: 2
**PERCHLORYL FLUORIDE**
SYNS: CHLORINE FLUORIDE OXIDE ◇ CHLORINE OXYFLUORIDE

OSHA PEL: (Transitional: TWA 3 ppm) TWA 3 ppm; STEL 6 ppm
ACGIH TLV: TWA 3 ppm; STEL 6 ppm

SAFETY PROFILE: Moderately toxic by inhalation route. While nonflammable, it supports combustion. Moderately explosive. When heated to decomposition it emits toxic fumes of $F^-$ and $Cl^-$.

PROP: Colorless, noncorrosive gas; characteristic sweet odor. Mp: −146°, bp: −46.8°, d: (liquid): 1.434. d: (gas): 0.637.

**PCJ400** HR: 1
**PERLITE**

OSHA PEL: TWA Total Dust: 15 $mg/m^3$; Respirable Fraction: 5 $mg/m^3$

ACGIH TLV: TWA (nuisance particulate) 10 $mg/m^3$ of total dust (when toxic impurities are not present, e.g., quartz < 1%).

SAFETY PROFILE: A nuisance dust.

PROP: Natural glass, amorphous mineral consisting of fused sodium potassium aluminum silicate, containing <1% quartz.

**PCL500** CAS: 79-21-0 $C_2H_4O_3$ HR: 3
**PEROXYACETIC ACID**
DOT: 2131
SYNS: ACETYL HYDROPEROXIDE ◇ ETHANEPEROXOIC ACID ◇ PERACETIC ACID (MAK)

DFG MAK: Very strong skin effects.
DOT Class: Organic Peroxide

SAFETY PROFILE: Poison by ingestion. Moderately toxic by inhalation and skin contact. A corrosive eye, skin, and mucous membrane irritant. Flammable. Severe explosion hazard. To fight fire, use water, foam, $CO_2$.

PROP: Not over 40% peracetic acid and not over 6% hydrogen peroxide. Colorless liquid; strong odor. Bp: 105°, explodes @ 110°, flash p: 105°F (OC), d: 1.15 @ 20°. Sol in water.

**PCS250** CAS: 8002-05-9 HR: 1
**PETROLEUM DISTILLATE**
DOT: 1268
SYN: NAPHTHA

OSHA PEL: (Transitional: TWA 500 ppm) TWA 400 ppm
DOT Class: Mildly toxic by inhalation. Combustible Liquid

SAFETY PROFILE: Flammable or combustible liquid.

**PCT250** CAS: 64475-85-0 HR: 3
**PETROLEUM SPIRITS**
DOT: 1115
SYNS: MINERAL SPIRITS ◇ STODDARD SOLVENT ◇ V.M. AND P. NAPHTHA

OSHA PEL: TWA 300 ppm
ACGIH TLV: TWA 300 ppm
NIOSH REL: TWA 350 $mg/m^3$; CL 1800 $mg/m^3$/15M

SAFETY PROFILE: Moderately toxic to humans. Mildly toxic by inhalation. Highly dangerous fire hazard. Explosive in the form of vapor. To fight fire, use foam, $CO_2$, dry chemical.

PROP: Volatile, clear, colorless and nonfluorescent liquid. Mp: <−73°, bp: 40-80°, ULC: 95-100, lel: 1.1%, uel: 5.9%, flash p: <0°F, d: 0.635-0.660, autoign temp: 550°F, vap d: 2.50.

**PCW250** CAS: 85-01-8 $C_{14}H_{10}$ HR: 3
**PHENANTHRENE**
SYN: PHENANTRIN

OSHA PEL: TWA 0.2 mg/m$^3$

SAFETY PROFILE: Moderately toxic by ingestion. An experimental neoplastigen by skin contact. A human skin photosensitizer. Combustible. To fight fire, use water, foam, $CO_2$, dry chemical.

PROP: Solid or monoclinic crystals. Mp: 100°, bp: 339°, d: 1.179 @ 25°, vap press: 1 mm @ 118.3°, vap d: 6.14. Insol in water; sol in $CS_2$, benzene and hot alc; very sol in ether.

**PDN750** CAS: 108-95-2 $C_6H_6O$ HR: 3
**PHENOL**
DOT: 1671/2312/2821
SYNS: CARBOLIC ACID ◇ HYDROXYBENZENE

OSHA PEL: TWA 5 ppm (skin)
ACGIH TLV: TWA 5 ppm (skin); BEI: 250 mg(total phenol)/g creatinine in urine at end of shift.
DFG MAK: 5 ppm (19 mg/m$^3$); BAT: 300 mg/L at end of shift.
NIOSH REL: TWA 20 mg/m$^3$; CL 60 mg/m$^3$/15M
DOT Class: Poison B

SAFETY PROFILE: Human poison by ingestion. Moderately toxic by skin contact. A severe eye and skin irritant. An experimental carcinogen. Combustible. To fight fire, use alcohol foam, $CO_2$, dry chemical.

PROP: White, crystalline mass which turns pink or red if not perfectly pure; burning taste, distinctive odor. Mp: 40.6°, bp: 181.9°, flash p: 175°F (CC), d: 1.072, autoign temp: 1319°F, vap press: 1 mm @ 40.1°, vap d: 3.24. Sol in water; misc in alc and ether.

**PDP250** CAS: 92-84-2 $C_{12}H_9NS$ HR: 3
**PHENOTHIAZINE**
SYNS: DIBENZO-1,4-THIAZINE ◇ THIODIPHENYLAMINE

OSHA PEL: TWA 5 mg/m$^3$ (skin)
ACGIH TLV: TWA 5 mg/m$^3$ (skin)

SAFETY PROFILE: Moderately toxic to humans by ingestion. Experimental reproductive effects. Can cause skin irritation and photosensitization. When heated to decomposition or on contact with acid or acid fumes it emits highly toxic fumes of $SO_x$ and $NO_x$.

PROP: Yellow, rhombic leaflets or diamond-shaped plates from toluene or butanol. Mp: 185.1°, sublimes at 130° at 1 mm, bp: 371°. Freely sol in benzene; sol in ether and hot acetic acid; sltly sol in alc and in mineral oils; practically insol in petroleum ether, chloroform, and water.

**PEY500** CAS: 106-50-3 $C_6H_8N_2$ HR: 3
**p-PHENYLENEDIAMINE**
DOT: 1673
SYNS: p-AMINOANILINE ◇ p-BENZENEDIAMINE ◇ C.I. 76060 ◇ PPD

OSHA PEL: TWA 0.1 mg/m$^3$ (skin)
ACGIH TLV: TWA 0.1 mg/m$^3$ (skin); (Proposed TWA 0.1 mg/m$^3$)
DFG MAK: 0.1 mg/m$^3$
DOT Class: ORM-A

SAFETY PROFILE: Poison by ingestion. A human skin irritant. Combustible. To fight fire, use water, $CO_2$, dry chemical.

PROP: White-sltly red crystals. Mp: 146°, flash p: 312°F, vap d: 3.72, bp: 267°. Sol in alc, chloroform, and ether.

**PFA850** CAS: 101-84-8 $C_{12}H_{10}O$ HR: 2
**PHENYL ETHER**
SYNS: BIPHENYL OXIDE ◇ DIPHENYL ETHER ◇ PHENOXYBENZENE

OSHA PEL: Vapor: TWA 1 ppm
ACGIH TLV: TWA 1 ppm; STEL 2 ppm (vapor)
DFG MAK: 1 ppm (7 mg/m$^3$)

SAFETY PROFILE: Moderately toxic by ingestion. A skin and eye irritant. Combustible. To fight fire, use water, foam, $CO_2$, dry chemical.

PROP: Colorless crystals, geranium odor. Mp: 28°, bp: 257°, flash p: 239°F, d: 1.0728 @ 20°, vap d: 5.86, autoign temp: 1148°F, lel: 0.8%, uel: 1.5%.

**PFA860** CAS: 8004-13-5 $C_{12}H_{10}•C_{12}H_{10}O$ HR: 3
**PHENYL ETHER-BIPHENYL MIXTURE**
SYNS: BIPHENYL, mixed with BIPHENYL OXIDE (3:7) ◇ DOWTHERM

OSHA PEL: Vapor: TWA 1 ppm

SAFETY PROFILE: Poison by inhalation. Moderately toxic by ingestion. A mild skin and eye irritant.

**PFH000** CAS: 122-60-1 $C_9H_{10}O_2$ HR: 3
**PHENYL GLYCYDYL ETHER**
SYNS: 1,2-EPOXY-3-PHENOXYPROPANE ◇ PGE ◇ PHENOL GLYCIDYL ETHER (MAK) ◇ PHENYL-2,3-EPOXYPROPYL ETHER

OSHA PEL: (Transitional: TWA 10 ppm) TWA 1 ppm
ACGIH TLV: TWA 1 ppm
DFG MAK: 1 ppm (6 mg/m$^3$), Suspected Carcinogen.
NIOSH REL: (Glycidyl Ethers) CL 5 mg/m$^3$/15M

SAFETY PROFILE: Moderately toxic by ingestion and skin contact. A severe eye and skin irritant. An experimental carcinogen. Experimental reproductive effects.

**PFI000** CAS: 100-63-0 $C_6H_8N_2$ HR: 3
**PHENYLHYDRAZINE**
DOT: 2572
SYNS: HYDRAZINE-BENZENE ◇ HYDRAZINOBENEZENE

OSHA PEL: (Transitional: TWA 5 ppm (skin)) TWA 5 ppm (skin); STEL 10 ppm
ACGIH TLV: TWA 5 ppm (skin); STEL 10 ppm; Suspected Human Carcinogen; (Proposed: TWA 0.1 ppm; Suspected Human Carcinogen)
DFG MAK: 5 ppm (22 $mg/m^3$), Suspected Carcinogen.
NIOSH REL: CL 0.6 $mg/m^3$/2H
DOT Class: Poison B

SAFETY PROFILE: Poison by ingestion. An experimental carcinogen. Experimental reproductive effects. Flammable. To fight fire, use alcohol foam. When heated to decomposition it emits highly toxic fumes of $NO_x$.

PROP: Yellow, monoclinic crystals or oil. Mp: 19.6°, bp: 243.5° (decomp), flash p: 192°F (CC), d: 1.0978 @ 20°/4°, vap press: 1 mm @ 71.8°, vap d: 3.7. Sltly sol in hot water; misc in alc, chloroform, ether, and benzene.

**PFI250** CAS: 59-88-1 $C_6H_8N_2 \cdot ClH$ HR: 3
**PHENYLHYDRAZINE HYDROCHLORIDE**
SYN: PHENYLHYDRAZINIUM CHLORIDE

NIOSH REL: CL 0.6 $mg/m^3$/2H

SAFETY PROFILE: Poison by ingestion. Experimental reproductive effects. When heated to decomposition it emits very toxic fumes of $NO_x$ and HCl.

PROP: Leaflet crystals from alc. Mp: 245°. Very sol in water; sol in alc; insol in ether.

**PFL850** CAS: 108-98-5 $C_6H_6S$ HR: 3
**PHENYL MERCAPTAN**
DOT: 2337
SYNS: BENZENETHIOL (DOT) ◇ THIOPHENOL (DOT)

OSHA PEL: TWA 0.5 ppm
ACGIH TLV: TWA 0.5 ppm
NIOSH REL: CL 0.5 $mg/m^3$/15M
DOT Class: Poison B

SAFETY PROFILE: Poison by ingestion, inhalation, and skin contact. A severe eye irritant. When heated to decomposition or on contact with acids it emits toxic fumes of $SO_x$.

PROP: Liquid, repulsive odor. Bp: 168.3°, d: 1.0728 @ 25°/4°.



**PFT500** CAS: 135-88-6 $C_{16}H_{13}N$ HR: 3
**N-PHENYL-β-NAPHTHYLAMINE**
SYNS: 2-ANILINONAPHTHALENE ◇ 2-NAPHTHYLPHENYLAMINE

ACGIH TLV: Suspected Human Carcinogen.
DFG MAK: Suspected Carcinogen.

SAFETY PROFILE: An experimental carcinogen. A suspected human carcinogen. Moderately toxic by ingestion. When heated to decomposition it emits toxic fumes of $NO_x$.

PROP: Rhombic crystals from methanol. Mp: 107-108°, bp: 395.5°. Insol in water; sol in hot benzene; very sol in hot alc and ether.

**PFV250** CAS: 638-21-1 $C_6H_7P$ HR: 3
**PHENYLPHOSPHINE**

OSHA PEL: CL 0.05 ppm
ACGIH TLV: CL 0.05 ppm

SAFETY PROFILE: Poison by inhalation. Ignites spontaneously in air. When heated to decomposition it emits toxic fumes of $PO_x$.

PROP: Needles from aq alc. Mp: 164-165°, bp: 305-308°. Insol in water, sol in alkali; very sol in alc and ether.

**PGS000** CAS: 298-02-2 $C_7H_{17}O_2PS_3$ HR: 3
**PHORATE**
SYN: O,O-DIETHYL-S-ETHYLTHIOMETHYL DITHIOPHOSPHONATE

OSHA PEL: TWA 0.05 mg/m$^3$ (skin); STEL 0.2 mg/m$^3$
ACGIH TLV: TWA 0.05 mg/m$^3$ (skin)

SAFETY PROFILE: Poison by ingestion and skin contact. Experimental reproductive effects. When heated to decomposition it emits toxic fumes of $PO_x$ and $SO_x$.

PROP: Liquid. Bp: 118-120° @ 0.8 mm, d: 1.156. @ 25°/4°. Insol in water; misc with carbon tetrachloride, dioxane, xylene.

**PGX000** CAS: 75-44-5 $CCl_2O$ HR: 3
**PHOSGENE**
DOT: 1076
SYNS: CARBON OXYCHLORIDE ◇ CARBONYL CHLORIDE ◇ CHLOROFORMYL CHLORIDE

OSHA PEL: TWA 0.1 ppm
ACGIH TLV: TWA 0.1 ppm
DFG MAK: 0.1 ppm (0.4 mg/m$^3$)
NIOSH REL: TWA 0.1 ppm; CL 0.2 ppm/15M
DOT Class: Poison A

SAFETY PROFILE: A human poison by inhalation. A severe eye, skin, and mucous membrane irritant. When heated to decomposition or on contact with water or steam it will react to produce toxic and corrosive fumes of $Cl^-$.

PROP: Colorless, poison gas or volatile liquid; odor of new mown hay or green corn. Mp: −118°, bp: 8.3°, d: 1.37 @ 20°, vap press: 1180 mm @ 20°, vap d: 3.4. Very sltly sol in water; very sol in benzene and acetic acid; decomp sltly in water.

**PGY000** CAS: 7803-51-2 $H_3P$ HR: 3
**PHOSPHINE**
DOT: 2199
SYNS: HYDROGEN PHOSPHIDE ◇ PHOSPHORUS TRIHYDRIDE

OSHA PEL: (Transitional: TWA 0.3 ppm) TWA 0.3 ppm; STEL 1 ppm
ACGIH TLV: TWA 0.3 ppm; STEL 1 ppm
DFG MAK: 0.1 ppm (0.15 mg/m$^3$)
DOT Class: Poison A

SAFETY PROFILE: A poison by inhalation. Very dangerous fire hazard by spontaneous chemical reaction. Moderately explosive. To fight fire, use $CO_2$, dry chemical or water spray. When heated to decomposition it emits highly toxic fumes of $PO_x$.

PROP: Colorless gas; foul odor of decaying fish. Mp: −132.5°, bp: −87.5°, d: 1.529 g/L @ 0°, autoign temp: 212°F, lel: 1%. Sltly sol in water.

**PHB250** CAS: 7664-38-2 $H_3O_4P$ HR: 3
**PHOSPHORIC ACID**
DOT: 1805
SYN: ORTHOPHOSPHORIC ACID

OSHA PEL: (Transitional: TWA 1 mg/m$^3$) TWA 1 mg/m$^3$; STEL 3 mg/m$^3$
ACGIH TLV: TWA 1 mg/m$^3$; STEL 3 mg/m$^3$
DOT Class: Corrosive Material

SAFETY PROFILE: Human poison. Moderately toxic by skin contact. A corrosive irritant to eyes, skin, and mucous membranes. When heated to decomposition it emits toxic fumes of $PO_x$.

PROP: Colorless liquid or rhombic crystals. Mp: 42.35°, loses 1/2$H_2O$ @ 213°, fp: 42.4°, d: 1.864 @ 25°, vap press: 0.0285 mm @ 20°. Misc with water and alc.

**PHO500** CAS: 7723-14-0 P HR: 3
**PHOSPHORUS (red)**
DOT: 1381

DFG MAK: 0.1 mg/m$^3$
DOT Class: Flammable Solid and Poison

SAFETY PROFILE: A human poison. Dangerous fire hazard. Moderate explosion hazard by chemical reaction. May explode on impact. To fight fire, use water. When heated to decomposition it emits toxic fumes of $PO_x$.

PROP: Reddish-brown powder. Bp: 280° (with ignition), mp: 590° @ 43 atm, d: 2.34, autoign temp: 500°F in air, vap d: 4.77.

**PHO750** CAS: 7723-14-0 $P_4$ HR: 3
**PHOSPHORUS (yellow)**

DOT: 1381/2447
SYN: WHITE PHOSPHORUS

OSHA PEL: TWA 0.1 mg/m$^3$
ACGIH TLV: TWA 0.1 mg/m$^3$
DOT Class: Flammable Solid and Poison

SAFETY PROFILE: Human poison by ingestion. Experimental reproductive effects. Dangerous fire hazard; ignites spontaneously in air. Very reactive. To fight fire, use water. When heated to decomposition it emits highly toxic fumes of $PO_x$.

PROP: Cubic crystals; colorless to yellow, wax-like solid. Mp: 44.1°, bp: 280°, flash p: spontaneously flammable in air, d: 1.82, autoign temp: 86°F, vap press: 1 mm @ 76.6°, vap d: 4.42.

**PHQ800** CAS: 10025-87-3 $Cl_3OP$ HR: 3
**PHOSPHORUS OXYCHLORIDE**
DOT: 1810
SYNS: PHOSPHORUS OXYTRICHLORIDE ◇ PHOSPHORYL CHLORIDE

OSHA PEL: TWA 0.1 ppm
ACGIH TLV: TWA 0.1 ppm
DFG MAK: 0.2 ppm (1 mg/m$^3$)
DOT Class: Corrosive

SAFETY PROFILE: Poison by inhalation and ingestion. A corrosive eye, skin, and mucous membrane irritant. Potentially explosive reaction with water evolves hydrogen chloride and phosphine which then ignites. When heated to decomposition it emits highly toxic fumes of $Cl^-$ and $PO_x$.

PROP: Colorless to sltly yellow, fuming liquid. Mp: 1.2°, bp: 105.1°, d: 1.685 @ 15.5°, vap press: 40 mm @ 27.3°, vap d: 5.3.

**PHR500** CAS: 10026-13-8 $Cl_5P$ HR: 3
**PHOSPHORUS PENTACHLORIDE**
DOT: 1806
SYNS: PHOSPHORIC CHLORIDE ◇ PHOSPHORUS PERCHLORIDE

OSHA PEL: TWA 1 mg/m$^3$
ACGIH TLV: TWA 0.85 mg/m$^3$
DFG MAK: 1 mg/m$^3$
DOT Class: Corrosive

SAFETY PROFILE: Poison by inhalation. Moderately toxic by ingestion. A severe eye, skin, and mucous membrane irritant. Corrosive to body tissues. Flammable by chemical reaction. To fight fire, use $CO_2$, dry chemical. Will react with water or steam to produce heat and toxic and corrosive fumes. When heated to decomposition it emits highly toxic fumes of $Cl^-$ and $PO_x$.

PROP: Yellowish-white, fuming, crystalline mass; pungent odor. Mp: (under press) 148° decomp, bp: subl @ 160°, d: 4.65 g/L @ 296°, vap press: 1 mm @ 55.5°.

**PHS000** CAS: 1314-80-3 $P_2S_5$ HR: 3

**PHOSPHORUS PENTASULFIDE**

DOT: 1340

SYNS: PHOSPHORIC SULFIDE ◇ SULFUR PHOSPHIDE ◇ THIOPHOSPHORIC ANHYDRIDE

OSHA PEL: (Transitional: TWA 1 $mg/m^3$) TWA 1 $mg/m^3$; STEL 3 $mg/m^3$
ACGIH TLV: TWA 1 $mg/m^3$; STEL 3 $mg/m^3$
DFG MAK: 1 $mg/m^3$
DOT Class: Flammable Solid

SAFETY PROFILE: A poison by ingestion. A severe eye and skin irritant. Dangerous fire hazard. Spontaneous heating in the presence of moisture. Moderate explosion hazard in solid form. Reacts with water, steam, or acids to produce toxic and flammable vapors. Incompatible with air, alcohols, and water. To fight fire use $CO_2$ snow, dry chemical or sand. When heated to decomposition it emits highly toxic fumes of $SO_x$ and $PO_x$.

PROP: Gray to yellow-green, crystalline, deliquescent mass. Bp: 514°, d: 2.09, autoign temp: 287°F. Mp: 286-290°.

**PHS250** CAS: 1314-56-3 $O_5P_2$ HR: 3

**PHOSPHORUS PENTOXIDE**

DOT: 1807

SYNS: DIPHOSPHORUS PENTOXIDE ◇ PHOSPHORIC ANHYDRIDE

DFG MAK: 1 $mg/m^3$
DOT Class: Corrosive Material

SAFETY PROFILE: Poison by inhalation. A corrosive irritant to the eyes, skin and mucous membranes. When heated to decomposition it emits toxic fumes of $PO_x$.

PROP: Deliquescent crystals. D: 2.30; mp: 340°, sublimes @ 360°.

**PHT275** CAS: 7719-12-2 $Cl_3P$ HR: 3

**PHOSPHORUS TRICHLORIDE**

DOT: 1809

SYNS: CHLORIDE of PHOSPHORUS ◇ PHOSPHORUS CHLORIDE

OSHA PEL: (Transitional: TWA 0.5 ppm) TWA 0.2 ppm; STEL 0.5 ppm
ACGIH TLV: TWA 0.2 ppm; STEL 0.5 ppm
DFG MAK: 0.5 ppm (3 $mg/m^3$)
DOT Class: Corrosive Material

SAFETY PROFILE: Poison by inhalation. Moderately toxic by ingestion. A corrosive irritant to skin, eyes, and mucous membranes. Will react with water, steam or acids to produce heat, toxic, and corrosive fumes. To fight fire, use $CO_2$, dry chemical. When heated to decomposition it emits highly toxic fumes of $Cl^-$ and $PO_x$.

PROP: Clear, colorless, fuming liquid. Mp: −111.8°, bp: 76°, d: 1.574 @ 21°, vap press: 100 mm @ 21°, vap d: 4.75. Decomp by water and alc; sol in benzene, chloroform, and ether.

**PHW750** CAS: 85-44-9 $C_8H_4O_3$ HR: 3
**PHTHALIC ANHYDRIDE**
DOT: 2214
SYNS: 1,2-BENZENEDICARBOXYLIC ACID ANHYDRIDE ◇ ESEN ◇ PHTHALIC ACID ANHYDRIDE

OSHA PEL: (Transitional: TWA 2 ppm) TWA 1 ppm
ACGIH TLV: TWA 1 ppm
DFG MAK: 5 $mg/m^3$
DOT Class: Corrosive

SAFETY PROFILE: Poison by ingestion. Experimental teratogenic effects. A corrosive eye, skin, and mucous membrane irritant. Combustible. Moderate explosion hazard. To fight fire, use $CO_2$, dry chemical.

PROP: White, crystalline needles. Mp: 131.2°, lel: 1.7%, uel: 10.4%, bp: 295° (sublimes), flash p: 305°F (CC), d: 1.527 @ 4°, autoign temp: 1058°F, vap press: 1 mm @ 96.5°, vap d: 5.10. Very sltly sol in water; sol in alc; sltly sol in ether.

**PHX550** CAS: 626-17-5 $C_8H_4N_2$ HR: 3
**m-PHTHALODINITRILE**
SYNS: 1,3-BENZENEDICARBONITRILE ◇ 1,3-DICYANOBENZENE ◇ IPN ◇ ISOPHTHALODINITRILE

OSHA PEL: TWA 5 $mg/m^3$
ACGIH TLV: TWA 5 $mg/m^3$

SAFETY PROFILE: Poison by ingestion. An eye irritant. When heated to decomposition it emits toxic fumes of $NO_x$ and $CN^-$.

PROP: Colorless crystals. Vap d: 4.42, mp: 138°, bp: subl. Insol in water; sol in benzene and acetone.

**PIB900** CAS: 1918-02-1 $C_6H_3Cl_3N_2O_2$ HR: 3
**PICLORAM**
SYNS: 4-AMINO-3,5,6-TRICHLOROPICOLINIC ACID ◇ ATCP ◇ TORDON

OSHA PEL: (Transitional: Total Dust: 15 $mg/m^3$; Respirable Fraction: 5 $mg/m^3$) TWA Total Dust: 10 $mg/m^3$; Respirable Fraction: 5 $mg/m^3$
ACGIH TLV: TWA 10 $mg/m^3$

SAFETY PROFILE: Moderately toxic by ingestion. An experimental carcinogen and teratogen. When heated to decomposition it emits very toxic fumes of $Cl^-$ and $NO_x$.

PROP: Crystals. Mp: 218°.

**PID000** CAS: 88-89-1 $C_6H_3N_3O_7$ HR: 3
**PICRIC ACID**
DOT: 0154

SYNS: CARBAZOTIC ACID ◇ C.I. 10305 ◇ 2-HYDROXY-1,3,5-TRINITROBENZENE ◇ 1,3,5-TRINITROPHENOL

OSHA PEL: TWA 0.1 mg/m$^3$ (skin)
ACGIH TLV: TWA 0.1 mg/m$^3$; STEL 0.3 mg/m$^3$ (skin) (Proposed: TWA 0.1 mg/m$^3$)
DFG MAK: 0.1 mg/m$^3$
DOT Class: Class A Explosive.

SAFETY PROFILE: Poison by ingestion. An irritant and allergen. Skin contact can cause local and systemic allergic reactions. Combustible. Very unstable; a severe explosion hazard.

PROP: Yellow crystals or yellow liquid; very bitter. Mp: 121.8°, bp: explodes > 300°, flash p: 302°F, d: 1.763, autoign temp: 572°F, vap d: 7.90.

**PIH175** CAS: 83-26-1 $C_{14}H_{14}O_3$ HR: 3
**PINDONE**
DOT: 2472
SYNS: PIVAL ◇ 2-PIVALYL-1,3-INDANDIONE

OSHA PEL: TWA 0.1 mg/m$^3$
ACGIH TLV: TWA 0.1 mg/m$^3$
DOT Class: Poison B

SAFETY PROFILE: Poison by ingestion.

PROP: Yellow crystals. Mp: 108°.

**PIK000** CAS: 142-64-3 $C_4H_{10}N_2 \cdot 2ClH$ HR: 2
**PIPERAZINE DIHYDROCHLORIDE**

OSHA PEL: TWA 5 mg/m$^3$
ACGIH TLV: TWA 5 mg/m$^3$

SAFETY PROFILE: Mildly toxic by ingestion. When heated to decomposition it emits very toxic fumes of $NO_x$ and HCl.

**PJD500** CAS: 7440-06-4 Pt HR: 3
**PLATINUM**
SYN: C.I. 77795

OSHA PEL: TWA (metal) 1 mg/m$^3$; (soluble salts as Pt) 0.002 mg/m$^3$
ACGIH TLV: TWA (metal) 1 mg/m$^3$; (soluble salts as Pt) 0.002 mg/m$^3$
DFG MAK: 0.002 mg/m$^3$

SAFETY PROFILE: An experimental tumorigen by implant route. Finely divided platinum is a powerful catalyst and can be dangerous to handle. Used catalysts are especially dangerous and may be explosive.

PROP: Silvery-white, malleable, ductile metal; stable in air. Mp: 1772°, bp: 3827°, d: 21.45 @ 20°.

**PJL750** CAS: 1336-36-3 HR: 3
**POLYCHLORINATED BIPHENYLS**
DOT: 2315
SYNS: AROCLOR ◇ CHLORINATED BIPHENYL ◇ PCB (DOT, USDA)

NIOSH REL: TWA (Polychlorinated Biphenyls) 0.001 $mg/m^3$
DFG MAK: Suspected Carcinogen.
DOT Class: ORM-E

SAFETY PROFILE: Moderately toxic by ingestion. Suspected human carcinogens. Experimental carcinogens. Experimental reproductive effects. Combustible. When heated to decomposition they emit highly toxic fumes of $Cl^-$.

PROP: A series of technical mixtures consisting of many isomers and compounds that vary from mobile oily liquids to white crystalline solids and hard noncrystalline resins. Bp: 340-375°, flash p: 383°F (COC), d: 1.44 @ 30°.

**PJM500** CAS: 53469-21-9 HR: 3
**POLYCHLORINATED BIPHENYL (AROCLOR 1242)**
SYNS: AROCLOR 1242 ◇ CHLORODIPHENYL (42% Cl) (OSHA) ◇ PCB's

OSHA PEL: TWA 1 $mg/m^3$ (skin)
ACGIH TLV: TWA 1 $mg/m^3$ (skin)
DFG MAK: 0.1 ppm (1 $mg/m^3$)
NIOSH REL: TWA (Polychlorinated Biphenyls) 0.001 $mg/m^3$

SAFETY PROFILE: Moderately toxic by ingestion. Suspected human carcinogen. Experimental reproductive effects. When heated to decomposition it emits toxic fumes of $Cl^-$.

**PJN000** CAS: 11097-69-1 HR: 3
**POLYCHLORINATED BIPHENYL (AROCLOR 1254)**
SYNS: AROCHLOR 1254 ◇ CHLORODIPHENYL (54% Cl) (OSHA) ◇ PCB's

OSHA PEL: TWA 0.5 $mg/m^3$ (skin)
ACGIH TLV: TWA 0.5 $mg/m^3$ (skin)
NIOSH REL: TWA (Polychlorinated Biphenyls) 0.001 $mg/m^3$

SAFETY PROFILE: Moderately toxic by ingestion. A suspected human carcinogen. Experimental carcinogenic, teratogenic and reproductive effects. When heated to decomposition it emits toxic fumes of $Cl^-$.

PROP: Composed of 11% tetra-, 49% penta-, 34% hexa- and 6% heptachlorobiphenyls.

**PKQ059** CAS: 9002-86-2 $(C_2H_3Cl)_n$ HR: 2
**POLYVINYL CHLORIDE**
SYNS: CHLOROETHYLENE POLYMER ◇ POLY(CHLOROETHYLENE) ◇ VINYL CHLORIDE POLYMER

DFG MAK: 6 $mg/m^3$ (dust)

SAFETY PROFILE: An experimental tumorigen by ingestion and implantation. Chronic inhalation of dusts can cause pulmonary damage, blood effects, and abnormal liver function. ''Meat wrappers asthma'' has resulted from the cutting of PVC films with a hot knife. Can cause allergic dermatitis. When heated to decomposition it emits toxic fumes of $Cl^-$ and phosgene.

PROP: Polymers with molecular weights ranging from 60,000-150,000. White powder. D: 1.406.

**PKS750** CAS: 65997-15-1 HR: 1
**PORTLAND CEMENT**
SYN: CEMENT, PORTLAND

OSHA PEL: (Transitional: TWA 50 mppcf) TWA Total Dust: 10 mg/$m^3$; Respirable Fraction: 5 mg/$m^3$
ACGIH TLV: TWA (nuisance particulate) 10 mg/$m^3$ of total dust (when toxic impurities are not present, e.g., quartz $< 1\%$).

SAFETY PROFILE: A nuisance dust. A skin irritant.

PROP: Fine gray powder composed of compounds of lime, aluminum, silica, and iron oxide. Small amounts of magnesia, sodium, potassium, chromium, and sulfur are also present in combined form. Containing less than 1% crystalline silica.

**PLC500** CAS: 151-50-8 CN•K HR: 3
**POTASSIUM CYANIDE**
DOT: 1680
SYNS: CYANIDE OF POTASSIUM ◇ HYDROCYANIC ACID, POTASSIUM SALT

OSHA PEL: TWA 5 mg(CN)/$m^3$
ACGIH TLV: TWA 5 mg(CN)/$m^3$ (skin)
DFG MAK: 5 mg/$m^3$
NIOSH REL: CL (Cyanide) 5 mg(CN)/$m^3$/10M
DOT Class: Poison B

SAFETY PROFILE: A deadly human poison by ingestion. Experimental teratogenic and reproductive effects. Reacts with acids or acid fumes to liberate deadly HCN. When heated to decomposition it emits very toxic fumes of $K_2O$, $CN^-$ and $NO_x$.

PROP: Colorless water soln; slt odor of bitter almonds.

**PLJ500** CAS: 1310-58-3 HKO HR: 3
**POTASSIUM HYDROXIDE**
DOT: 1813/1814
SYNS: CAUSTIC POTASH ◇ LYE ◇ POTASSIUM HYDRATE (DOT)

OSHA PEL: CL 2 mg/$m^3$
ACGIH TLV: CL 2 mg/$m^3$
DOT Class: Corrosive Material

SAFETY PROFILE: Poison by ingestion. Very corrosive to the eyes, skin, and mucous membranes. Violent, exothermic reaction with water. When heated to decomposition it emits toxic fumes of $K_2O$.

PROP: White, deliquescent pieces, lumps or sticks having crystalline fracture. Mp: 360°±7°, bp: 1320°, d: 2.044. Sol in water and alc.

**PLP000** CAS: 7722-64-7 $MnO_4 \cdot K$ HR: 3
**POTASSIUM PERMANGANATE**
DOT: 1490
SYNS: C.I. 77755 ◇ PERMANGANATE OF POTASH (DOT)

OSHA PEL: CL 5 mg(Mn)/$m^3$
ACGIH TLV: TWA 5 mg(Mn)/$m^3$
DOT Class: Oxidizer

SAFETY PROFILE: A human poison by ingestion. Experimental reproductive effects. Flammable by chemical reaction. A dangerous explosion hazard. When heated to decomposition it emits toxic fumes of $K_2O$.

PROP: Dark purple crystals with a blue metallic sheen; sweetish astringent taste. Mp: decomp @ <240°, d: 2.703.

**PMJ750** CAS: 74-98-6 $C_3H_8$ HR: 3
**PROPANE**
DOT: 1075/1978
SYNS: DIMETHYLMETHANE ◇ PROPYL HYDRIDE

OSHA PEL: TWA 1000 ppm
ACGIH TLV: Asphyxiant
DFG MAK: 1000 ppm (1800 mg/$m^3$)
DOT Class: Flammable Gas

SAFETY PROFILE: An asphyxiant. Flammable gas. Explosive in the form of vapor. To fight fire, stop flow of gas.

PROP: Colorless gas. Bp: −42.1°, lel: 2.3%, uel: 9.5%, fp: −187.1°, flash p: −156°F, d: 0.5852 @ −44.5°/4°, autoign temp: 842°F, vap d: 1.56. Sol in water, alc, and ether.

**PML400** CAS: 1120-71-4 $C_3H_6O_3S$ HR: 3
**PROPANE SULTONE**
SYNS: 3-HYDROXY-1-PROPANESULPHONIC ACID SULFONE ◇ 1,3-PROPANE SULTONE (MAK)

ACGIH TLV: Suspected Human Carcinogen.
DFG MAK: Animal Carcinogen, Suspected Human Carcinogen.

SAFETY PROFILE: Moderately toxic by skin contact. An experimental carcinogen and teratogen. Experimental reproductive effects. A skin irritant. When heated to decomposition it emits toxic fumes of $SO_x$.

**PML500** CAS: 107-03-9 $C_3H_8S$ HR: 2
**PROPANETHIOL**
DOT: 2704
SYNS: 3-MERCAPTOPROPANOL ◇ PROPYL MERCAPTAN

NIOSH REL: CL 0.5 ppm/15M
DOT Class: Label: Flammable Liquid.

SAFETY PROFILE: Moderately toxic by ingestion. A severe eye irritant. Very dangerous fire hazard. When heated to decomposition it emits toxic fumes of $SO_x$.

PROP: Flash p: −4°F.

**PMN450** CAS: 107-19-7 $C_3H_4O$ HR: 3
**PROPARGYL ALCOHOL**
SYNS: ETHYNYLCARBINOL ◇ ETHYNYLMETHANOL ◇ 2-PROPYN-1-OL

OSHA PEL: TWA 1 ppm (skin)
ACGIH TLV: TWA 1 ppm (skin)
DFG MAK: 2 ppm (5 $mg/m^3$)
DOT Class: Flammable Liquid

SAFETY PROFILE: Poison by ingestion and skin contact. Moderately toxic by inhalation. A skin and mucous membrane irritant. Dangerous fire hazard. To fight fire, use foam, $CO_2$, dry chemical.

PROP: Moderately volatile liquid; geranium odor. D: 0.9715 @ 20°/4°, mp: −48° to −52°, bp: 114-115°, flash p: 33°C (97°F) (OC), vap press: 11.6 mm @ 20°, vap d: 1.93.

**PMT100** CAS: 57-57-8 $C_3H_4O_2$ HR: 3
**β-PROPIOLACTONE**
SYNS: BPL ◇ 2-OXETANONE ◇ β-PROPIONOLACTONE

OSHA: Carcinogen.
ACGIH TLV: TWA 0.5 ppm; Suspected Human Carcinogen.
DFG MAK: Animal Carcinogen, Suspected Human Carcinogen.

SAFETY PROFILE: Poison by inhalation. A human carcinogen.

**PMU750** CAS: 79-09-4 $C_3H_6O_2$ HR: 3
**PROPIONIC ACID**
DOT: 1848
SYNS: CARBOXYETHANE ◇ ETHANECARBOXYLIC ACID

OSHA PEL: TWA 10 ppm
ACGIH TLV: TWA 10 ppm
DFG MAK: 10 ppm (30 $mg/m^3$)
DOT Class: Corrosive Material

SAFETY PROFILE: Moderately toxic by ingestion and skin contact. A corrosive irritant to eyes, skin, and mucous membranes. Flammable liquid. To fight fire, use alcohol foam.

PROP: Oily liquid; pungent, disagreeable, rancid odor. D: 0.998 @ 15°/4°, mp: −21.5°, bp: 141.1°, vap press: 10 mm @ 39.7°, vap d: 2.56, autoign temp: 955°F. Misc in water, alc, ether, and chloroform.

**PMV750** CAS: 107-12-0 $C_3H_5N$ HR: 3
**PROPIONONITRILE**
DOT: 2404
SYNS: CYANOETHANE ◇ ETHYL CYANIDE

NIOSH REL: TWA (Nitriles) 14 $mg/m^3$
DOT Class: Flammable Liquid

SAFETY PROFILE: Poison by ingestion and skin contact. Experimental teratogenic effects. An eye irritant. Dangerous fire hazard. To fight fire, use water spray, foam, mist, $CO_2$, dry chemical. When heated to decomposition it emits toxic fumes of $NO_x$ and $CN^-$.

PROP: Colorless liquid; ethereal odor. Bp: 97.1°, d: 0.783 @ 21°/4°, vap d: 1.9, flash p: 36°F, lel: 3.1%; mp: −91.8°. Misc with alc and ether.

**PMY300** CAS: 114-26-1 $C_{11}H_{15}NO_3$ HR: 3
**PROPOXUR**
SYNS: APROCARB ◇ o-ISOPROPOXYPHENYL METHYLCARBAMATE ◇ BAYGON

OSHA PEL: TWA 0.5 $mg/m^3$
ACGIH TLV: TWA 0.5 $mg/m^3$
DFG MAK: 2 $mg/m^3$

SAFETY PROFILE: A poison by ingestion. Moderately toxic by inhalation and skin contact. Experimental reproductive effects. When heated to decomposition it emits toxic fumes of $NO_x$.

PROP: A white to tan, crystalline solid. Sltly sol in water; sol in all polar organic solvents.

**PNC250** CAS: 109-60-4 $C_5H_{10}O_2$ HR: 3
**n-PROPYL ACETATE**
DOT: 1276
SYNS: ACETIC ACID, n-PROPYL ESTER ◇ 1-ACETOXYPROPANE

OSHA PEL: (Transitional: TWA 200 ppm) TWA 200 ppm; STEL 250 ppm
ACGIH TLV: TWA 200 ppm; STEL 250 ppm
DFG MAK: 200 ppm (840 $mg/m^3$)
DOT Class: Flammable Liquid

SAFETY PROFILE: Mildly toxic by ingestion and inhalation. A skin irritant. Dangerous fire hazard; explosive in the form of vapor. To fight fire, use alcohol foam, $CO_2$, dry chemical.

PROP: Clear, colorless liquid; pleasant odor. Mp: −92.5°, bp: 101.6°, flash p: 58°F, lel: 2.0%, uel: 8.0%, d: 0.887, autoign temp: 842°F, vap press: 40 mm @ 28.8°, vap d: 3.52. Misc with alc and ether; sol in water.

**PND000** CAS: 71-23-8 $C_3H_8O$ HR: 3
**n-PROPYL ALCOHOL**
DOT: 1274
SYNS: ETHYL CARBINOL ◇ 1-HYDROXYPROPANE ◇ n-PROPANOL

OSHA PEL: (Transitional: TWA 200 ppm) TWA 200 ppm; STEL 250 ppm

ACGIH TLV: TWA 200 ppm; STEL 250 ppm (skin)
DOT Class: Flammable Liquid

SAFETY PROFILE: Moderately toxic by inhalation and ingestion. An experimental carcinogen. A skin and severe eye irritant. Dangerous fire hazard. Explosive in the form of vapor. To fight fire, use alcohol foam, $CO_2$, dry chemical.

PROP: Clear liquid; alcohol-like odor. Mp: −127°, bp: 97.19°, flash p: 59°F (CC), ULC: 55-60, d: 0.8044 @ 20°/4°, lel: 2.1%, uel: 13.5%, autoign temp: 824°F, vap press: 10 mm @ 14.7°, vap d: 2.07. Misc in water, alc, and ether.

**PNJ400** CAS: 78-87-5 $C_3H_6Cl_2$ HR: 2
**PROPYLENE DICHLORIDE**
SYN: 1,2-DICHLOROPROPANE

OSHA PEL: (Transitional: TWA 75 ppm) TWA 75 ppm; STEL 110 ppm
ACGIH TLV: TWA 75 ppm; STEL 110 ppm
DFG MAK: 75 ppm (350 mg/m$^3$)
DOT Class: Flammable Liquid

SAFETY PROFILE: Moderately toxic by inhalation and ingestion. An eye irritant. Can cause dermatitis. A very dangerous fire hazard. To fight fire, use water, foam, $CO_2$, dry chemical. When heated to decomposition it emits toxic fumes of $Cl^-$.

PROP: Colorless liquid. Bp: 96.8°, flash p: 60°F, d: 1.1593 @ 20°/20°, vap press: 40 mm @ 19.4°, vap d: 3.9, autoign temp: 1035°F, lel: 3.4%, uel: 14.5%.

**PNL000** CAS: 6423-43-4 $C_3H_6N_2O_6$ HR: 3
**PROPYLENE GLYCOL DINITRATE**
SYNS: PGDN ◇ PROPYLENE GLYCOL-1,2-DINITRATE

OSHA PEL: TWA 0.05 ppm
ACGIH TLV: TWA 0.05 ppm (skin)
DFG MAK: 0.05 ppm (0.3 mg/m$^3$)

SAFETY PROFILE: Poison by ingestion. An eye irritant. When heated to decomposition it emits toxic fumes of $NO_x$.

**PNL250** CAS: 107-98-2 $C_4H_{10}O_2$ HR: 2
**PROPYLENE GLYCOL MONOMETHYL ETHER**
SYNS: METHOXY ETHER of PROPYLENE GLYCOL ◇ 1-METHOXY-2-PROPANOL ◇ PROPYLENE GLYCOL METHYL ETHER

OSHA PEL: TWA 100 ppm; STEL: 150 ppm
ACGIH TLV: TWA 100 ppm; STEL: 150 ppm
DFG MAK: 100 ppm (375 mg/m$^3$)

SAFETY PROFILE: Mildly toxic by ingestion, inhalation, and skin contact. A skin and eye irritant. An experimental teratogen. Many glycol ethers have dangerous human reproductive effects. Very dangerous fire hazard. To fight fire, use foam, $CO_2$, dry chemical.

PROP: Colorless liquid. Mp: −96.7°, bp: 120°, flash p: 100°F, d: 0.919 @ 25°/25°.

**PNL400** CAS: 75-55-8 $C_3H_7N$ HR: 3
**PROPYLENE IMINE**
DOT: 1921
SYNS: 2-METHYLAZIRIDINE ◇ METHYLETHYLENIMINE

OSHA PEL: TWA 2 ppm (skin)
ACGIH TLV: TWA 2 ppm (skin), Suspected Human Carcinogen.
DFG MAK: Animal Carcinogen, Suspected Human Carcinogen.
DOT Class: Flammable Liquid

SAFETY PROFILE: Poison by ingestion and skin contact. Moderately toxic by inhalation. An experimental carcinogen. Severe eye irritant. A very dangerous fire hazard. Polymerizes explosively on exposure to acids or acid fumes. When heated to decomposition it emits toxic fumes of $NO_x$.

PROP: Liquid. Vap d: 2.0, flash p: 14°F.

**PNL600** CAS: 75-56-9 $C_3H_6O$ HR: 3
**PROPYLENE OXIDE**
DOT: 1280
SYNS: 1,2-EPOXYPROPANE ◇ METHYL ETHYLENE OXIDE ◇ METHYL OXIRANE

OSHA PEL: (Transitional: TWA 100 ppm) TWA 20 ppm
ACGIH TLV: TWA 20 ppm
DFG MAK: Animal Carcinogen, Suspected Human Carcinogen.
DOT Class: Flammable Liquid

SAFETY PROFILE: Moderately toxic by ingestion, inhalation, and skin contact. An experimental carcinogen. A suspected human carcinogen. Experimental reproductive effects. A severe skin and eye irritant. A very dangerous fire and explosion hazard when exposed to heat or flame. To fight fire, use alcohol foam, $CO_2$, dry chemical.

PROP: Colorless liquid; ethereal odor. Bp: 33.9°, lel: 2.8%, uel: 37%, fp: −104.4°, flash p: −35°F (TOC), d: 0.8304 @ 20°/20°, vap press: 400 mm @ 17.8°, vap d: 2.0. Sol in water, alc, and ether.

**PNQ500** CAS: 627-13-4 $C_3H_7NO_3$ HR: 3
**n-PROPYL NITRATE**
DOT: 1865

OSHA PEL: (Transitional: TWA 25 ppm) TWA 25 ppm; STEL 40 ppm
ACGIH TLV: TWA 25 ppm; STEL 40 ppm
DFG MAK: 25 ppm (110 mg/$m^3$)
DOT Class: Flammable Liquid

SAFETY PROFILE: Dangerous fire hazard. A shock-sensitive explosive. When heated to decomposition it emits toxic fumes of $NO_x$.

PROP: Pale yellow liquid; sickly odor. Bp: 110.5°, d: 1.054 @ 20°/4°, flash p: 68°F, autoign temp: 347°F (in air), lel: 2%, uel: 100%. Very sltly sol in water; sol in alc and ether.

**PON250** CAS: 129-00-0 $C_{16}H_{10}$ HR: 3
**PYRENE**
SYNS: BENZO(def)PHENANTHRENE ◇ β-PYRINE

OSHA PEL: TWA 0.2 mg/m$^3$

SAFETY PROFILE: Poison by inhalation. Moderately toxic by ingestion. A skin irritant.

PROP: Colorless solid, solutions have a slight blue color. Mp: 156°, d: 1.271 @ 23°, bp: 404°. Insol in water, fairly sol in organic solvents.

**POO250** CAS: 8003-34-7 HR: 3
**PYRETHRINS**
DOT: 9184
SYNS: CINERIN I or II ◇ PYRETHRUM (ACGIH)

OSHA PEL: TWA 5 mg/m$^3$
ACGIH TLV: TWA 5 mg/m$^3$
DFG MAK: 5 mg/m$^3$
DOT Class: ORM-E

SAFETY PROFILE: Moderately toxic to humans by ingestion. Experimental reproductive effects. An allergen. Combustible.

PROP: Viscous liquid. Bp: 170° @ 0.1 mm (decomp).

**POP250** CAS: 110-86-1 $C_5H_5N$ HR: 3
**PYRIDINE**
DOT: 1282
SYN: AZABENZENE

OSHA PEL: TWA 5 ppm
ACGIH TLV: TWA 5 ppm
DFG MAK: 5 ppm (15 mg/m$^3$)
DOT Class: Flammable Liquid

SAFETY PROFILE: Moderately toxic by ingestion and skin contact. A skin and severe eye irritant. Dangerous fire hazard. Severe explosion hazard in the form of vapor. Incompatible with oxidizing materials. To fight fire, use alcohol foam. When heated to decomposition it emits highly toxic fumes of $NO_x$.

PROP: Colorless liquid; sharp, penetrating, empyreumatic odor; burning taste. Bp: 115.3°, lel: 1.8%, uel: 12.4%, fp: −42°, flash p: 68°F (CC), d: 0.982, autoign temp: 900°F, vap press: 10 mm @ 13.2°, vap d: 2.73. Volatile with steam; misc with water, alc, and ether.

**QQS200** CAS: 106-51-4 $C_6H_4O_2$ HR: 3
**QUINONE**
SYNS: BENZOQUINONE (DOT) ◇ 1,4-CYCLOHEXADIENEDIONE

OSHA PEL: TWA 0.1 ppm
ACGIH TLV: TWA 0.1 ppm
DFG MAK: 0.1 ppm (0.4 mg/$m^3$)
DOT Class: Poison B

SAFETY PROFILE: Poison by ingestion. An experimental tumorigen by skin contact. Causes severe damage to the skin and mucous membranes on contact with it in the solid state, in solution, or in the form of condensed vapors.

PROP: Yellow crystals; characteristic irritating odor. Mp: 115.7°, bp: sublimes, d: 1.318 @ 20°/4°.

**REA000** CAS: 108-46-3 $C_6H_6O_2$ HR: 3
**RESORCINOL**
DOT: 2876
SYNS: m-BENZENEDIOL ◇ C.I. 76505 ◇ 1,3-DIHYDROXYBENZENE ◇ m-HYDROQUINONE

OSHA PEL: TWA 10 ppm; STEL 20 ppm
ACGIH TLV: TWA 10 ppm; STEL 20 ppm
DOT Class: ORM-E

SAFETY PROFILE: Human poison by ingestion. Moderately toxic experimentally by skin contact. An experimental tumorigen. A skin and severe eye irritant. Combustible. To fight fire, use water, $CO_2$, dry chemical.

PROP: Very white crystals, become pink on exposure to light when not perfectly pure; unpleasant sweet taste. Mp: 110°, bp: 280.5°, flash p: 261°F (CC), d: 1.285 @ 15°, autoign temp: 1126°F, vap press: 1 mm @ 108.4°, vap d: 3.79. Very sol in alc, ether, and glycerol; sltly sol in chloroform; sol in water.

**RHF000** CAS: 7440-16-6 Rh HR: 2
**RHODIUM**

OSHA PEL: TWA Metal, Fume, Insoluble Compounds: 0.1 mg(Rh)/$m^3$; Soluble Compounds: 0.001 mg(Rh)/$m^3$
ACGIH TLV: TWA (Metal) 1 mg/$m^3$, (insoluble compounds as Rh) 1 mg/$m^3$, (soluble compounds as Rh) 0.01 mg/$m^3$

SAFETY PROFILE: It may be a sensitizer but not to the same extent as platinum. Most rhodium compounds have only moderate toxicity by ingestion. Flammable.

PROP: A silvery-white, metallic element. Mp: 1966°, bp: 3727°, d: 2.41 @ 20°.

**RMA500** CAS: 299-84-3 $C_8H_8Cl_3O_3PS$ HR: 3
**RONNEL**
SYNS: DIMETHYL TRICHLOROPHENYL THIOPHOSPHATE ◇ FENCHLOROPHOS

OSHA PEL: (Transitional: TWA 15 $mg/m^3$) TWA 10 $mg/m^3$
ACGIH TLV: TWA 10 $mg/m^3$

SAFETY PROFILE: Poison by ingestion. Moderately toxic by skin contact. An experimental teratogen. Experimental reproductive effects. When heated to decomposition it emits very toxic fumes of $Cl^-$, $PO_x$, and $SO_x$.

PROP: White powder. Mp: 41°, vap press: $8 \times 10^{-4}$ mm.

**RNZ000** CAS: 83-79-4 $C_{23}H_{22}O_6$ HR: 3
**ROTENONE**

OSHA PEL: TWA 5 $mg/m^3$
ACGIH TLV: TWA 5 $mg/m^3$
DFG MAK: 5 $mg/m^3$

SAFETY PROFILE: Human poison by ingestion. An experimental neoplastigen and teratogen. Experimental reproductive effects. An skin and eye irritant.

PROP: Orthorhombic plates. Mp: 165-166° (dimorphic form mp: 185-186°). D: 1.27 @ 20°. Almost insol in water; sol in alc, acetone, carbon tetrachloride, chloroform, ether and other organic solvents. Decomp on exposure to light and air.

**ROU000** HR: 3
**RUBBER SOLVENT**
SYNS: LACQUER DILUENT ◇ NAPHTHA

ACGIH TLV: TWA 400 ppm
NIOSH REL: TWA (Petroleum Solvent) 350 $mg/m^3$; CL 1800 $mg/m^3$/15M

SAFETY PROFILE: Mildly toxic by inhalation. A very dangerous fire hazard. Explosive in the form of vapor. To fight fire, use foam, alcohol foam.

PROP: A petroleum cut distilling between 38 to 149°C consisting chiefly of $C_5$ to $C_9$ aliphatic hydrocarbons. Flash p: −40°F (varies with manufacturer), lel: 1.0%, uel: 7.0%, d: $<1$, bp: 100-280°F. Insol in water.

**SBO500** CAS: 7782-49-2 Se HR: 3
**SELENIUM**
DOT: 2658
SYN: C.I. 77805

OSHA PEL: TWA 0.2 mg(Se)/$m^3$
ACGIH TLV: TWA 0.2 mg(Se)/$m^3$
DFG MAK: 0.1 $mg/m^3$
DOT Class: Poison B

SAFETY PROFILE: Poison by inhalation. An experimental tumorigen and teratogen. Experimental reproductive effects. When heated to decomposition it emits toxic fumes of Se.

PROP: Steel gray, nonmetallic element. Mp: 170-217°, bp: 690°, d: 4.81-4.26, vap press: 1 mm @ 356°. Insol in $H_2O$ and alc; very sltly sol in ether.

**SBS000** CAS: 7783-79-1 $F_6Se$ HR: 3
**SELENIUM HEXAFLUORIDE**
DOT: 2194
SYN: SELENIUM FLUORIDE

OSHA PEL: TWA 0.05 ppm (Se)
ACGIH TLV: TWA 0.05 ppm (Se)
DOT Class: Poison A

SAFETY PROFILE: Poison by inhalation. When heated to decomposition it emits very toxic fumes of $F^-$ and Se.

PROP: Colorless gas. Mp: −39° (subl @ −40.6°), bp: −34.5°, d: 3.25 @ −25°.

**SCH000** CAS: 7631-86-9 $O_2Si$ HR: 1
**SILICA, AMORPHOUS FUMED**
SYNS: AMORPHOUS SILICA DUST ◇ COLLOIDAL SILICA ◇ FUMED SILICA ◇ SILICA AEROGEL ◇ SILICON DIOXIDE (FCC)

OSHA PEL: (Transitional: TWA 80 $mg/m^3/\%SiO_2$) TWA 6 $mg/m^3$
ACGIH TLV: (Proposed: TWA 2 $mg/m^3$ (Respirable Dust))

SAFETY PROFILE: Moderately toxic by ingestion. Much less toxic than crystalline forms. Does not cause silicosis.

PROP: A finely powdered microcellular silica foam with minimum $SiO_2$ content of 89.5%. Insol in water; sol in hydrofluoric acid.

**SCI000** CAS: 7631-86-9 $O_2Si$ HR: 1
**SILICA, AMORPHOUS HYDRATED**
SYNS: SILICA AEROGEL ◇ SILICA GEL ◇ SILICIC ACID

OSHA PEL: (Transitional: TWA 80 $mg/m^3/\%SiO_2$) TWA 6 $mg/m^3$
ACGIH TLV: TWA (nuisance particulate) 10 $mg/m^3$ of total dust (when toxic impurities are not present, e.g., quartz $< 1\%$).

SAFETY PROFILE: Some deposits contain small amounts of crystalline quartz which is therefore fibrogenic. When diatomaceous earth is calcined (with or without fluxing agents) some silica is converted to cristobalite and is therefore fibrogenic.

**SCI500** $SiO_2$ HR: 3
**SILICA (CRYSTALLINE)**
SYNS: CHALCEDONY ◇ CRISTOBALITE ◇ PURE QUARTZ ◇ SAND ◇ SILICA FLOUR ◇ SILICON DIOXIDE ◇ TRIDYMITE ◇ TRIPOLI

OSHA PEL: (Transitional: TWA Respirable Fraction: 10 $mg/m^3/2(\%SiO_2+2)$; Total Dust: 30 $mg/m^3/2(\%SiO_2+2)$) Respirable Fraction: TWA 0.05 $mg/m^3$

ACGIH TLV: TWA Respirable Fraction: 0.05 $mg/m^3$
DFG MAK: 0.15 $mg/m^3$
NIOSH REL: TWA 50 $\mu g/m^3$

SAFETY PROFILE: Moderately toxic as an acute irritating dust. Prolonged inhalation of dusts containing free silica may result in development of a disabling pulmonary fibrosis known as silicosis.

PROP: Transparent, tasteless crystals or amorphous powder. Mp: 1710°, bp: 2230°, d (amorphous): 2.2, d (crystalline): 2.6, vap press: 10 mm @ 1732°. Practically insol in water or acids. Dissolves readily in HF, forming silicon tetrafluoride.

**SCJ000** CAS: 14464-46-1 $O_2Si$ HR: 3
**SILICA, CRYSTALLINE-CRISTOBALITE**
SYNS: CALCINED DIATOMITE ◇ CRISTOBALITE

OSHA PEL: (Transitional: TWA Respirable Fraction: (10 $mg/m^3$/2(%$SiO_2$+2); Total Dust: 30 $mg/m^3$/2(%$SiO_2$+2)) TWA Respirable Fraction: 0.05 $mg/m^3$
ACGIH TLV: TWA Respirable Fraction: 0.05 $mg/m^3$
DFG MAK: 0.15 $mg/m^3$
NIOSH REL: TWA 50 $\mu g/m^3$

SAFETY PROFILE: An experimental carcinogen and tumorigen. About twice as toxic as silica in causing silicosis.

PROP: White, cubic-system crystals formed from quartz at temperatures above 1000°C.

**SCJ500** CAS: 14808-60-7 $O_2Si$ HR: 3
**SILICA, CRYSTALLINE-QUARTZ**
SYNS: CHALCEDONY ◇ CHERTS ◇ QUARTZ ◇ SILICIC ANHYDRIDE

OSHA PEL: (Transitional: TWA Respirable Fraction: 10 $mg/m^3$/2(%$SiO_2$+2); Total Dust: 30 $mg/m^3$/2(%$SiO_2$+2)) TWA Respirable Fraction: 0.1 $mg/m^3$
ACGIH TLV: TWA Respirable Fraction: 0.1 $mg/m^3$
DFG MAK: 0.15 $mg/m^3$
NIOSH REL: TWA 50 $\mu g/m^3$; 3000000 fibers/$m^3$

SAFETY PROFILE: An experimental carcinogen. Causes silicosis.

PROP: Mp: 1710°, bp: 2230°, d: 2.6.

**SCK000** CAS: 15468-32-3 $O_2Si$ HR: 3
**SILICA, CRYSTALLINE-TRIDYMITE**
SYN: TRIDYMITE

OSHA PEL: (Transitional: TWA Respirable: 10 $mg/m^3$/2(%$SiO_2$+2); Total Dust: TWA 30 $mg/m^3$/2(%$SiO_2$+2)) TWA 0.05 $mg/m^3$
ACGIH TLV: TWA Respirable Fraction: 0.05 $mg/m^3$
DFG MAK: 0.15 $mg/m^3$
NIOSH REL: TWA 50 $\mu g/m^3$

SAFETY PROFILE: An experimental tumorigen. About twice as toxic as silica in causing silicosis.

PROP: White or colorless platelets or orthorhombic (crystals) formed from quartz @ temperatures >870°.

**SCK600** CAS: 60676-86-0 $O_2Si$ HR: 3
**SILICA, FUSED**
SYNS: AMORPHOUS FUSED SILICA ◇ FUSED QUARTZ ◇ FUSED SILICA (ACGIH) ◇ QUARTZ GLASS ◇ SILICA, AMORPHOUS FUSED

OSHA PEL: (Transitional: TWA Respirable: 10 $mg/m^3/2(\%SiO_2+2)$; Total Dust: TWA 30 $mg/m^3/2(\%SiO_2+2)$) TWA 0.1 $mg/m^3$
ACGIH TLV: TWA Respirable Fraction: 0.1 $mg/m^3$

SAFETY PROFILE: Prolonged inhalation may result in development of a disabling pulmonary fibrosis known as silicosis.

PROP: Made up of spherical submicroscopic particles under 0.1 micron in size.

**SCL000** CAS: 7699-41-4 $H_2O_3Si$ HR: 1
**SILICA, GEL and AMORPHOUS-PRECIPITATED**
SYNS: METASILICIC ACID ◇ PRECIPITATED SILICA ◇ SILICA GEL ◇ SILICIC ACID

OSHA PEL: (Transitional: TWA 80 $mg/m^3/\%SiO_2$) TWA 6 $mg/m^3$
ACGIH TLV: TWA (nuisance particulate) 10 $mg/m^3$ of total dust (when toxic impurities are not present, e.g., quartz < 1%).

SAFETY PROFILE: An eye irritant and nuisance dust.

**SCN000** $3MgO \cdot 4SiO_2 \cdot H_2O$ HR: 2
**SILICATE SOAPSTONE**
SYN: SOAPSTONE

OSHA PEL: (Transitional: TWA 20 mppcf) TWA Total Dust: 6 $mg/m^3$; Respirable Fraction: 3 $mg/m^3$
ACGIH TLV: TWA Respirable Fraction: 3 $mg/m^3$; 6 $mg/m^3$ of total dust (when toxic impurities are not present, e.g., quartz < 1%).

SAFETY PROFILE: Less toxic than quartz.

PROP: Contains less than 1% crystalline silica.

**SCP000** CAS: 7440-21-3 Si HR: 3
**SILICON**
DOT: 1346

OSHA PEL: (Transitional: TWA Total Dust: 15 $mg/m^3$; Respirable Fraction: 5 $mg/m^3$) TWA Total Dust: 10 $mg/m^3$ of total; Respirable Fraction: 5 $mg/m^3$
ACGIH TLV: TWA (nuisance particulate) 10 $mg/m^3$ of total dust (when toxic impurities are not present, e.g., quartz < 1%).
DOT Class: Flammable Solid.

SAFETY PROFILE: Does not occur freely in nature, but is found as silicon dioxide (silica) and as various silicates. Elemental silicon is flammable when exposed to flame or by chemical reaction with oxidizers. When heated it will react with water or steam to produce $H_2$.

PROP: Cubic, steel-gray crystals or dark brown powder. Mp: 1420°, bp: 2600°, d: 2.42 or 2.3 @ 20°, vap press: 1 mm @ 1724°. Almost insol in water; sol in molten alkali oxides.

**SCQ000** CAS: 409-21-2 CSi HR: 3
**SILICON CARBIDE**
SYNS: CARBORUNDUM ◇ SILICON MONOCARBIDE

OSHA PEL: (Transitional: TWA Total Dust: 15 mg/m$^3$; Respirable Fraction: 5 mg/m$^3$) TWA Total Dust: 10 mg/m$^3$; Respirable Fraction: 5 mg/m$^3$
ACGIH TLV: TWA (nuisance particulate) 10 mg/m$^3$ of total dust (when toxic impurities are not present, e.g., quartz $< 1\%$).
DFG MAK: 4 mg/m$^3$0

SAFETY PROFILE: An experimental neoplastigen.

PROP: Bluish-black, iridescent crystals. Mp: 2600°, bp: subl $>$ 2000°, decomp @ 2210°, d: 3.17.

**SDH575** CAS: 7803-62-5 $H_4Si$ HR: 3
**SILICON TETRAHYDRIDE**
DOT: 2203
SYNS: MONOSILANE ◇ SILANE

OSHA PEL: TWA 5 ppm
ACGIH TLV: TWA 5 ppm
DOT Class: Flammable Gas

SAFETY PROFILE: Mildly toxic by inhalation. Silanes are irritating to skin, eyes and mucous membranes. Easily ignited in air. Ignites in oxygen. It may self-explode. When heated to decomposition it burns or explodes.

PROP: Gas with repulsive odor; slowly decomp by water. D: 0.68 @ −185°, mp: −185°, bp: 112°, fp: −200°.

**SDI500** CAS: 7440-22-4 Ag HR: 2
**SILVER**
SYNS: ARGENTUM ◇ C.I. 77820

OSHA PEL: Metal, Dust, and Fume: TWA 0.01 mg/m$^3$
ACGIH TLV: TWA (metal) 0.1 mg/m$^3$, (soluble compounds as Ag) 0.01 mg/m$^3$
DFG MAK: 0.01 mg/m$^3$

SAFETY PROFILE: An experimental tumorigen. Flammable in the form of dust.

PROP: Soft, ductile, malleable, lustrous, white metal. Mp: 961.93°, bp: 2212°, d: 10.50 @ 20°.

**SFA000** CAS: 26628-22-8 $N_3Na$ HR: 3
**SODIUM AZIDE**
DOT: 1687

OSHA PEL: As $NH_3$: CL 0.1 ppm; As $NaN_3$: Cl 0.3 mg/m$^3$ (skin)
ACGIH TLV: CL 0.3 mg/m$^3$
DFG MAK: 0.07 ppm (0.2 mg/m$^3$)
DOT Class: Poison B

SAFETY PROFILE: Poison by ingestion and skin contact. An experimental tumorigen. An unstable explosive sensitive to impact. When heated to decomposition it emits very toxic fumes of $NO_x$ and $Na_2O$.

PROP: Colorless, hexagonal crystals. Mp: decomp, d: 1.846. Insol in ether; sol in liquid ammonia.

**SFE000** CAS: 7631-90-5 $HO_3S \cdot Na$ HR: 3
**SODIUM BISULFITE**
DOT: 2693
SYNS: SODIUM ACID SULFITE ◇ SODIUM HYDROGEN SULFITE

OSHA PEL: TWA 5 mg/m$^3$
ACGIH TLV: TWA 5 mg/m$^3$
DOT Class: ORM-B

SAFETY PROFILE: Moderately toxic by ingestion. A corrosive irritant to skin, eyes, and mucous membranes. An allergen. When heated to decomposition it emits toxic fumes of $SO_x$ and $Na_2O$.

PROP: White, crystalline powder; odor of sulfur dioxide, disagreeable taste. D: 1.48, mp: decomp. Very sol in hot or cold water; sltly sol in alc.

**SFE500** CAS: 1303-96-4 $B_4O_7 \cdot 2Na$ HR: D
**SODIUM BORATE**
SYNS: BORATES, TETRA, SODIUM SALT, ANHYDROUS (OSHA, ACGIH) ◇ SODIUM BORATE ANHYDROUS

OSHA PEL: 10 mg/m$^3$ (anhydrous, decahydrate, pentahydrate)
ACGIH TLV: TWA 1 mg/m$^3$

SAFETY PROFILE: Experimental reproductive effects. When heated to decomposition it emits toxic fumes of $Na_2O$ and B.

PROP: White crystals. Mp: 741°, bp: 1575° (decomp), d: 2.367. Slowly soluble in water.

**SFF000** CAS: 1303-96-4 $B_4O_7 \cdot 2Na \cdot 10H_2O$ HR: 3
**SODIUM BORATE DECAHYDRATE**
SYNS: BORATES, TETRA, SODIUM SALT, ANHYDROUS (OSHA, ACGIH) ◇ BORAX (8CI) ◇ BORAX DECAHYDRATE ◇ SODIUM TETRABORATE DECAHYDRATE

OSHA PEL: TWA 10 mg/m$^3$
ACGIH TLV: TWA 5 mg/m$^3$

SAFETY PROFILE: Moderately toxic to humans by ingestion. Experimental reproductive effects. When heated to decomposition it emits toxic fumes of $Na_2O$, B.

PROP: Hard, odorless crystals, granules or crystalline powder. D: 1.73, mp: 75° (when rapidly heated).

**SGA500** CAS: 143-33-9 CNNa HR: 3
**SODIUM CYANIDE**
DOT: 1689
SYNS: CYANIDE of SODIUM ◇ CYANOGRAN ◇ HYDROCYANIC ACID, SODIUM SALT

OSHA PEL: TWA 5 mg(CN)/$m^3$ (skin)
ACGIH TLV: TWA 5 mg(CN)/$m^3$ (skin)
DFG MAK: 5 mg(CN)/$m^3$
NIOSH REL: CL 5 mg(CN)/$m^3$/10M
DOT Class: Poison B

SAFETY PROFILE: A deadly human poison by ingestion. An experimental teratogen. Experimental reproductive effects. Flammable by chemical reaction with heat, moisture, and acid.

PROP: White, deliquescent, crystalline powder. Mp: 563.7°, bp: 1496°, vap press: 1 mm @ 817°.

**SHF500** CAS: 7681-49-4 FNa HR: 3
**SODIUM FLUORIDE**
DOT: 1690
SYNS: DISODIUM DIFLUORIDE ◇ SODIUM MONOFLUORIDE

OSHA PEL: TWA 2.5 mg(F)/$m^3$
ACGIH TLV: TWA 2.5 mg(F)/$m^3$
NIOSH REL: TWA (Inorganic Fluorides) 2.5 mg(F)/$m^3$
DOT Class: ORM-B

SAFETY PROFILE: Human poison by ingestion. Experimental poison by skin contact. An experimental tumorigen and teratogen. A corrosive irritant to skin, eyes, and mucous membranes. Experimental reproductive effects. When heated to decomposition it emits toxic fumes of $F^-$ and $Na_2O$.

PROP: Clear, lustrous crystals or white powder or balls. Mp: 993°, bp: 1700°, d: 2 @ 41°, vap press: 1 mm @ 1077°.

**SHG500** CAS: 62-74-8 $C_2H_2FO_2 \cdot Na$ HR: 3
**SODIUM FLUOROACETATE**
DOT: 2629
SYNS: 1080 ◇ COMPOUND NO. 1080

OSHA PEL: (Transitional: TWA 0.05 mg/$m^3$ (skin)) TWA 0.05 mg/$m^3$ (skin); STEL 0.15 mg/$m^3$ (skin)
ACGIH TLV: TWA 0.05 mg/$m^3$ (skin); STEL 0.15 mg/$m^3$ (skin)
DFG MAK: 0.05 mg/$m^3$
DOT Class: Poison B: Label: Poison.

SAFETY PROFILE: A deadly human poison by ingestion. When heated to decomposition it emits highly toxic fumes of $Na_2O$ and $F^-$.

PROP: Fine, white powder. Sol in water.

**SHS000** CAS: 1310-73-2 HNaO HR: 3
**SODIUM HYDROXIDE**
DOT: 1823/1824
SYNS: CAUSTIC SODA ◇ LYE (DOT) ◇ SODA LYE ◇ SODIUM HYDRATE (DOT)

OSHA PEL: (Transitional: TLV 2 $mg/m^3$) CL 2 $mg/m^3$
ACGIH TLV: Cl 2 $mg/m^3$
DFG MAK: 2 $mg/m^3$
NIOSH REL: (Sodium Hydroxide) CL 2 $mg/m^3$/15M
DOT Class: Corrosive Material

SAFETY PROFILE: Moderately toxic by ingestion. A corrosive irritant to skin, eyes, and mucous membranes. Dangerous material to handle. When heated to decomposition it emits toxic fumes of $Na_2O$.

PROP: White pieces, lumps, or sticks. Mp: 318.4°, bp: 1390°, d: 2.120 @ 20°/4°, vap press: 1 mm @ 739°. Deliquescent, sol in water and alc.

**SII000** CAS: 7681-57-4 $O_5S_2 \cdot 2Na$ HR: 3
**SODIUM METABISULFITE**
DOT: 2693
SYNS: SODIUM METABOSULPHITE ◇ SODIUM PYROSULFITE

OSHA PEL: TWA 5 $mg/m^3$
ACGIH TLV: TWA 5 $mg/m^3$
DOT Class: ORM-B

SAFETY PROFILE: Experimental reproductive effects. When heated to decomposition it emits toxic fumes of $SO_x$ and $Na_2O$.

PROP: Colorless crystals or white to yellowish powder; odor of sulfur dioxide. Sol in water; sltly sol in alc.

**SLJ500** CAS: 9005-25-8 HR: 1
**STARCH DUST**

OSHA PEL: Total Dust: 15 $mg/m^3$; Respirable Fraction: 5 $mg/m^3$
ACGIH TLV: TWA (nuisance particulate) 10 $mg/m^3$ of total dust (when toxic impurities are not present, e.g., quartz $< 1\%$).

SAFETY PROFILE: A nuisance dust. An allergen. Flammable and moderately explosive.

**SLQ000** CAS: 7803-52-3 $H_3Sb$ HR: 3
**STIBINE**
DOT: 2676
SYNS: ANTIMONY HYDRIDE ◇ ANTIMONY TRIHYDRIDE ◇ HYDROGEN ANTIMONIDE

OSHA PEL: TWA 0.1 ppm
ACGIH TLV: TWA 0.1 ppm
DFG MAK: 0.1 ppm (0.5 $mg/m^3$)
DOT Class: Poison A

SAFETY PROFILE: Poison by inhalation. Flammable. When heated to decomposition it emits toxic fumes of Sb.

PROP: Colorless gas, disagreeable odor. Mp: −88°, bp: −18.4°, d: 2.204 g/mL @ bp. Gas is sltly sol in water, very sol in alc, carbon disulfide, and organic solvents.

**SLU500** CAS: 8052-41-3 HR: 3
**STODDARD SOLVENT**
SYNS: NAPHTHA SAFETY SOLVENT ◇ WHITE SPIRITS

OSHA PEL: (Transitional: TWA 500 ppm) TWA 100 ppm
ACGIH TLV: TWA 100 ppm
NIOSH REL: TWA 350 $mg/m^3$; CL 1800 $mg/m^3$/15M

SAFETY PROFILE: Mildly toxic by inhalation. A human eye irritant. Flammable. Explosive in the form of vapor. To fight fire, use foam, $CO_2$, dry chemical.

PROP: Clear, colorless liquid. Composed of 85% nonane and 15% trimethyl benzene. Bp: 220-300°, flash p: 100-110°F, lel: 1.1%, uel: 6%, autoign temp: 450°F, d: 1.0. Insol in water; misc with abs alc, benzene, ether, and chloroform.

**SMH000** CAS: 7789-06-2 $CrO_4 \cdot Sr$ HR: 3
**STRONTIUM CHROMATE (1:1)**
DOT: 9149
SYNS: C.I. PIGMENT YELLOW 32 ◇ STRONTIUM YELLOW

OSHA PEL: CL 0.1 $mg(CrO_3)/m^3$
ACGIH TLV: TWA 0.05 $mg(Cr)/m^3$; Confirmed Human Carcinogen; (Proposed: TWA 0.001 $mg(Cr)/m^3$; Suspected Human Carcinogen)
DFG TRK: 0.1 $mg/m^3$; Animal Carcinogen, Suspected Human Carcinogen.
NIOSH REL: TWA 0.0001 $mg(Cr(VI))/m^3$
DOT Class: ORM-E

SAFETY PROFILE: Moderately toxic by ingestion. A human carcinogen. An experimental carcinogen.

PROP: Monoclinic, yellow crystals. D: 3.895 @ 15°.

**SMN500** CAS: 57-24-9 $C_{21}H_{22}N_2O_2$ HR: 3
**STRYCHNINE**
DOT: 1692

OSHA PEL: TWA 0.15 $mg/m^3$
ACGIH TLV: TWA 0.15 $mg/m^3$
DFG MAK: 0.15 $mg/m^3$
DOT Class: Poison B

SAFETY PROFILE: Human poison by ingestion. Experimental reproductive effects. An allergen. When heated to decomposition it emits toxic fumes of $NO_x$.

PROP: Hard, white, crystalline alkaloid; very bitter taste. Mp: 268°, bp: 270°, d: 1.359 @ 18°.

**SMQ000** CAS: 100-42-5 $C_8H_8$ HR: 3
**STYRENE**
DOT: 2055
SYNS: ETHENYLBENZENE ◇ STYRENE MONOMER (ACGIH) ◇ VINYLBENZENE

OSHA PEL: (Transitional: TWA 100 ppm; CL 200; Pk 600/5M/3H) TWA 50 ppm; STEL: 100 ppm
ACGIH TLV: TWA 50 ppm; STEL: 100 ppm (skin); BEI: 1 g(mandelic acid)/L in urine at end of shift; 40 ppb styrene in mixed-exhaled air prior to shift; 18 ppm styrene in mixed-exhaled air during shift; 0.55 mg/L styrene in blood end of shift; 0.02 mg/L styrene in blood prior to shift.
DFG MAK: 20 ppm (85 mg/m$^3$); BAT: 2g/L of mandelic acid in urine at end of shift.
NIOSH REL: (Styrene) TWA 50 ppm; CL 100 ppm
DOT Class: Flammable Liquid

SAFETY PROFILE: Experimental poison by ingestion and inhalation. A suspected human carcinogen. An experimental carcinogen and teratogen. Experimental reproductive effects. A skin and eye irritant. A very dangerous fire hazard. Explosive in the form of vapor. To fight fire, use foam, $CO_2$, dry chemical.

PROP: Colorless, refractive, oily liquid. Mp: −31°, bp: 146°, lel: 1.1%, uel: 6.1%, flash p: 88°F, d: 0.9074 @ 20°/4°, autoign temp: 914°F, vap d: 3.6, fp: −33°, ULC: 40-50. Very sltly sol in water; misc in alc and ether.

**SNE000** CAS: 110-61-2 $C_4H_4N_2$ HR: 3
**SUCCINONITRILE**
SYNS: 1,4-BUTANEDINITRILE ◇ ETHYLENE CYANIDE

NIOSH REL: (Nitriles) TWA 20 mg/m$^3$

SAFETY PROFILE: Poison by ingestion. An experimental teratogen. Experimental reproductive effects. Combustible. To fight fire, use alcohol foam, $CO_2$, dry chemical. When heated to decomposition, or on contact with acid or acid fumes, it emits highly toxic fumes of $NO_x$ and $CN^-$.

PROP: Colorless, odorless, waxy material. Mp: 58.1°, bp: 267°, flash p: 270°F, d: 1.022 @ 25°, vap press: 2 mm @ 100°, vap d: 2.1. Sltly sol in ether, water, and alc; sol in acetone.

**SNH000** CAS: 57-50-1 $C_{12}H_{22}O_{11}$ HR: 1
**SUCROSE**
SYNS: BEET SUGAR ◇ CANE SUGAR ◇ SACCHAROSE ◇ SACCHARUM ◇ SUGAR

OSHA PEL: TWA Total Dust: 15 $mg/m^3$; Respirable Fraction: 5 $mg/m^3$
ACGIH TLV: TWA (nuisance particulate) 10 $mg/m^3$ of total dust (when toxic impurities are not present, e.g., quartz < 1%).

SAFETY PROFILE: Mildly toxic by ingestion. An experimental teratogen. Experimental reproductive effects.

PROP: White crystals; sweet taste. D: 1.587 @ 25°/4°, mp: 170-186° (decomp). Sol in water, alc; insol in ether.

**SOD100** CAS: 3689-24-5 $C_8H_{20}O_5P_2S_2$ HR: 3
**SULFOTEP**
DOT: 1704
SYNS: BIS-O,O-DIETHYLPHOSPHOROTHIONIC ANHYDRIDE ◇ ETHYL THIOPYROPHOSPHATE ◇ TEDP (OSHA, MAK) ◇ TETRAETHYL DITHIOPYROPHOSPHATE

OSHA PEL: TWA 0.2 $mg/m^3$ (skin)
ACGIH TLV: TWA 0.2 $mg/m^3$ (skin)
DFG MAK: 0.015 ppm (0.2 $mg/m^3$)
DOT Class: Poison B

SAFETY PROFILE: Poison by ingestion, skin contact, and inhalation. When heated to decomposition it emits toxic fumes of $PO_x$ and $SO_x$.

PROP: A liquid almost insol in water.

**SOH500** CAS: 7446-09-5 $O_2S$ HR: 3
**SULFUR DIOXIDE**
DOT: 1079
SYNS: BISULFITE ◇ SULFUROUS ACID ANHYDRIDE ◇ SULFUROUS ANHYDRIDE

OSHA PEL: (Transitional: TWA 5 ppm) TWA 2 ppm; STEL 5 ppm
ACGIH TLV: TWA 2 ppm; STEL 5 ppm
DFG MAK: 2 ppm (5 $mg/m^3$)
NIOSH REL: (Sulfur Dioxide) TWA 0.5 ppm
DOT Class: Nonflammable Gas

SAFETY PROFILE: A poison gas. An experimental tumorigen and teratogen. Experimental reproductive effects. A corrosive irritant to eyes, skin, and mucous membranes. A nonflammable Gas. When heated to decomposition it emits toxic fumes of $SO_x$.

PROP: Colorless gas or liquid under pressure; pungent odor. Mp: −75.5°, bp: −10.0°, d (liquid): 1.434 @ 0°, vap d: 2.264 @ 0°, vap press: 2538 mm @ 21.1°. Sol in water.

**SOI000** CAS: 2551-62-4 $F_6S$ HR: 1
**SULFUR HEXAFLUORIDE**
DOT: 1080
SYN: SULFUR FLUORIDE

OSHA PEL: TWA 1000 ppm
ACGIH TLV: TWA 1000 ppm

DFG MAK: 1000 ppm (6000 mg/m$^3$)
DOT Class: Nonflammable Gas

SAFETY PROFILE: It may act as a simple asphyxiant in high concentrations and when pure. May explode. When heated to it decomposition emits highly toxic fumes of $F^-$ and $SO_x$.

PROP: Colorless gas. Mp: −51° (subl @ −64°), vap d: 6.602, d (liquid): 1.67 @ −100°.

**SOI500** CAS: 7664-93-9 $H_2O_4S$ HR: 3
**SULFURIC ACID**
DOT: 1830/1832
SYNS: OIL of VITRIOL (DOT) ◇ SULPHURIC ACID

OSHA PEL: TWA 1 mg/m$^3$
ACGIH TLV: TWA 1 mg/m$^3$; STEL 3 ppm
DFG MAK: 1 mg/m$^3$
NIOSH REL: (Sulfuric Acid) TWA 1 mg/m$^3$
DOT Class: Corrosive Material

SAFETY PROFILE: Human poison. A severe eye irritant. Extremely irritating, corrosive, and toxic to tissue. A very powerful, acidic oxidizer which can ignite or explode on contact with many materials. When heated to decomposition it emits toxic fumes of $SO_x$.

PROP: Colorless oily liquid; odorless. Mp: 10.49°, d: 1.834, vap press: 1 mm @ 145.8°, bp: 290°, decomp @ 340°. Misc with water and alc, liberating great heat.

**SON510** CAS: 10025-67-9 $Cl_2S_2$ HR: 3
**SULFUR MONOCHLORIDE**
DOT: 1828
SYNS: CHLORIDE of SULFUR (DOT) ◇ SULFUR DICHLORIDE ◇ SULFUR CHLORIDE

OSHA PEL: (Transitional: TWA 1 ppm) CL 1 ppm
ACGIH TLV: CL 1 ppm
DFG MAK: 1 ppm (6 mg/m$^3$)
DOT Class: Corrosive Material

SAFETY PROFILE: Poison by ingestion and inhalation. A fuming, corrosive liquid very irritating to skin, eyes, and mucous membranes. It decomposes on contact with water to form the highly irritating hydrogen chloride, thiosulfuric acid, and sulfur. Combustible. Will react with water or steam to produce heat and toxic and corrosive fumes. To fight fire, use $CO_2$, dry chemical. When heated to decomposition it emits highly toxic fumes of $Cl^-$ and $SO_x$.

PROP: Amber to yellowish-red, oily, fuming liquid; penetrating odor. mp: −80°, bp: 138.0°, flash p: 245°F (CC), d: 1.6885 @ 15.5°/15.5°, autoign temp: 453°F, vap press: 10 mm @ 27.5°, vap d: 4.66. Decomp in water.

**SOQ450** CAS: 5714-22-7 $F_{10}S_2$ HR: 3
**SULFUR PENTAFLUORIDE**
SYN: SULFUR DECAFLUORIDE

OSHA PEL: (Transitional: TWA 0.25 mg/m$^3$) CL 0.01 ppm
ACGIH TLV: CL 0.01 ppm
DFG MAK: 0.025 ppm (0.25 mg/m$^3$)

SAFETY PROFILE: Moderately toxic by inhalation. When heated to decomposition it emits very toxic fumes of $F^-$ and $SO_x$.

**SOR000** CAS: 7783-60-0 $F_4S$ HR: 3
**SULFUR TETRAFLUORIDE**
DOT: 2418
SYN: TETRAFLUOROSULFURANE

OSHA PEL: CL 0.1 ppm
ACGIH TLV: CL 0.1 ppm
NIOSH REL: (Inorganic Fluorides) TWA 2.5 mg(F)/m$^3$
DOT Class: Poison A

SAFETY PROFILE: Poison by inhalation. A powerful irritant. Will react with water, steam or acids to yield toxic and corrosive fumes. When heated to decomposition it emits very toxic fumes of $F^-$ and $SO_x$.

PROP: Gas. Bp: −40°, mp: −124°.

**SOU500** CAS: 2699-79-8 $F_2O_2S$ HR: 3
**SULFURYL FLUORIDE**
DOT: 2191
SYN: SULFURIC OXYFLUORIDE

OSHA PEL: (Transitional: TWA 5 ppm) TWA 5 ppm; STEL 10 ppm
ACGIH TLV: TWA 5 ppm; STEL 10 ppm
DOT Class: Nonflammable Gas

SAFETY PROFILE: Poison by ingestion. Mildly toxic by inhalation. When heated to decomposition it emits very toxic fumes of $F^-$ and $SO_x$.

PROP: Colorless gas. Mp: −137°, bp: −55°, d: 3.72 g/L.

**SOU625** CAS: 35400-43-2 $C_{12}H_{19}O_2PS_3$ HR: 3
**SULPROFOS**
SYNS: O-ETHYL-O-(4-(METHYLTHIO)PHENYL)PHOSPHORODITHIOIC ACID-S-PROPYL ESTER ◇ HELOTHION

OSHA PEL: TWA 1 mg/m$^3$
ACGIH TLV: TWA 1 mg/m$^3$

SAFETY PROFILE: Poison by ingestion. Moderately toxic by skin contact. When heated to decomposition it emits very toxic fumes of $PO_x$ and $SO_x$.

**TAA100** CAS: 93-76-5 $C_8H_5Cl_3O_3$ HR: 3
**2,4,5-T**
DOT: 2765
SYN: 2,4,5-TRICHLOROPHENOXYACETIC ACID

OSHA PEL: TWA 10 mg/m$^3$
ACGIH TLV: TWA 10 mg/m$^3$
DFG MAK: 10 mg/m$^3$
DOT Class: ORM-A

SAFETY PROFILE: Poison by ingestion. An experimental neoplastigen and teratogen. Experimental reproductive effects. When heated to decomposition it emits toxic fumes of $Cl^-$.

PROP: Crystals; light tan solid. Mp: 151-153°. Usually has 2,3,7,8-TCDD as a minor component.

**TAB750** CAS: 14807-96-6 $H_2O_3Si \cdot 3/4Mg$ HR: 3
**TALC (powder)**
SYNS: ASBESTINE ◇ C.I. 77718 ◇ TALCUM

OSHA PEL: (Transitional: TWA 20 mppcf (containing no asbestos fibers)) TWA 2 mg/m$^3$
ACGIH TLV: TWA 2 mg/m$^3$, respirable dust (use asbestos TLV if asbestos fibers are present)
DFG MAK: 2 mg/m$^3$

SAFETY PROFILE: Talc with less than 1 % asbestos is regarded as a nuisance dust. An experimental tumorigen. A human skin irritant. Prolonged or repeated exposure can produce a form of pulmonary fibrosis (talc pneumoconiosis) which may be due to asbestos content.

PROP: White to grayish-white, fine powder; odorless and tasteless. Insol in water, cold acids, or alkalies. Contains less than 1% crystalline silica.

**TAE750** CAS: 7440-25-7 Ta HR: 3
**TANTALUM**

OSHA PEL: TWA 5 mg/m$^3$
ACGIH TLV: TWA 5 mg/m$^3$
DFG MAK: 5 mg/m$^3$

SAFETY PROFILE: An experimental tumorigen. The dry powder ignites spontaneously in air.

PROP: Gray, very hard, malleable, ductile metal. Mp: 2996°; bp: 5429°; d: 16.69. Insol in water.

**TAI000** CAS: 1746-01-6 $C_{12}H_4Cl_4O_2$ HR: 3
**TCDD**
SYNS: TCDBD ◇ 2,3,6,7-TETRACHLORODIBENZO-p-DIOXIN ◇ TETRADIOXIN

DFG MAK: Animal Carcinogen, Suspected Human Carcinogen.
NIOSH REL: (Dioxin) Reduce to lowest feasible level.

SAFETY PROFILE: A deadly experimental poison by ingestion and skin contact. An experimental carcinogen and teratogen. Causes allergic dermatitis. Experimental reproductive effects. An eye irritant. When heated to decomposition it emits toxic fumes of $Cl^-$.

PROP: Colorless needles. Mp: 305°.

**TAJ000** CAS: 13494-80-9 Te HR: 3
**TELLURIUM**

OSHA PEL: TWA 0.1 mg(Te)/$m^3$
ACGIH TLV: TWA 0.1 mg(Te)/$m^3$
DFG MAK: 0.1 mg/$m^3$

SAFETY PROFILE: Poison by ingestion. An experimental teratogen. Experimental reproductive effects. Respiratory tract irritant. When heated to decomposition it emits toxic fumes of Te.

PROP: Silvery-white quite brittle, metallic, lustrous element. Mp: 449.5°; bp: 989.8°; d: 6.24 @ 20°; vap press: 1 mm @ 520°. Insol in water, benzene, carbon disulfide.

**TAK250** CAS: 7783-80-4 $F_6Te$ HR: 3
**TELLURIUM HEXAFLUORIDE**
DOT: 2195

OSHA PEL: TWA 0.02 ppm
ACGIH TLV: TWA 0.02 ppm
DOT Class: Poison A

SAFETY PROFILE: Poison by inhalation. Human skin (systemic) effects. When heated to decomposition it emits very toxic fumes of $F^-$ and Te.

PROP: Colorless gas, repulsive odor. Mp: −37.6°; bp: −38.9° (subl); d: (solid) 4.006 @ −191°; (liquid) 2.499 @ −10°.

**TAL250** CAS: 3383-96-8 $C_{16}H_{20}O_6P_2S_3$ HR: 3
**TEMEPHOS**
SYN: ABATE

OSHA PEL: (Transitional: Total Dust: 15 mg/$m^3$; Respirable Fraction: 5 mg/$m^3$) TWA Total Dust: 10 mg/$m^3$; Respirable Fraction: 5 mg/$m^3$
ACGIH TLV: TWA 10 mg/$m^3$

SAFETY PROFILE: Poison by ingestion. Moderately toxic by skin contact. An experimental teratogen. A skin irritant. When heated to decomposition it emits toxic fumes of $PO_x$ and $SO_x$.

PROP: White crystals. Mp: 30°.

**TAL570** CAS: 107-49-3 $C_8H_{20}O_7P_2$ HR: 3
**TEPP**
DOT: 2783
SYNS: DIPHOSPHORIC ACID THETRAETHYL ESTER ◇ TETRAETHYL PYROPHOSPHATE

OSHA PEL: TWA 0.05 $mg/m^3$ (skin)
ACGIH TLV: TWA 0.004 $mg/m^3$ (skin)
DFG MAK: 0.0005 ppm (0.05 $mg/m^3$)
DOT Class: Poison B

SAFETY PROFILE: Human poison by ingestion. Experimental poison by skin contact. When heated to decomposition it emits toxic fumes of $PO_x$.

PROP: Water-white to amber hygroscopic liquid. D: 1.20.

**TBC600** CAS: 26140-60-3 $C_{18}H_{14}$ HR: 2
**TERPHENYLS**
SYNS: DIPHENYLBENZENE ◇ TRIPHENYL

OSHA PEL: (Transitional: TWA CL 1 ppm) CL 0.5 ppm
ACGIH TLV: TWA CL 0.5 ppm

SAFETY PROFILE: Moderately toxic by ingestion. Combustible. To fight fire, use water, $CO_2$, dry chemical.

**TBC620** CAS: 92-06-8 $C_{18}H_{14}$ HR: 2
**m-TERPHENYL**
SYNS: m-DIPHENYLBENZENE ◇ ISODIPHENYLBENZENE

OSHA PEL: (Transitional: TWA CL 1 ppm) CL 0.5 ppm
ACGIH TLV: TWA CL 0.5 ppm

SAFETY PROFILE: Moderately toxic by ingestion. Combustible. To fight fire, use water, $CO_2$, dry chemical.

**TBC640** CAS: 84-15-1 $C_{18}H_{14}$ HR: 2
**o-TERPHENYL**
SYN: 1,2-DIPHENYLBENZENE

OSHA PEL: (Transitional: TWA CL 1 ppm) CL 0.5 ppm
ACGIH TLV: TWA CL 0.5 ppm

SAFETY PROFILE: Moderately toxic by ingestion. Combustible. To fight fire, use water, $CO_2$, dry chemical.

**TBC750** CAS: 92-94-4 $C_{18}H_{14}$ HR: 2
**p-TERPHENYL**
SYNS: p-DIPHENYLBENZENE ◇ 4-PHENYLDIPHENYL

OSHA PEL: (Transitional: TWA CL 1 ppm) CL 0.5 ppm
ACGIH TLV: TWA CL 0.5 ppm

SAFETY PROFILE: Moderately toxic by ingestion. Combustible. To fight fire, use water, $CO_2$, dry chemical.

PROP: Leaves or needles. D: 1.234 @ 0/4°, mp: 212-213°, bp: 276°, flash p: 405°F (OC), vap d: 7.95. Sol in hot benzene; very sol in hot alc, sltly sol in ether.

**TBP000** CAS: 76-11-9 $C_2Cl_4F_2$ HR: 1
**1,1,1,2-TETRACHLORO-2,2-DIFLUOROETHANE**
SYN: HALOCARBON 112a

OSHA PEL: TWA 500 ppm
ACGIH TLV: TWA 500 ppm
DFG MAK: 1000 ppm (8340 $mg/m^3$)

SAFETY PROFILE: Mildly toxic by inhalation. When heated to decomposition it emits very toxic fumes of $Cl^-$ and $F^-$.

**TBP050** CAS: 76-12-0 $C_2Cl_4F_2$ HR: 2
**1,1,2,2-TETRACHLORO-1,2-DIFLUOROETHANE** ◇ SYNS: 1,2-DIFLUORO-1,1,2,2-TETRACHLOROETHANE ◇ FREON 112

OSHA PEL: TWA 500 ppm
ACGIH TLV: TWA 500 ppm
DFG MAK: 500 ppm (4170 $mg/m^3$)

SAFETY PROFILE: Moderately toxic by ingestion. Mildly toxic by inhalation. A skin and eye irritant. When heated to decomposition it emits toxic fumes of $F^-$ and $Cl^-$.

PROP: Liquid. Bp: 92.8°, d: 1.6447 @ 25°, vap d: 7.03.

**TBP250** CAS: 31242-94-1 $C_{12}H_6Cl_4O$ HR: 3
**TETRACHLORODIPHENYL OXIDE**
SYN: PHENYL ETHER TETRACHLORO

OSHA PEL: TWA 0.5 $mg/m^3$

SAFETY PROFILE: Poison by ingestion. When heated to decomposition it emits toxic fumes of $Cl^-$.

**TBP750** CAS: 25322-20-7 $C_2H_2Cl_4$ HR: 3
**TETRACHLOROETHANE**
DOT: 1702

NIOSH REL: (Tetrachloroethane) Reduce to lowest feasible level
DOT Class: ORM-A

SAFETY PROFILE: Poison by ingestion and inhalation. Experimental reproductive effects. When heated to decomposition it emits toxic fumes of $Cl^-$.

**TBQ100** CAS: 79-34-5 $C_2H_2Cl_4$ HR: 3
**1,1,2,2-TETRACHLOROETHANE**
DOT: 1702
SYNS: ACETYLENE TETRACHLORIDE ◇ TCE

OSHA PEL: (Transitional: TWA 5 ppm (skin)) TWA 1 ppm (skin)
ACGIH TLV: TWA 1 ppm (skin)
DFG MAK: 1 ppm (7 $mg/m^3$); Suspected Carcinogen.
NIOSH REL: (1,1,2,2-Tetrachlorethane) Reduce to lowest level.
DOT Class: IMO: Poison B

SAFETY PROFILE: Poison by inhalation and ingestion. An experimental tumorigen and carcinogen. A strong irritant of eyes and mucous membranes. When heated to decomposition it emits toxic fumes of $Cl^-$.

PROP: Heavy, colorless, mobile liquid; chloroform-like odor. Mp: −43.8°, bp: 146.4°, d: 1.600 @ 20°/4°.

**TBR000** CAS: 1335-88-2 $C_{10}H_4Cl_4$ HR: 3
**TETRACHLORONAPHTHALENE**
SYN: HALOWAX

OSHA PEL: TWA 2 mg/m$^3$
ACGIH TLV: TWA 2 mg/m$^3$

SAFETY PROFILE: Probably a poison. When heated to decomposition it emits highly toxic fumes of $Cl^-$.

PROP: Crystals. Mp: 182°.

**TCF000** CAS: 78-00-2 $C_8H_{20}Pb$ HR: 3
**TETRAETHYL LEAD**
DOT: 1649
SYNS: TEL ◇ TETRAETHYLPLUMBANE

OSHA PEL: TWA 0.075 mg(Pb)/m$^3$ (skin)
ACGIH TLV: TWA 0.1 mg(Pb)/m$^3$ (skin)
DFG MAK: 0.01 ppm (0.075 mg/m$^3$)
DOT Class: Poison B

SAFETY PROFILE: Human poison. Moderately toxic by inhalation and skin contact. An experimental carcinogen and teratogen. Experimental reproductive effects. May cause lead exposure intoxication by skin contact. Flammable. To fight fire, use dry chemical, $CO_2$, mist, foam. When heated to decomposition it emits toxic fumes of Pb.

PROP: Colorless, oily liquid; pleasant characteristic odor. Mp: 125-150°, bp: 198-202° with decomp, d: 1.659 @ 18°, vap press: 1 mm @ 38.4°, flash p: 200°F.

**TCR750** CAS: 109-99-9 $C_4H_8O$ HR: 3
**TETRAHYDROFURAN**
DOT: 2056
SYNS: BUTYLENE OXIDE ◇ DIETHYLENE OXIDE ◇ 1,4-EPOXYBUTANE ◇ THF

OSHA PEL: (Transitional: TWA 200 ppm) TWA 200 ppm; STEL 250 ppm
ACGIH TLV: TWA 200 ppm; STEL 250 ppm
DFG MAK: 200 ppm (590 mg/m$^3$)
DOT Class: Flammable Liquid

SAFETY PROFILE: Moderately toxic by ingestion. Mildly toxic by inhalation. Irritant to eyes and mucous membranes. Flammable liquid. Explosive in the form of vapor. Unstabilized tetrahydrofuran forms thermally explosive peroxides on exposure to air. To fight fire, use foam, dry chemical, $CO_2$.

PROP: Colorless, mobile liquid; ether-like odor. Bp: 65.4°, flash p: 1.4°F (TCC), lel: 1.8%, uel: 11.8%, fp: −108.5°, d: 0.888 @ 20/4°, vap press: 114 mm @ 15°, vap d: 2.5, autoign temp: 610°F. Misc with water, alc, ketones, esters, ethers, and hydrocarbons.

**TDR500** CAS: 75-74-1 $C_4H_{12}Pb$ HR: 3
**TETRAMETHYL LEAD**
SYNS: TETRAMETHYLPLUMBANE ◇ TML

OSHA PEL: TWA 0.075 mg(Pb)/m$^3$ (skin)
ACGIH TLV: TWA 0.15 mg(Pb)/m$^3$ (skin)
DFG MAK: 0.01 ppm (0.075 mg/m$^3$)

SAFETY PROFILE: Poison by ingestion. Moderately toxic by skin contact. An experimental teratogen. Experimental reproductive effects. A very dangerous fire hazard. Moderate explosion hazard in the form of vapor. To fight fire, use water, foam, $CO_2$, dry chemical. When heated to decomposition it emits toxic fumes of Pb.

PROP: Colorless liquid. Mp: −18°F, lel: 1.8%, bp: 110°, d: 1.99, vap d: 9.2, flash p: 100°F.

**TDW250** CAS: 3333-52-6 $C_8H_{12}N_2$ HR: 3
**TETRAMETHYLSUCCINONITRILE**

OSHA PEL: TWA 0.5 ppm (skin)
ACGIH TLV: TWA 0.5 ppm (skin)
DFG MAK: 0.5 ppm (3 mg/m$^3$)
NIOSH REL: (Nitriles) CL 6 mg/m$^3$/15M

SAFETY PROFILE: Poison by ingestion. An experimental teratogen. Experimental reproductive effects. A human skin irritant and allergen. Absorbed by skin. When heated to decomposition it emits toxic fumes of $CN^-$ and $NO_x$.

PROP: Crystallizes in plates; almost no odor. Mp: 169° (sublimes).

**TDY250** CAS: 509-14-8 $CN_4O_8$ HR: 3
**TETRANITROMETHANE**
DOT: 1510

OSHA PEL: TWA 1 ppm
ACGIH TLV: TWA 1 ppm
DFG MAK: 1 ppm (8 mg/m$^3$)
DOT Class: Oxidizer

SAFETY PROFILE: Poison by ingestion and inhalation. Irritating to the skin, eyes, mucous membranes, and respiratory passages. A very dangerous fire hazard. A severe explosion hazard. When heated to decomposition it emits highly toxic fumes of $NO_x$.

PROP: Colorless or yellow liquid. Mp: 13°, bp: 125.7°, d: 1.650 @ 13°, vap press: 10 mm @ 22.7°. Insol in water; very sol in alc and ether.

**TEE500** CAS: 7722-88-5 $O_7P_2 \cdot 4Na$ HR: 3
**TETRASODIUM PYROPHOSPHATE**
SYNS: SODIUM PYROPHOSPHATE (FCC) ◇ TSPP

OSHA PEL: TWA 5 mg/m$^3$
ACGIH TLV: TWA 5 mg/m$^3$

SAFETY PROFILE: Poison by ingestion. When heated to decomposition it emits toxic fumes of $PO_x$ and $Na_2O$.

PROP: White crystalline powder. Mp: 988°, d: 2.534. Sol in water; insol in alc.

**TEG250** CAS: 479-45-8 $C_7H_5N_5O_8$ HR: 3
**TETRYL**
DOT: 0208
SYNS: PICRYLMETHYLNITRAMINE ◇ TRINITROPHENYLMETHYLNITRAMINE

OSHA PEL: (Transitional: TWA 1.5 mg/m$^3$ (skin)) TWA 0.1 mg/m$^3$ (skin)
ACGIH TLV: TWA 1.5 mg/m$^3$
DOT Class: Class A Explosive

SAFETY PROFILE: An irritant, sensitizer, and allergen. A dangerous fire and explosion hazard. A high explosive sensitive to shock, friction or heat.

PROP: Yellow, monoclinic crystals. Mp: 130°, bp: explodes @ 187°, d: 1.57 @ 19°.

**TEI000** CAS: 7440-28-0 Tl HR: 3
**THALLIUM**

OSHA PEL: TWA 0.1 mg(Tl)/m$^3$ (skin)
ACGIH TLV: TWA 0.1 mg(Tl)/m$^3$ (skin)
DFG MAK: 0.1 mg/m$^3$

SAFETY PROFILE: Human poison. Flammable in the form of dust. When heated to decomposition it emits toxic fumes of Tl.

PROP: Bluish-white, soft, malleable metal. Mp: 303.5°, bp: 1457°, d: 11.85 @ 20°, vap press: 1 mm @ 825°.

**TFC600** CAS: 96-69-5 $C_{22}H_{30}O_2S$ HR: 3
**4,4′-THIOBIS(6-tert-BUTYL-m-CRESOL)**
SYNS: 4,4′-THIOBIS(2-tert-BUTYL-5-METHYLPHENOL)

OSHA PEL: (Transitional: TWA Total Dust: 15 mg/m$^3$; Respirable Fraction: 5 mg/m$^3$) TWA Total Dust: 10 mg/m$^3$; Respirable Fraction: 5 mg/m$^3$
ACGIH TLV: TWA 10 mg/m$^3$

SAFETY PROFILE: Probably a poison by ingestion and inhalation. When heated to decomposition it emits highly toxic fumes of $SO_x$.

PROP: Light gray to tan powder. Mp: 150°, d: 1.10.

**TFD750** CAS: 91-71-4 $C_{21}H_{17}AsN_2O_5S_2$ HR: 3
**THIOCARBAMIZINE**
SYN: 4-CARBAMIDOPHENYL BIS(o-CARBOXYPHENYLTHIO)ARSENITE

OSHA PEL: TWA 0.5 mg(As)/$m^3$
ACGIH TLV: TWA 0.2 mg(As)/$m^3$

SAFETY PROFILE: Moderately toxic by ingestion. When heated to decomposition it emits very toxic fumes of As, $SO_x$, and $NO_x$.

**TFI000** CAS: 139-65-1 $C_{12}H_{12}N_2S$ HR: 3
**4,4′-THIODIANILINE**
SYN: BIS(p-AMINOPHENYL)SULFIDE

DFG MAK: Animal Carcinogen, Suspected Human Carcinogen.

SAFETY PROFILE: Moderately toxic by ingestion. An experimental carcinogen and tumorigen. May be a human carcinogen. Experimental reproductive effects. When heated to decomposition it emits very toxic fumes of $NO_x$ and $SO_x$.

PROP: Needles. Mp: 108°.

**TFJ100** CAS: 68-11-1 $C_2H_4O_2S$ HR: 3
**THIOGLYCOLIC ACID**
DOT: 1940
SYN: MERCAPTOACETIC ACID

OSHA PEL: TWA 1 ppm (skin)
ACGIH TLV: TWA 1 ppm (skin)
DOT Class: Corrosive Material

SAFETY PROFILE: Poison by ingestion and skin contact. A corrosive irritant to skin, eyes, and mucous membranes. When heated to decomposition it emits toxic fumes of $SO_x$.

PROP: Liquid, strong odor. Mp: −16.5°, bp: 108° @ 15 mm. Misc with water alc, ether, chloroform, and benzene.

**TFL000** CAS: 7719-09-7 $Cl_2OS$ HR: 3
**THIONYL CHLORIDE**
DOT: 1836
SYNS: SULFINYL CHLORIDE ◇ SULFUROUS OXYCHLORIDE

OSHA PEL: CL 1 ppm
ACGIH TLV: CL 1 ppm
DOT Class: Corrosive Material

SAFETY PROFILE: Moderately toxic by inhalation. Violent reaction with water releases hydrogen chloride and sulfur dioxide. A corrosive irritant which causes burns to the skin and eyes. When heated to decomposition it emits toxic fumes of $SO_x$ and $Cl^-$.

PROP: Colorless to yellow to red liquid; suffocating odor. Mp: −105°, bp: 78.8° @ 746 mm, d: 1.640 @ 15.5/15.5°, vap press: 100 mm @ 21.4°. Misc with benzene, chloroform, and carbon tetrachloride.

**TFS350** CAS: 137-26-8 $C_6H_{12}N_2S_4$ HR: 3
**THIRAM**
DOT: 2771
SYNS: BIS(DIMETHYLTHIOCARBAMOYL) DISULFIDE ◇ METHYL THIRAM ◇ METHYL TUADS ◇ TERAMETHYL THIURAM DISULFIDE

OSHA PEL: TWA 5 mg/m$^3$
ACGIH TLV: TWA 5 mg/m$^3$ (Proposed: TWA 1 mg/m$^3$)
DFG MAK: 5 mg/m$^3$
DOT Class: ORM-A

SAFETY PROFILE: Poison by ingestion. An experimental carcinogen and teratogen. A mild allergen and irritant. When heated to decomposition emits toxic fumes of $SO_x$.

PROP: Crystals. Mp: 156°, d: 1.30, bp: 129° @ 20 mm. Insol in water; sol in alc, ether, acetone, and chloroform.

**TGB250** CAS: 7440-31-5 Sn HR: 3
**TIN and COMPOUNDS**

OSHA PEL: Organic Compounds: TWA 0.1 mg(Sn)/m$^3$ (skin); Inorganic Compounds (except oxides): TWA 2 mg/m$^3$
ACGIH TLV: TWA metal, oxide and inorganic compounds (except $SnH_4$) as Sn 2 mg/m$^3$; organic compounds 0.1 mg/m$^3$ (skin) (Proposed: TWA 0.1 mg(Sn)/m$^3$; STEL 0.2 mg(Sn)/m$^3$ (skin))
DFG MAK: Inorganic 2 mg/m$^3$, organic 0.1 mg/m$^3$
NIOSH REL: (Organotin Compounds) TWA 0.1 mg(Sn)/m$^3$

SAFETY PROFILE: An experimental tumorigen. Combustible.

PROP: Cubic, gray, crystalline metallic element. Mp: 231.9°, stabilizes <18°, d: 7.31, vap press: 1 mm @ 1492°, bp: 2507°.

**TGG760** CAS: 13463-67-7 $O_2Ti$ HR: 3
**TITANIUM DIOXIDE**
SYNS: C.I. 77891 ◇ RUTILE ◇ TITANIUM OXIDE

OSHA PEL: (Transitional: TWA Total Dust: 15 mg/m$^3$; Respirable Fraction: 5 mg/m$^3$) TWA Total Dust: 10 mg/m$^3$; Respirable Fraction: 5 mg/m$^3$
ACGIH TLV: TWA (nuisance particulate) 10 mg/m$^3$ of total dust (when toxic impurities are not present, e.g., quartz $< 1\%$).
DFG MAK: 6 mg/m$^3$

SAFETY PROFILE: An experimental carcinogen. A human skin irritant. A nuisance dust.

PROP: White amorphous powder. Mp: 1860° (decomp), d: 4.26. Insol in water, hydrochloric acid, dilute sulfuric acid, and alc.

**TGJ750** CAS: 119-93-7 $C_{14}H_{16}N_2$ HR: 3
**o-TOLIDINE**
SYNS: C.I. 37230 ◇ 4,4'-DIAMINO-3,3'-DIMETHYLDIPHENYL

ACGIH TLV: Suspected Human Carcinogen
DFG MAK: Animal Carcinogen, Suspected Human Carcinogen.
NIOSH REL: (o-Toluidine) CL 0.02 mg/m$^3$/60M; avoid skin contact.

SAFETY PROFILE: Moderately toxic by ingestion. An experimental carcinogen. When heated to decomposition it emits toxic fumes of $NO_x$.

PROP: White to reddish crystals. Mp: 129-131°C. Very sltly sol in water; sol in alc, ether, and acetic acid.

**TGK750** CAS: 108-88-3 $C_7H_8$ HR: 3
**TOLUENE**
DOT: 1294
SYNS: METHYLBENZENE ◇ TOLUOL (DOT)

OSHA PEL: (Transitional: TWA 200 ppm; CL 300 ppm; Pk 500 ppm/10M/8H) TWA 100 ppm; STEL 150 ppm
ACGIH TLV: TWA 100 ppm; STEL 150 ppm; BEI: 1 mg(toluene)/L in venous blood end of shift; 20 ppm toluene in end-exhaled air during shift.
DFG MAK: 100 ppm (380 mg/m$^3$); BAT: 340 μg/dl in blood at end of shift.
NIOSH REL: (Toluene) TWA 100 ppm; CL 200 ppm/10M
DOT Class: Flammable Liquid

SAFETY PROFILE: Mildly toxic by inhalation. An experimental teratogen. Experimental reproductive effects. An experimental skin and severe eye irritant. Flammable liquid. Explosive in the form of vapor. To fight fire, use foam, $CO_2$, dry chemical.

PROP: Colorless liquid; benzol-like odor. Mp: −95, bp: 110.4°, flash p: 40°F (CC), ULC: 75-80, lel: 1.27%, uel: 7%, d: 0.866 @ 20°/4°, autoign temp: 996°F, vap press: 36.7 mm @ 30°, vap d: 3.14. Insol in water; sol in acetone; misc in abs alc, ether, and chloroform.

**TGL750** CAS: 95-80-7 $C_7H_{10}N_2$ HR: 3
**TOLUENE-2,4-DIAMINE**
DOT: 1709
SYNS: 3-AMINO-p-TOLUIDINE ◇ C.I. 76035 ◇ 2,4-DIAMINOTOLUENE

DFG MAK: Animal Carcinogen, Suspected Human Carcinogen.
DOT Class: Poison B

SAFETY PROFILE: Poison by ingestion. An experimental carcinogen. Experimental reproductive effects. A skin and eye irritant. When heated to decomposition it emits toxic fumes of $NO_x$.

PROP: Prisms. Mp: 99°, bp: 280°, vap press: 1 mm @ 106.5°.

**TGM750** CAS: 584-84-9 $C_9H_6N_2O_2$ HR: 3

**TOLUENE-2,4-DIISOCYANATE**

SYNS: 2,4-DIISOCYANATOTOLUENE ◇ ISOCYANIC ACID, METHYLPHENYLENE ESTER ◇ TDI ◇ 2,4-TOLUENEDIISOCYANATE

OSHA PEL: (Transitional: CL 0.02 ppm) TWA 0.005 ppm; STEL 0.02 ppm
ACGIH TLV: TWA 0.005 ppm; STEL 0.02 ppm
DFG MAK: 0.01 ppm (0.07 mg/m$^3$
NIOSH REL: (Diisocyanates) TWA 0.005 ppm; CL 0.02 ppm/10M
DOT Class: Poison B

SAFETY PROFILE: Poison by ingestion and inhalation. An experimental carcinogen. A severe skin and eye irritant. Combustible. Explosive in the form of vapor. To fight fire, use dry chemical, $CO_2$. When heated to decomposition it emits highly toxic fumes of $NO_x$.

PROP: Clear, faintly yellow liquid; sharp, pungent odor. Mp: 19.5-21.5°, d (liquid): 1.2244 @ 20/4°, bp: 251°, flash p: 270°F (OC), vap d: 6.0, lel: 0.9%, uel: 9.5%. Misc with alc (decomp), ether, acetone, carbon tetrachloride, benzene, chlorobenzene, kerosene, and alcolive oil.

**TGM800** CAS: 91-08-7 $C_9H_6N_2O_2$ HR: 3

**TOLUENE-2,6-DIISOCYANATE**

SYNS: 2,6-DIISOCYANATOTOLUENE ◇ 2-METHYL-m-PHENYLENE ISOCYANATE ◇ 2,6-TDI ◇ 2,6-TOLUENE DIISOCYANATE

DFG MAK: 0.01 ppm (0.07 mg/m$^3$)
NIOSH REL: (Diisocyanates) TWA 0.005 ppm; CL 0.02 ppm/10M

SAFETY PROFILE: Poison by ingestion and inhalation. When heated to decomposition it emits toxic fumes of $NO_x$.

**TGQ500** CAS: 108-44-1 $C_7H_9N$ HR: 3

**m-TOLUIDINE**

DOT: 1708
SYNS: 3-AMINO-1-METHYLBENZENE ◇ m-METHYLANILINE

OSHA PEL: TWA 2 ppm (skin)
ACGIH TLV: TWA 2 ppm (skin)

SAFETY PROFILE: Poison by ingestion. A skin and eye irritant. Flammable. To fight fire, use foam, $CO_2$, dry chemical. When heated to decomposition it emits highly toxic fumes of $NO_x$.

PROP: Colorless liquid. Mp: −50.5°, bp: 203.3°, d: 0.989 @ 20/4°, vap press: 1 mm @ 41°, vap d: 3.90. Sltly sol in water; sol in alc, and ether.

**TGQ750** CAS: 95-53-4 $C_7H_9N$ HR: 3

**o-TOLUIDINE**

DOT: 1708
SYNS: 1-AMINO-2-METHYLBENZENE ◇ o-AMINOTOLUENE ◇ C.I. 37077

OSHA PEL: TWA 5 ppm (skin)
ACGIH TLV: TWA 2 ppm (skin); Suspected Human Carcinogen.
DFG MAK: Animal Carcinogen, Suspected Human Carcinogen.
DOT Class: Poison B

SAFETY PROFILE: Poison by ingestion. Moderately toxic by skin contact. An experimental neoplastigen. A skin and eye irritant. Flammable. To fight fire, use foam, $CO_2$, dry chemical. When heated to decomposition it emits highly toxic fumes of $NO_x$.

PROP: Colorless liquid. Mp: −16.3°, bp: 200-202°, ULC: 20-25, flash p: 185° (CC), d: 1.004 @ 20/4°, autoign temp: 900°F, vap press: 1 mm @ 44°, vap d: 3.69. Sltly sol in water and dilute acid; sol in alc and ether.

**TGR000** CAS: 106-49-0 $C_7H_9N$ HR: 3
**p-TOLUIDINE**
DOT: 1708
SYNS: p-AMINOTOLUENE ◇ C.I. 37107 ◇ 4-METHYLANILINE

OSHA PEL: TWA 2 ppm (skin)
ACGIH TLV: TWA 2 ppm (skin); Suspected Human Carcinogen.

SAFETY PROFILE: Poison by ingestion. A severe skin and eye irritant. Flammable. To fight fire, use foam, $CO_2$, dry chemical. When heated to decomposition it emits highly toxic fumes of $NO_x$.

PROP: Colorless leaflets. Mp: 44.5°, bp: 200.4°, flash p: 188°F (CC), d: 1.046 @ 20/4°, autoign temp: 900°F, vap press: 1 mm @ 42°, vap d: 3.90. Sol in water, dilute acid and $CS_2$; very sol in alc and ether.

**TIA250** CAS: 126-73-8 $C_{12}H_{27}O_4P$ HR: 3
**TRIBUTYL PHOSPHATE**

OSHA PEL: (Transitional: TWA 5 mg/m$^3$) TWA 0.2 ppm
ACGIH TLV: TWA 0.2 ppm

SAFETY PROFILE: Moderately toxic by ingestion and inhalation. A skin, eye, and mucous membrane irritant. Combustible. To fight fire, use $CO_2$, dry chemical, fog, mist. When heated to decomposition it emits toxic fumes of $PO_x$.

PROP: Colorless odorless liquid. Bp: 289° (decomp), mp: $<-80°$, flash p: 295°F (COC), d: 0.982 @ 20°, vap d: 9.20. Sol in water; misc in alc, ether.

**TII250** CAS: 76-03-9 $C_2HCl_3O_2$ HR: 3
**TRICHLOROACETIC ACID**
DOT: 1839/2564

OSHA PEL: TWA 1 ppm
ACGIH TLV: TWA 1 ppm
DOT Class: Corrosive Material

SAFETY PROFILE: A corrosive irritant to skin, eyes and mucous membranes. When heated to decomposition it emits toxic fumes of $Cl^-$ and $Na_2O$.

PROP: Colorless, rhombic, deliquescent crystals. Bp: 197.5°, fp: 57.7°, flash p: none, d: 1.6298 @ 61°/4°, vap press: 1 mm @ 51.0°.

**TIK250** CAS: 120-82-1 $C_6H_3Cl_3$ HR: 3
**1,2,4-TRICHLOROBENZENE**
DOT: 2321

OSHA PEL: CL 5 ppm
ACGIH TLV: CL 5 ppm
DFG MAK: 5 ppm (40 $mg/m^3$)
DOT Class: Poison B

SAFETY PROFILE: Poison by ingestion. An experimental teratogen. Experimental reproductive effects. A skin irritant. Combustible. To fight fire, use water, foam, $CO_2$, dry chemical. When heated to decomposition it emits toxic fumes of $Cl^-$.

PROP: Colorless liquid. Mp: 17°, bp: 213°, flash p: 230°F (CC), d: 1.454 @ 25°/25°, vap press: 1 mm @ 38.4°, vap d: 6.26. Sol in water.

**TIL360** CAS: 2431-50-7 $C_4H_5C_{13}$ HR: 3
**2,3,4-TRICHLOROBUTENE-1**

DFG MAK: Animal Carcinogen; Suspected Human Carcinogen.

SAFETY PROFILE: Poison by ingestion. When heated to decomposition it emits toxic fumes of $Cl^-$.

**TIN000** CAS: 79-00-5 $C_2H_3Cl_3$ HR: 3
**1,1,2-TRICHLOROETHANE**
SYNS: ETHANE TRICHLORIDE ◇ VINYL TRICHLORIDE

OSHA PEL: TWA 10 ppm (skin)
ACGIH TLV: TWA 10 ppm (skin)
DFG MAK: 10 ppm (55 $mg/m^3$); Suspected Carcinogen.

SAFETY PROFILE: Poison by ingestion. Moderately toxic by inhalation and skin contact. An experimental carcinogen. An eye and severe skin irritant. When heated to decomposition it emits toxic fumes of $Cl^-$.

PROP: Liquid; pleasant odor. Bp: 114°, fp: −35°, d: 1.4416 @ 20°/4°, vap press: 40 mm @ 35.2°.

**TIO750** CAS: 79-01-6 $C_2HCl_3$ HR: 3
**TRICHLOROETHYLENE**
DOT: 1710
SYNS: ACETYLENE TRICHLORIDE ◇ ETHYLENE TRICHLORIDE

OSHA PEL: (Transitional: TWA 100 ppm; CL 200 ppm; Pk 300 ppm/5M/2H)TWA 50 ppm; STEL 200 ppm
ACGIH TLV: TWA 50 ppm; STEL 200 ppm; BEI: 320 mg(trichloroethanol)/g creatinine in urine at end of shift; 0.5 ppm trichloroethylene in end-exhaled air prior to shift and end of work week.

DFG MAK: Suspected Carcinogen; 50 ppm (270 mg/m$^3$); BAT: 500 µg/dL in blood at end of shift or work week.
NIOSH REL: (Trichloroethylene) TWA 250 ppm; (Waste Anesthetic Gases) CL 2 ppm/1H
DOT Class: ORM-A

SAFETY PROFILE: Mildly toxic to humans by ingestion and inhalation. An experimental carcinogen and teratogen. Experimental reproductive effects. An eye and severe skin irritant. High concentrations of trichloroethylene vapor in high-temperature air can be made to burn mildly if plied with a strong flame. When heated to decomposition it emits toxic fumes of $Cl^-$.

PROP: Clear, colorless, mobile liquid; characteristic sweet odor of chloroform. D: 1.4649 @ 20°/4°, bp: 86.7°, flash p: 89.6°F (but practically nonflammable), lel: 12.5%, uel: 90% @ > 30°, mp: −73°, fp: −86.8°, autoign temp: 788°F, vap press: 100 mm @ 32°, vap d: 4.53, refr index: 1.477 @ 20°. Immiscible with water; misc with alc, ether, acetone, and carbon tetrachloride.

**TIP500** CAS: 75-69-4 $CCl_3F$ HR: 2
**TRICHLOROFLUOROMETHANE**
SYNS: FLUOROTRICHLOROMETHANE (OSHA) ◇ FREON 11

OSHA PEL: (Transitional: TWA 1000 ppm) CL 1000 ppm
ACGIH TLV: CL 1000 ppm
DFG MAK: 1000 ppm (5600 mg/m$^3$)

SAFETY PROFILE: High concentrations cause narcosis and anesthesia. When heated to decomposition it emits highly toxic fumes of $F^-$ and $Cl^-$.

PROP: Colorless liquid. Mp: −111°, bp: 24.1°, d: 1.484 @ 17.2°.

**TIT500** CAS: 1321-65-9 $C_{10}H_5Cl_3$ HR: 3
**TRICHLORONAPHTHALENE**
SYNS: HALOWAX ◇ NIBREN WAX ◇ SEEKAY WAX

OSHA PEL: TWA 5 mg/m$^3$ (skin)
ACGIH TLV: TWA 5 mg/m$^3$ (skin)
DFG MAK: 5 mg/m$^3$

SAFETY PROFILE: A poison.

PROP: A white solid.

**TJB600** CAS: 96-18-4 $C_3H_5Cl_3$ HR: 3
**1,2,3-TRICHLOROPROPANE**
SYNS: ALLYL TRICHLORIDE ◇ GLYCEROL TRICHLOROHYDRIN ◇ TRICHLOROHYDRIN

OSHA PEL: (Transitional: TWA 50 ppm) TWA 10 ppm
ACGIH TLV: TWA 10 ppm (skin)
DFG MAK: 50 ppm (300 mg/m$^3$)

SAFETY PROFILE: Poison by ingestion. Moderately toxic by inhalation and skin contact. Experimental reproductive effects. A skin and severe eye irritant. Moderately flammable. When heated to decomposition it emits toxic fumes of $Cl^-$. To fight fire, use water (as a blanket), spray, mist, dry chemical.

PROP: Bp: 142°, d: 1.414 @ 20°/20°, flash p: 180°F (OC).

**TJO000** CAS: 121-44-8 $C_6H_{15}N$ HR: 2
**TRIETHYLAMINE**
DOT: 1296
SYNS: N,N-DIETHYLETHANAMINE ◇ TEN

OSHA PEL: (Transitional: TWA 25 ppm) TWA 10 ppm; STEL 15 ppm
ACGIH TLV: TWA 10 ppm; STEL 15 ppm
DFG MAK: 10 ppm (40 mg/m$^3$)
DOT Class: Flammable Liquid

SAFETY PROFILE: Moderately toxic by ingestion and skin contact. Experimental reproductive effects. A skin and severe eye irritant. A very dangerous fire hazard. Explosive in the form of vapor. To fight fire, use $CO_2$, dry chemical, alcohol foam. When heated to decomposition it emits toxic fumes of $NO_x$.

PROP: Colorless liquid; ammonia odor. Mp: −114.8°, bp: 89.5°, flash p: 20°F (OC), d: 0.7255 @ 25/4°, vap d: 3.48, lel: 1.2%, uel: 8.0%. Misc in water, alc, and ether.

**TJY100** CAS: 75-63-8 $CBrF_3$ HR: 1
**TRIFLUOROBROMOMETHANE**
DOT: 1009
SYNS: BROMOFLUOROFORM ◇ FREON 13B1

OSHA PEL: TWA 1000 ppm
ACGIH TLV: TWA 1000 ppm
DFG MAK: 1000 ppm (6100 mg/m$^3$)
DOT Class: Nonflammable Gas

SAFETY PROFILE: Mildly toxic by inhalation. When heated to decomposition it emits toxic fumes of $F^-$ and $Br^-$.

**TKB250** CAS: 406-90-6 $C_4H_5F_3O$ HR: 3
**2,2,2-TRIFLUOROETHYL VINYL ETHER**
SYNS: FLUOROXENE ◇ (2,2,2-TRIFLUOROETHOXY)ETHENE

NIOSH REL: (Waste Anesthetic Gases) CL 2 ppm/1H

SAFETY PROFILE: Human poison by ingestion. Experimental poison by inhalation. An experimental teratogen. Experimental reproductive effects. When heated to decomposition it emits toxic fumes of $F^-$.

**TKV000** CAS: 552-30-7 $C_9H_4O_5$ HR: 1
**TRIMELLITIC ANHYDRIDE**
SYNS: 1,2,4-BENZENETRICARBOXYLIC ANHYDRIDE ◇ TMAN ◇ TRIMELLIC ACID ANHYDRIDE

OSHA PEL: TWA 0.005 ppm
ACGIH TLV: TWA 0.005 ppm
DFG MAK: 0.005 ppm (0.04 mg/m$^3$)

SAFETY PROFILE: Mildly toxic by ingestion. Irritant to lungs and air passages. May be a powerful allergen.

PROP: Crystals. Mp: 162°, bp: 240-245° @ 14 mm. Sol in acetone, ethyl acetate, and dimethylformamide.

**TLD500** CAS: 75-50-3 $C_3H_9N$ HR: 3
**TRIMETHYLAMINE**
DOT: 1083/1297
SYNS: N,N-DIMETHYLMETHANAMINE ◇ TMA

OSHA PEL: TWA 10 ppm; STEL 15 ppm
ACGIH TLV: TWA 10 ppm; STEL 15 ppm
DOT Class: Flammable Gas

SAFETY PROFILE: Mildly toxic by inhalation. A very dangerous fire hazard. Self-reactive. Moderately explosive in the form of vapor. To fight fire, stop flow of gas. When heated to decomposition it emits toxic fumes of $NO_x$.

PROP: Colorless gas; pungent, fishy, ammoniacal odor, saline taste. Bp: 2.87°, lel: 2%, uel: 11.6%, fp: −117.1°, d: 0.662 @ −5°, autoign temp: 374°F, vap d: 2.0, flash p: 20°F (CC). Misc with alc; sol in ether, benzene, toluene, xylene, and chloroform.

**TLG250** CAS: 137-17-7 $C_9H_{13}N$ HR: 3
**2,4,5-TRIMETHYLANILINE**
SYNS: 1-AMINO-2,4,5-TRIMETHYLBENZENE ◇ PSEUDOCUMIDINE

DFG MAK: Animal Carcinogen, Suspected Human Carcinogen.

SAFETY PROFILE: Moderately toxic by ingestion. An experimental carcinogen and tumorigen. When heated to decomposition it emits toxic fumes of $NO_x$.

**TLL250** CAS: 25551-13-7 $C_9H_{12}$ HR: 1
**TRIMETHYL BENZENE**

OSHA PEL: TWA 25 ppm
ACGIH TLV: TWA 25 ppm

SAFETY PROFILE: Mildly toxic by ingestion. A skin and eye irritant. Flammable.

**TLY500** CAS: 540-84-1 $C_8H_{18}$ HR: 3
**2,2,4-TRIMETHYLPENTANE**
DOT: 1262
SYNS: ISOBUTYLTRIMETHYLETHANE ◇ ISOOCTANE (DOT)

NIOSH REL: TWA (Alkanes) 350 mg/m$^3$
DOT Class: Flammable Liquid

SAFETY PROFILE: High concentrations can cause narcosis. A very dangerous fire hazard. Explosive in the form of vapor. To fight fire, use $CO_2$, dry chemical.

PROP: Clear liquid; odor of gasoline. Bp: 99.2°, fp: −116°, flash p: 10°F, d: 0.692 @ 20/4°, autoign temp: 779°F, vap press: 40.6 mm @ 21°, vap d: 3.93, lel: 1.1%, uel: 6.0%.

**TMD250** CAS: 512-56-1 $C_3H_9O_4P$ HR: 3
**TRIMETHYL PHOSPHATE**
SYNS: METHYL PHOSPHATE ◇ TMP

DFG MAK: Suspected Carcinogen.

SAFETY PROFILE: Moderately toxic by ingestion and skin contact. An experimental carcinogen and teratogen. Experimental reproductive effects. When heated to decomposition it emits toxic fumes of $PO_x$.

PROP: Liquid. D: 1.97 @ 19.5/0°, bp: 197.2°. Sol in alc, water, and ether.

**TMD500** CAS: 121-45-9 $C_3H_9O_3P$ HR: 2
**TRIMETHYL PHOSPHITE**
DOT: 2329
SYN: TRIMETHOXYPHOSPHINE

OSHA PEL: TWA 2 ppm
ACGIH TLV: TWA 2 ppm
DOT Class: Flammable or Combustible Liquid

SAFETY PROFILE: Moderately toxic by ingestion and skin contact. An experimental teratogen. Experimental reproductive effects. A severe skin and eye irritant. Flammable. To fight fire, use water, foam, fog, $CO_2$. When heated to decomposition it emits toxic fumes of $PO_x$.

PROP: Colorless liquid. D: 1.046 @ 20/4°, vap d: 4.3, bp: 232-234°F, flash p: 130°F (OC). Insol in water; sol in hexane, benzene, acetone, alc, ether, carbon tetrachloride, and kerosene.

**TMM250** CAS: 129-79-3 $C_{13}H_5N_3O_7$ HR: 3
**2,4,7-TRINITROFLUOREN-9-ONE**
SYN: 2,4,7-TRINITROFLUORENONE (MAK)

DFG MAK: Suspected Carcinogen.

SAFETY PROFILE: Mildly toxic by ingestion. An experimental tumorigen. A skin and eye irritant. When heated to decomposition it emits highly toxic fumes of $NO_x$.

**TMN000** CAS: 75321-19-6 $C_{16}H_7N_3O_6$ HR: 3
**1,3,6-TRINITROPYRENE**

DFG MAK: Suspected Carcinogen.

SAFETY PROFILE: When heated to decomposition it emits toxic fumes of $NO_x$.

**TMN490** CAS: 118-96-7 $C_7H_5N_3O_6$ HR: 3
**2,4,6-TRINITROTOLUENE**
DOT: 0209
SYNS: ENTSUFON ◇ TNT

OSHA PEL: (Transitional: TWA 1.5 $mg/m^3$ (skin)) TWA 0.5 $mg/m^3$ (skin)
ACGIH TLV: TWA 0.5 $mg/m^3$ (skin)
DFG MAK: 0.01 ppm (0.1 $mg/m^3$)
DOT Class: Class A Explosive

SAFETY PROFILE: Moderately toxic by ingestion. Experimental reproductive effects. A skin irritant. Flammable. Moderate explosion hazard; will detonate under strong shock. When heated to decomposition it emits highly toxic fumes of $NO_x$.

PROP: Colorless, monoclinic crystals. Mp: 80.7°, bp: 240° (explodes), flash p: explodes, d: 1.654. Sol in hot water, alc, and ether.

**TMO600** CAS: 78-30-8 $C_{21}H_{21}O_4P$ HR: 3
**TRIORTHOCRESYL PHOSPHATE**
SYNS: o-CRESYL PHOSPHATE ◇ o-TOLYL PHOSPHATE ◇ TOCP

OSHA PEL: (Transitional: TWA 0.1 $mg/m^3$) TWA 0.1 $mg/m^3$ (skin)
ACGIH TLV: TWA 0.1 $mg/m^3$ (skin)

SAFETY PROFILE: Moderately toxic by ingestion. Combustible. To fight fire, use $CO_2$, dry chemical. When heated to decomposition it emits highly toxic fumes of $PO_x$.

PROP: Colorless liquid. Mp: −25 to −30°, bp: 410° (slt decomp), flash p: 437°F, d: 1.17, autoign temp: 725°F, vap d: 12.7. Insol in water; sol in alc and ether.

**TMQ500** CAS: 603-34-9 $C_{18}H_{15}N$ HR: 2
**TRIPHENYL AMINE**
SYN: N,N-DIPHENYLANILINE

OSHA PEL: TWA 5 $mg/m^3$
ACGIH TLV: TWA 5 $mg/m^3$

SAFETY PROFILE: Moderately toxic by ingestion. When heated to decomposition it emits toxic fumes of $NO_x$.

PROP: Crystals. D: 0.774 @ 0/0°, mp: 127°, bp: 365°.

**TMT750** CAS: 115-86-6 $C_{18}H_{15}O_4P$ HR: 3
**TRIPHENYL PHOSPHATE**

OSHA PEL: TWA 3 $mg/m^3$
ACGIH TLV: TWA 3 $mg/m^3$

SAFETY PROFILE: Moderately toxic by ingestion. Combustible. To fight fire, use $CO_2$, dry chemical. When heated to decomposition it emits toxic fumes of $PO_x$.

PROP: Colorless, odorless, crystalline solid. Mp: 49-50°, bp: 245° @ 11 mm, flash p: 428°F (CC), d: 1.268 @ 60°, vap press: 1 mm @ 193.5°. Insol in water; sol in alc, benzene, ether, chloroform, and acetone.

**TMX500** CAS: 1317-95-9 HR: 3
**TRIPOLI**

OSHA PEL: (Transitional: TWA Respirable: 10 $mg/m^3/2(\%SiO_2+2)$; Total Dust: TWA 30 $mg/m^3/2(\%SiO_2+2)$) TWA 0.1 $mg/m^3$
ACGIH TLV: TWA 0.1 $mg/m^3$ (of contained respirable quartz dust)

SAFETY PROFILE: Prolonged inhalation of dusts containing free silica may result in development of a disabling pulmonary fibrosis known as silicosis.

PROP: Finely granulated white or gray siliceous rock. A form of crystalline silica.

**TOA750** CAS: 7440-33-7 W HR: 2
**TUNGSTEN and COMPOUNDS**
SYN: WOLFRAM

OSHA PEL: TWA (insoluble compounds) 5 $mg(W)/m^3$; STEL 10 $mg(W)/m^3$; (soluble compounds) 1 $mg(W)/m^3$; STEL 3 $mg(W)/m^3$
ACGIH TLV: TWA (insoluble compounds) 5 $mg(W)/m^3$; STEL 10 $mg(W)/m^3$; (soluble compounds) 1 $mg(W)/m^3$; STEL 3 $mg(W)/m^3$
NIOSH REL: (Tungsten, Insoluble) TWA 5 $mg(W)/m^3$

SAFETY PROFILE: Mildly toxic. An experimental teratogen. Experimental reproductive effects. A skin and eye irritant. Tungsten compounds are considered somewhat more toxic than those of molybdenum. Flammable in the form of dust. The powdered metal may ignite on contact with air or oxidants.

PROP: A steely-gray to white, cuttable, forgeable, and spinnable metallic element. Mp: 3410°, d: 19.3 @ 20°, bp: 5900°.

**TOD750** CAS: 8006-64-2 HR: 3
**TURPENTINE**
DOT: 1299
SYN: SPIRITS of TURPENTINE

OSHA PEL: TWA 100 ppm
ACGIH TLV: TWA 100 ppm
DFG MAK: 100 ppm (560 $mg/m^3$)
DOT Class: Flammable Liquid

SAFETY PROFILE: Moderately toxic to humans by ingestion. An experimental tumorigen. A human eye irritant. Irritating to skin and mucous membranes. A very dangerous fire hazard. Moderate explosion hazard in the form of vapor. To fight fire, use foam, $CO_2$, dry chemical.

PROP: Colorless liquid, characteristic odor. Bp: 154-170°, lel: 0.8%, flash p: 95°F (CC), d: 0.854-0.868 @ 25/25°, autoign temp: 488°F, vap d: 4.84, ULC: 40-50.

**UNS000** CAS: 7440-61-1 U HR: 3
**URANIUM**

OSHA PEL: (Transitional: TWA Soluble Compounds: 0.05 $mg(U)/m^3$; Insoluble Compounds 0.25 $mg(U)/m^3$) TWA Soluble Compounds: 0.05 $mg(U)/m^3$; Insoluble Compounds 0.2 $mg(U)/m^3$; STEL 0.6 $mg(U)/m^3$
ACGIH TLV: TWA 0.2 $mg(U)/m^3$; STEL 0.6 $mg(U)/m^3$
DFG MAK: 0.25 $mg/m^3$
DOT Class: Radioactive Material

SAFETY PROFILE: A highly toxic element on an acute basis. The permissible levels for soluble compounds are based on chemical toxicity, while the permissible body level for insoluble compounds is based on radiotoxicity. A very dangerous fire hazard in the form of a solid or dust. It can react violently with air, water.

PROP: A heavy, silvery-white, malleable, ductile, softer-than-steel, metallic, and radioactive element. Mp: 1132°, bp: 3818°, d: 18.95 (ca).

**UVA000** CAS: 51-79-6 $C_3H_7NO_2$ HR: 3
**URETHANE**
DOT: 2979
SYNS: ETHYL CARBAMATE ◇ ETHYL URETHANE

DFG MAK: Animal Carcinogen, Suspected Human Carcinogen.

SAFETY PROFILE: Moderately toxic by ingestion. An experimental carcinogen. A transplacental carcinogen. Experimental reproductive effects. A powerful teratogen in mice. When heated it emits toxic fumes of $NO_x$.

PROP: Colorless, odorless crystals. Mp: 49°, bp: 184°, d: 0.9862, vap press: 10 mm @ 77.8°, vap d: 3.07. Very sol in water, alc, and ether.

**VAG000** CAS: 110-62-3 $C_5H_{10}O$ HR: 2
**n-VALERALDEHYDE**
DOT: 2058
SYNS: PENTANAL ◇ VALERIC ACID ALDEHYDE

OSHA PEL: TWA 50 ppm
ACGIH TLV: TWA 50 ppm
DOT Class: Flammable Liquid

SAFETY PROFILE: Moderately toxic by ingestion. Mildly toxic by inhalation and skin contact. A severe eye and skin irritant. A very dangerous fire hazard.

PROP: Liquid. Flash p: 53.6°F, bp: 102-103°, d: 0.8095 @ 20/4°. Very sltly sol in water; misc with organic solvents.

**VCP000** CAS: 7440-62-2 V HR: 3
**VANADIUM**

OSHA PEL: (Transitional: Respirable Dust: Cl 0.5 $mg(V_2O_5)/m^3$; Fume: Cl 0.1 $mg(V_2O_5)/m^3$) Respirable Dust and Fume: TWA 0.05 $mg(V_2O_5)/m^3$
NIOSH REL: TWA 1.0 $mg(V)/m^3$

SAFETY PROFILE: An experimental tumorigen. Flammable dust. When heated to decomposition it emits toxic fumes of $VO_x$.

PROP: A bright, white, soft, ductile metal; sltly radioactive. Bp: 3000°, d: 6.11 @ 18.7°, mp: 1917°. Insol in water.

**VDU000** CAS: 1314-62-1 $O_5V_2$ HR: 3
**VANADIUM PENTOXIDE (dust)**
DOT: 2862
SYNS: C.I. 77938 ◇ VANADIC ANHYDRIDE ◇ VANADIUM DUST AND FUME (ACGIH)

OSHA PEL: (Transitional: Respirable Dust: Cl 0.5 $mg(V_2O_5)/m^3$; Fume: Cl 0.1 $mg(V_2O_5)/m^3$) Respirable Dust and Fume: TWA 0.05 $mg(V_2O_5)/m^3$
ACGIH TLV: TWA 0.05 $mg(V_2O_5)/m^3$
DFG MAK: (fine dust) 0.05 $mg/m^3$
NIOSH REL: (Vanadium Compounds) CL 0.05 $mg(V)/m^3$/15M
DOT Class: ORM-E

SAFETY PROFILE: Poison by ingestion and inhalation. An experimental teratogen. Experimental reproductive effects. A respiratory irritant. When heated to decomposition it emits toxic fumes of $VO_x$.

PROP: Yellow to red, crystalline powder. Mp: 690°, bp: decomp @ 1750°, d: 3.357 @ 18°.

**VDZ000** CAS: 1314-62-1 $O_5V_2$ HR: 3
**VANADIUM PENTOXIDE (fume)**
SYN: VANADIUM DUST AND FUME (ACGIH)

OSHA PEL: (Transitional: Fume: Cl 0.1 $mg(V_2O_5)/m^3$) Fume: TWA 0.05 $mg(V_2O_5)/m^3$
ACGIH TLV: TWA 0.05 $mg(V_2O_5)/m^3$
NIOSH REL: (Vanadium Compound) CL 0.05 $mg(V)/m^3$/15M

SAFETY PROFILE: A poison. When heated to decomposition it emits toxic fumes of $VO_x$.

**VLU250** CAS: 108-05-4 $C_4H_6O_2$ HR: 3
**VINYL ACETATE**
DOT: 1301
SYNS: ACETIC ACID VINYL ESTER ◇ ETHENYL ACETATE ◇ VAC

OSHA PEL: TWA 10 ppm; STEL 20 ppm
ACGIH TLV: TWA 10 ppm; STEL 20 ppm

DFG MAK: 10 ppm (35 $mg/m^3$)
NIOSH REL: (Vinyl Acetate) CL 15 $mg/m^3$/15M
DOT Class: Label: Flammable Liquid.

SAFETY PROFILE: An experimental carcinogen. Moderately toxic by ingestion and inhalation. A skin and eye irritant. Highly dangerous fire hazard. Reacts with air or water to form peroxides.

PROP: Colorless, mobile liquid; polymerizes to solid on exposure to light. Mp: −92.8°, bp: 73°, flash p: 18°F, d: 0.9335 @ 20°, autoign temp: 800°F, vap press: 100 mm @ 21.5°, lel: 2.6%, uel: 13.4%, vap d: 3.0. Misc in alc and ether. Somewhat sol in water.

**VMP000** CAS: 593-60-2 $C_2H_3Br$ HR: 3
**VINYL BROMIDE**
DOT: 1085
SYNS: BROMOETHENE ◇ BROMOETHYLENE

OSHA PEL: TWA 5 ppm
ACGIH TLV: TWA 5 ppm; Suspected Human Carcinogen.
DFG MAK: Human Carcinogen.
NIOSH REL: (Vinyl Bromide) Lowest Detectable Level
DOT Class: Flammable Gas

SAFETY PROFILE: Moderately toxic by ingestion. An experimental carcinogen. A very dangerous fire hazard. To fight fire, use $CO_2$, dry chemical or water spray. When heated to decomposition it emits toxic fumes of $Br^-$.

PROP: A gas. Mp: −138°, bp: 15.6°, d: 1.51. Insol in water; misc in alc and ether.

**VNP000** CAS: 75-01-4 $C_2H_3Cl$ HR: 3
**VINYL CHLORIDE**
DOT: 1086
SYNS: CHLORETHYLENE ◇ MONOCHLOROETHYLENE (DOT) ◇ VC ◇ VINYL CHLORIDE MONOMER

OSHA PEL: TWA 1 ppm; CL 5 ppm/15M; Cancer Suspect Agent
ACGIH TLV: TWA 5 ppm; Human Carcinogen.
DFG TRK: Existing installations: 3 ppm, Human Carcinogen; Others: 2 ppm.
NIOSH REL: (Vinyl Chloride) Lowest Detectable Level
DOT Class: Flammable Gas

SAFETY PROFILE: Poison by inhalation. Moderately toxic by ingestion. A human carcinogen. A severe irritant to skin, eyes, and mucous membranes. Causes skin burns by rapid evaporation and consequent freezing. A very dangerous fire hazard. A severe explosion hazard in the form of vapor. To fight fire, stop flow of gas. When heated to decomposition it emits highly toxic fumes of $Cl^-$.

PROP: Colorless liquid or gas (when inhibited); faintly sweet odor. Mp: −160°; bp: −13.9°, lel: 4%, uel: 22%; flash p: 17.6°F (COC), fp: −159.7°, d (liquid): 0.9195 @ 15/4°, vap press: 2600 mm @ 25°, vap d: 2.15, autoign temp: 882°F. Sltly sol in water; sol in alc; very sol in ether.

**VOA000** CAS: 106-87-6 $C_8H_{12}O_2$ HR: 3
**VINYL CYCLOHEXENE DIOXIDE**
SYNS: 1-EPOXYETHYL-3,4-EPOXYCYCLOHEXANE ◇ VINYL CYCLOHEXENE DIEPOXIDE ◇ 4-VINYL-1-CYCLOHEXENE DIOXIDE (MAK)

OSHA PEL: TWA 10 ppm (skin)
ACGIH TLV: TWA 10 ppm (skin); Suspected Human Carcinogen.
DFG MAK: Animal Carcinogen, Suspected Human Carcinogen.

SAFETY PROFILE: Poison. Moderately toxic by ingestion and skin contact routes. An experimental carcinogen. A severe skin irritant. Combustible when exposed to heat or flame. To fight fire, use water, foam, dry chemical.

PROP: Colorless liquid. D: 1.098 @ 20/20°, bp: 227°, flash p: 230°F.

**VPA000** CAS: 75-02-5 $CH_2:CHF$ HR: 3
**VINYL FLUORIDE**
DOT: 1860
SYNS: FLUOROETHENE ◇ FLUOROETHYLENE

NIOSH REL: (Vinyl Chloride) TWA 1 ppm; CL 5 ppm/15M
DOT Class: Label: Flammable Gas.

SAFETY PROFILE: A poison. A very dangerous fire hazard. To fight fire, stop flow of gas. When heated to decomposition it emits toxic fumes of $F^-$.

PROP: Colorless gas. Mp: −160.5°, bp: −72°, lel: 2.6%, uel: 21.7%. Insol in water; sol in alc and ether.

**VPK000** CAS: 75-35-4 $C_2H_2Cl_2$ HR: 3
**VINYLIDENE CHLORIDE**
SYNS: 1,1-DICHLOROETHYLENE ◇ VDC ◇ VINYLIDENE DICHLORIDE

OSHA PEL: TWA 1 ppm
ACGIH TLV: TWA 5 ppm; STEL 20 ppm
DFG MAK: Suspected Carcinogen.

SAFETY PROFILE: Poison by inhalation and ingestion. An experimental carcinogen and teratogen. Experimental reproductive effects. A very dangerous fire hazard. Moderately explosive in the form of gas. To fight fire, use alcohol foam, $CO_2$, dry chemical. When heated to decomposition it emits toxic fumes of $Cl^-$.

PROP: Colorless, volatile liquid. Bp: 31.6°, lel: 7.3%, uel: 16.0%, fp: −122°, flash p: 0°F (OC), d: 1.213 @ 20°/4°, autoign temp: 1058°F.

**VPP000** CAS: 75-38-7 $C_2H_2F_2$ HR: 3
**VINYLIDENE FLUORIDE**
DOT: 1959
SYNS: 1,1-DIFLUOROETHYLENE (DOT, MAK) ◇ HALOCARBON 1132A ◇ VDF

DFG MAK: Suspected Carcinogen.
DOT Class: Flammable Gas

SAFETY PROFILE: Mildly toxic by inhalation. An experimental neoplastigen. A very dangerous fire hazard. Explosive in the form of vapor. To fight fire, stop flow of gas. When heated to decomposition it emits toxic fumes of $F^-$.

PROP: Colorless gas. Bp: $<-70°$, lel: 5.5%, uel: 21.3%.

**VQK650** CAS: 25013-15-4 $C_9H_{10}$ HR: 2
**VINYL TOLUENE**
SYN: METHYL STYRENE

OSHA PEL: TWA 100 ppm
ACGIH TLV: TWA 50 ppm; STEL 100 ppm
DOT Class: Flammable or Combustible Liquid

SAFETY PROFILE: Moderately toxic by ingestion and inhalation. An experimental teratogen. Experimental reproductive effects. A skin and eye irritant. Flammable.

**WAT200** CAS: 81-81-2 $C_{19}H_{16}O_4$ HR: 3
**WARFARIN**
SYNS: 3-(α-ACETONYLBENZYL)-4-HYDROXYCOUMARIN ◇ COUMADIN

OSHA PEL: TWA 0.1 mg/m$^3$
ACGIH TLV: TWA 0.1 mg/m$^3$
DFG MAK: 0.5 mg/m$^3$

SAFETY PROFILE: A human poison by ingestion. Poison by inhalation. Moderately toxic by skin contact. Causes human teratogenic and reproductive effects.

PROP: Colorless crystals; odorless and tasteless. Mp: 161°. Sol in acetone and dioxane; sltly sol in methanol and ethanol; very sol in alkaline solution; insol in water and benzene.

**WBJ000** HR: 3
**WELDING FUMES**

ACGIH TLV: TWA 5 mg/m$^3$

SAFETY PROFILE: Toxicity of the fumes depends on what is being welded and the fluxes used. Common contaminants are cadmium, lead, $NO_x$, CO, fluoride fumes, $Cl^-$, $PO_x$, $SO_x$, Iron fume, and zinc oxide. When oily surfaces are welded, toxic fumes can be liberated. Carbon monoxide may evolved if the welding torch is improperly ignited.

**XGS000** CAS: 1330-20-7 $C_8H_{10}$ HR: 2
**XYLENE**
DOT: 1307
SYNS: DIMETHYLBENZENE ◇ METHYL TOLUENE ◇ XYLOL (DOT)

OSHA PEL: (Transitional: TWA 100 ppm) TWA 100 ppm; STEL 150 ppm
ACGIH TLV: TWA 100 ppm; STEL 150 ppm; BEI: 1.5 g(methyl hippuric acids)/g creatinine in urine end of shift.
DFG MAK: (all isomers) 100 ppm (440 mg/m$^3$); BAT: 150 μg/dL in blood at end of shift.
NIOSH REL: (Xylene) TWA 100 ppm; CL 200 ppm/10M
DOT Class: Flammable Liquid

SAFETY PROFILE: Mildly toxic by ingestion and inhalation. An experimental teratogen. Experimental reproductive effects. A skin and severe eye irritant. A very dangerous fire hazard. To fight fire, use foam, $CO_2$, dry chemical.

PROP: A clear liquid. Bp: 138.5°, flash p: 100°F (TOC), d: 0.864 @ 20°/4°, vap press: 6.72 mm @ 21°. Composition: as nonaromatics 0.07%, toluene 14%, ethyl benzene 19.27%, p-xylene 7.84%, m-xylene 65.01%, o-xylene 7.63%, C9 and aromatics 0.04% (TXAPA9 33,543,75).

**XHA000** CAS: 108-38-3 $C_8H_{10}$ HR: 3
**m-XYLENE**
DOT: 1307
SYNS: m-DIMETHYLBENZENE ◇ m-XYLOL (DOT)

OSHA PEL: (Transitional: TWA 100 ppm) TWA 100 ppm; STEL 150 ppm
ACGIH TLV: TWA 100 ppm; STEL 150 ppm; BEI: methyl hippuric acids in urine end of shift 1.5 g/g creatinine
NIOSH REL: (Xylene) TWA 100 ppm; CL 200 ppm/10M
DOT Class: Flammable or Combustible Liquid

SAFETY PROFILE: Mildly toxic by ingestion, skin contact and inhalation. An experimental teratogen. A severe skin irritant. A very dangerous fire hazard. Explosive in the form of vapor. To fight fire, use foam, $CO_2$, dry chemical.

PROP: Colorless liquid. Mp: −47.9°, bp: 139°, lel: 1.1%, uel: 7.0%, flash p: 77°F, d: 0.864 @ 20/4°, vap press: 10 mm @ 28.3°, vap d: 3.66, autoign temp: 986°F. Insol in water; misc with alc, ether, and some organic solvents.

**XHJ000** CAS: 95-47-6 $C_8H_{10}$ HR: 3
**o-XYLENE**
DOT: 1307
SYNS: o-DIMETHYLBENZENE ◇ o-XYLOL (DOT)

OSHA PEL: (Transitional: TWA 100 ppm) TWA 100 ppm; STEL 150 ppm
ACGIH TLV: TWA 100 ppm; STEL 150 ppm; BEI: methyl hippuric acids in urine end of shift 1.5 g/g creatinine
NIOSH REL: (Xylene) TWA 100 ppm; CL 200 ppm/10M
DOT Class: Flammable or Combustible Liquid

SAFETY PROFILE: Mildly toxic by ingestion and inhalation. An experimental teratogen. A very dangerous fire hazard. Explosive in the form of vapor. To fight fire, use foam, $CO_2$, dry chemical. Incompatible with oxidizing materials.

PROP: Colorless liquid. D: 0.880 @ 20/4°, mp: −25.2°, bp: 144.4°, flash p: 62.6°F. lel: 1.0%, uel: 6.0%. Insol in water; misc in abs alc and ether.

**XHS000** CAS: 106-42-3 $C_8H_{10}$ HR: 3
**p-XYLENE**
DOT: 1307
SYNS: p-DIMETHYLBENZENE ◇ p-METHYLTOLUENE ◇ p-XYLOL (DOT)

OSHA PEL: (Transitional: TWA 100 ppm) TWA 100 ppm; STEL 150 ppm
ACGIH TLV: TWA 100 ppm; STEL 150 ppm; BEI: methyl hippuric acids in urine end of shift 1.5 g/g creatinine
NIOSH REL: (Xylene) TWA 100 ppm; CL 200 ppm/10M
DOT Class: Flammable Liquid

SAFETY PROFILE: Mildly toxic by ingestion and inhalation. An experimental teratogen. A very dangerous fire hazard. Explosive in the form of vapor. To fight fire, use foam, $CO_2$, dry chemical.

PROP: Clear plates. Bp: 138.3°, lel: 1.1%, uel: 7.0%, flash p: 77°F (CC), d: 0.8611 @ 20/4°, vap press: 10 mm @ 27.3°, vap d: 3.66, autoign temp: 986°F, mp: 13-14°. Insol in water; sol in alc, ether, and organic solvents.

**XHS800** CAS: 1477-55-0 $C_8H_{12}N_2$ HR: 2
**m-XYLENE-α,α′-DIAMINE**
SYNS: MXDA ◇ m-PHENYLENEBIS(METHYLAMINE)

OSHA PEL: TWA CL 0.1 $mg/m^3$ (skin)
ACGIH TLV: TWA CL 0.1 $mg/m^3$ (skin)

SAFETY PROFILE: Moderately toxic by skin contact and ingestion. A severe skin and eye irritant. When heated to decomposition it emits toxic fumes of $NO_x$.

**XIJ000** CAS: 3634-83-1 $C_{10}H_8N_2O_2$ HR: 2
**m-XYLENE-α,α′-DIISOCYANATE**

NIOSH REL: (Diisocyanates) TWA 0.005 ppm; CL 0.02 ppm/10M

SAFETY PROFILE: Mildly toxic by ingestion. A severe skin and eye irritant. When heated to decomposition it emits very toxic fumes of $NO_x$.

**XMA000** CAS: 1300-73-8 $C_8H_{11}N$ HR: 3
**XYLIDINE**
DOT: 1711
SYNS: AMINODIMETHYLBENZENE ◇ DIMETHYLPHENYLAMINE

OSHA PEL: (Transitional: TWA 5 ppm (skin)) TWA 0.2 ppm (skin)
ACGIH TLV: TWA 2 ppm (Proposed: TWA 0.5 ppm (skin); Suspected Human Carcinogen.)
DFG MAK: (all isomers except 2,4-xylidene) 5 ppm (25 mg/$m^3$)
DOT Class: Poison B

SAFETY PROFILE: Moderately toxic by ingestion. Twice as toxic as aniline. Combustible. To fight fire, use foam, $CO_2$, dry chemical. When heated to decomposition emits toxic fumes of $NO_x$.

PROP: Usually liquid (except for o-4-xylidine). Bp: 213-226°, flash p: 206° (CC), d: 0.97-0.99, vap d: 4.17. Sltly sol in water; sol in alc.

**XMS000** CAS: 95-68-1 $C_8H_{11}N$ HR: 3
**2,4-XYLIDINE**
DOT: 1711
SYNS: 1-AMINO-2,4-DIMETHYLBENZENE ◇ 2-METHYL-p-TOLUIDINE ◇ 2,4-XYLIDENE (MAK) ◇ m-XYLIDINE (DOT)

DFG MAK: 5 ppm (25 mg/$m^3$); Suspected Carcinogen.
DOT Class: Poison B

SAFETY PROFILE: Poison by ingestion. When heated to decomposition it emits toxic fumes of $NO_x$.

PROP: Liquid. Bp: 214°, mp: 16°, d: 0.978 @ 19.6/4°. Very sltly sol in water.

**YEJ000** CAS: 7440-65-5 Y HR: 2
**YTTRIUM**

PROP: Hexagonal, gray-black, metallic, rare earth element. Mp: 1509°, bp: 3200°, d: 4.472.
SYN: YTTRIUM-89

OSHA PEL: TWA 1 mg(Y)/$m^3$
ACGIH TLV: TWA 1 mg(Y)/$m^3$
DFG MAK: 5 mg(Y)/$m^3$

SAFETY PROFILE: As a lanthanon, it may have an anticoagulant effect on the blood. Flammable in the form of dust when reacted with air or halogens.

**ZFA000** CAS: 7646-85-7 $Cl_2Zn$ HR: 3
**ZINC CHLORIDE**
DOT: 1840/2331
SYNS: TINNING GLUX (DOT) ◇ ZINC DICHLORIDE

OSHA PEL: Fume: (Transitional: TWA 1 mg/$m^3$) TWA 1 mg/$m^3$; STEL 2 mg/$m^3$
ACGIH TLV: TWA 1 mg/$m^3$; STEL 2 mg/$m^3$ (fume)
DOT Class: Corrosive Material

SAFETY PROFILE: Poison by ingestion. An experimental tumorigen and teratogen. Experimental reproductive effects. A corrosive irritant to skin, eyes, and mucous membranes. The fumes are highly toxic. When heated to decomposition it emits toxic fumes of $Cl^-$ and ZnO.

PROP: Odorless, cubic, white, deliquescent crystals. Mp: 290°, bp: 732°, d: 2.91 @ 25°, vap press: 1 mm @ 428°.

**ZFJ100** CAS: 13530-65-9 $CrH_2O_4 \cdot Zn$ HR: 3
**ZINC CHROMATE**
SYNS: CHROMIUM ZINC OXIDE ◇ C.I. 77955 ◇ ZINC CHROME YELLOW

OSHA PEL: (Transitional: 1 $mg(CrO_3)/10m^3$) CL 0.1 $mg(CrO_3)/m^3$
ACGIH TLV: TWA 0.01 $mg(Cr)/M^3$; Confirmed Human Carcinogen
DFG TRK: 0.1 $mg/m^3$; Human Carcinogen.
NIOSH REL: (Chromium (VI)) TWA 0.001 $mg(Cr(VI))/m^3$

SAFETY PROFILE: A human carcinogen by inhalation with tumors of the lung. An experimental tumorigen.

**ZFJ120** CAS: 37300-23-5 $CrO_4 \cdot Zn \cdot H_4O_2Zn \cdot CrO_3$ HR: 3
**ZINC CHROMATE with ZINC HYDROXIDE and CHROMIUM OXIDE (9:1)**
SYN: ZINC YELLOW

OSHA PEL: (Transitional: 1 $mg(CrO_3)/10m^3$) CL 0.1 $mg(CrO_3)/m^3$
ACGIH TLV: TWA 0.01 $mg(Cr)/M^3$; Confirmed Human Carcinogen
DFG TRK: 0.1 $mg/m^3$; Human Carcinogen.
NIOSH REL: (Chromium (VI)) TWA 0.001 $mg(Cr(VI))/m^3$

SAFETY PROFILE: A human carcinogen by inhalation with tumors of the lung.

**ZJA000** CAS: 22323-45-1 $7ZnO \cdot 2HgO \cdot 2CrO_3 \cdot 7H_2O$ HR: 2
**ZINC MERCURY CHROMATE COMPLEX**
SYNS: CHROMIC ACID, MERCURY ZINC COMPLEX ◇ MERCURY ZINC CHROMATE COMPLEX

OSHA PEL: (Transitional: 1 $mg/10m^3$) CL 0.1 $mg(CrO_3)/m^3$
ACGIH TLV: TWA 0.05 $mg(Cr)/m^3$, Confirmed Human Carcinogen.
DFG MAK: Animal Carcinogen, Suspected Human Carcinogen.
NIOSH REL: TWA 0.025 $mg(Cr(VI))/m^3$; CL 0.05/15M; 0.05 $mg(Hg)/m^3$

SAFETY PROFILE: Moderately toxic by ingestion. Chromate salts are carcinogens. When heated to decomposition it emits very toxic fumes of Hg and ZnO.

**ZKA000** CAS: 1314-13-2 OZn HR: 3
**ZINC OXIDE**
SYNS: CHINESE WHITE ◇ C.I. 77947 ◇ FLOWERS OF ZINC ◇ ZINC WHITE

OSHA PEL: Fume: (Transitional: TWA 5 $mg/m^3$) TWA 5 $mg/m^3$; STEL 10 $mg/m^3$; Dust: (Transitional: Total Dust: 15 $mg/m^3$; Respirable Fraction: 5 $mg/m^3$;) TWA Total Dust: 10 $mg/m^3$; Respirable Fraction: 5 $mg/m^3$

ACGIH TLV: Fume: TWA 5 mg/m$^3$; STEL 10 mg/m$^3$; Dust: 10 mg/m$^3$ of total dust (when toxic impurities are not present, e.g., quartz $< 1\%$).
DFG MAK: 5 mg/m$^3$
NIOSH REL: TWA (Zinc Oxide) 5 mg/m$^3$; CL 15 mg/m$^3$/15M

SAFETY PROFILE: An experimental teratogen. Human systemic effects by inhalation of freshly formed fumes: metal fume fever with chills, fever, tightness of chest, and cough. A skin and eye irritant. When heated to decomposition it emits toxic fumes of ZnO.

PROP: Odorless, white or yellowish powder. Mp: >1800°, d: 5.47. Insol in water and alc; sol in dil acetic or mineral acids and ammonia.

**ZMS000** CAS: 557-05-1 $Zn(C_{18}H_{35}O_2)_2$ HR: 3
**ZINC STEARATE**
SYNS: OCTADECANOIC ACID, ZINC SALT ◇ STEARIC ACID, ZINC SALT ◇ ZINC OCTADECANOATE

OSHA PEL: (Transitional: TWA Total Dust: 15 mg/m$^3$; Respirable Fraction: 5 mg/m$^3$) TWA Total Dust: 10 mg/m$^3$; Respirable Fraction: 5 mg/m$^3$
ACGIH TLV: TWA 10 mg/m$^3$ of total dust when toxic impurities are not present, e.g., quartz $< 1\%$

SAFETY PROFILE: Inhalation of zinc stearate has been reported as causing pulmonary fibrosis. A nuisance dust. Combustible. To fight fire, use water, foam, $CO_2$, dry chemical. When heated to decomposition it emits toxic fumes of ZnO.

PROP: White powder. Mp: 130°, flash p: 530°F (OC), autoign temp: 790°F. Insol in water, alc, and ether; sol in benzene. Decomp by dil acids.

**ZOA000** CAS: 7440-67-7 Zr HR: 3
**ZIRCONIUM and COMPOUNDS**
DOT: 1308/1358/2008/2009/2858

OSHA PEL: (Transitional: TWA 5 mg(Zr)/m$^3$) TWA 5 mg(Zr)/m$^3$; STEL 10 mg(Zr)/m$^3$
ACGIH TLV: TWA 5 mg(Zr)/m$^3$; STEL 10 mg(Zr)/m$^3$
DFG MAK: 5 mg(Zr)/m$^3$
DOT Class: Flammable Solid

SAFETY PROFILE: Most zirconium compounds in common use are insoluble and considered inert. The metal is a very dangerous fire hazard in the form of dust and may ignite spontaneously. A dangerous explosion hazard in the form of dust by chemical reaction with air, and other substances. To fight fire, use special mixtures, dry chemical, salt or dry sand.

PROP: A grayish-white, lustrous, very sltly radioactive, metallic element. Mp: 1852°, bp: 3577°, d: 6.506 @ 20°.

# Appendix I
# Synonym Cross-reference

1080 see SHG500
2-AAF see FDR000
AAT see AIC250
ABATE see TAL250
ABSOLUTE ETHANOL see EFU000
ACETALDEHYDE see AAG250
ACETAMIDE see AAI000
2-ACETAMINOFLUORENE see FDR000
ACETDIMETHYLAMIDE see DOO800
ACETIC ACID see AAT250
ACETIC ACID AMIDE see AAI000
ACETIC ACID, AMYL ESTER see AOD750
ACETIC ACID, ANHYDRIDE see AAX500
ACETIC ACID-2-BUTOXY ESTER see BPV000
ACETIC ACID n-BUTYL ESTER see BPU750
ACETIC ACID-tert-BUTYL ESTER see BPV100
ACETIC ACID DIMETHYLAMIDE see DOO800
ACETIC ACID-1,1-DIMETHYLETHYL ESTER see BPV100
ACETIC ACID-4,6-DINITRO-o-CRESYL ESTER see AAU250
ACETIC ACID METHYL ESTER see MFW100
ACETIC ACID, n-PROPYL ESTER see PNC250
ACETIC ACID VINYL ESTER see VLU250
ACETIC ALDEHYDE see AAG250
ACETIC ANHYDRIDE see AAX500
ACETIC ETHER see EFR000
ACETIC OXIDE see AAX500
ACETIMIDIC ACID see AAI000
ACETONE see ABC750
ACETONE CYANOHYDRIN (DOT) see MLC750
ACETONITRILE see ABE500
3-(α-ACETONYLBENZYL)-4-HYDROXYCOUMARIN see WAT200
ACETONYL CHLORIDE see CDN200
2-ACETOXYBENZOIC ACID see ADA725
ACETOXYETHANE see EFR000
2-ACETOXYPENTANE see AOD735
1-ACETOXYPROPANE see PNC250
2-ACETOXYPROPANE see INE100
2-ACETYLAMINOFLUORENE (OSHA) see FDR000
ACETYL ANHYDRIDE see AAX500
ACETYLEN see ACI750

ACETYLENE see ACI750
ACETYLENE BLACK see CBT750
ACETYLENE DICHLORIDE see DFI100
trans-ACETYLENE DICHLORIDE see ACK000
ACETYLENE TETRABROMIDE see ACK250
ACETYLENE TETRACHLORIDE see TBQ100
ACETYLENE TRICHLORIDE see TIO750
ACETYL HYDROPEROXIDE see PCL500
ACETYL PEROXIDE see ACV500
ACETYLSALICYLIC ACID see ADA725
ACROLEIC ACID see ADS750
ACROLEIN see ADR000
ACRYLALDEHYDE see ADR000
ACRYLAMIDE see ADS250
ACRYLIC ACID see ADS750
ACRYLIC ACID n-BUTYL ESTER (MAK) see BPW100
ACRYLIC ACID ETHYL ESTER see EFT000
ACRYLIC ACID METHYL ESTER (MAK) see MGA500
ACRYLIC AMIDE see ADS250
ACRYLONITRILE see ADX500
ACTINOLITE ASBESTOS see ARM250
ADIPIC ACID DINITRILE see AER250
ADIPONITRILE see AER250
AGE see AGH150
ALDRIN see AFK250
ALLYL ALCOHOL see AFV500
ALLYL CHLORIDE see AGB250
ALLYL-2,3-EPOXYPROPYL ETHER see AGH150
ALLYL GLYCIDYL ETHER see AGH150
ALLYL PROPYL DISULFIDE see AGR500
ALLYL TRICHLORIDE see TJB600
ALPHANAPHTHYL THIOUREA see AQN635
ALTOX see AFK250
α-ALUMINA (OSHA) see AHE250
ALUMINUM and COMPOUNDS see AGX000
ALUMINUM OXIDE see EAL100
ALUMINUM OXIDE (2–3) see AHE250
ALUMINUM POWDER see AGX000
AMATIN see HCC500
AMINIC ACID see FNA000
p-AMINOANILINE see PEY500
o-AMINOANISOLE see AOV900
p-AMINOANISOLE see AOW000
2-AMINO-5-AZOTOLUENE see AIC250
o-AMINOAZOTOLUENE (MAK) see AIC250
AMINOBENZENE see AOQ000
4-(4-AMINOBENZYL)ANILINE see MJQ000
1-AMINOBUTANE see BPX750
2-AMINOBUTANE see BPY000
4-AMINO-6-tert-BUTYL-3-METHYLTHIO-as-TRIAZIN-5-ONE see MQR275
2-AMINO-4-CHLOROTOLUENE see CLK225

BERYLLIUM and COMPOUNDS see BFO750
BGE see BRK750
α-BHC see BBQ000
β-BHC see BBR000
γ-BHC see BBQ500
BHT (FOOD GRADE) see BFW750
BICYCLOPENTADIENE see DGW000
BINITROBENZENE see DUQ200
BIPHENYL see BGE000
BIPHENYLAMINE see AJS100
4-BIPHENYLAMINE see AJS100
BIPHENYL, mixed with BIPHENYL OXIDE (3:7) see PFA860
4,4′-BIPHENYLDIAMINE see BBX000
(1,1′-BIPHENYL)-4,4′-DIAMINE SULFATE (1:1) see BBY000
BIPHENYL OXIDE see PFA850
BIS(p-AMINOPHENYL)METHANE see MJQ000
BIS(p-AMINOPHENYL)SULFIDE see TFI000
BIS(2-CHLOROETHYL) ETHER see DFJ050
BIS(β-CHLOROETHYL)METHYLAMINE see BIE250
BIS(2-CHLOROETHYL)SULFIDE see BIH250
BIS(CHLOROMETHYL) ETHER see BIK000
BIS-CME see BIK000
BISCYCLOPENTADIENYLIRON see FBC000
BIS-O,O-DIETHYLPHOSPHOROTHIONIC ANHYDRIDE see SOD100
BIS(N,N-DIETHYLTHIOCARBAMOYL)DISULPHIDE see DXH250
4,4′-BIS(DIMETHYLAMINO)BENZOPHENONE see MQS500
BIS(DIMETHYLTHIOCARBAMOYL) DISULFIDE see TFS350
BIS(2,3-EPOXYPROPYL)ETHER see DKM200
BIS(2-HYDROXY ETHYL)AMINE see DHF000
2,2-BIS(HYDROXYMETHYL)-1,3-PROPANEDIOL see PBB750
BIS(4-ISOCYANATOCYCLOHEXYL)METHANE see MJM600
BIS(p-ISOCYANATOPHENYL)METHANE see MJP400
1,1-BIS(p-METHOXYPHENYL)-2,2,2-TRICHLOROETHANE see MEI450
BISULFITE see SOH500
BITUMEN (MAK) see ARO500
BLADAN see EEH600
BLASTING GELATIN (DOT) see NGY000
BLUE ASBESTOS (DOT) see ARM250
BORATES, TETRA, SODIUM SALT, ANHYDROUS (OSHA, ACGIH) see SFE500, SFF000
BORAX (8CI) see SFF000
BORAX DECAHYDRATE see SFF000
2-BORNANONE see CBA750
BOROETHANE see DDI450
BORON BROMIDE see BMG400
BORON FLUORIDE see BMG700
BORON HYDRIDE see DDI450
BORON OXIDE see BMG000
BORON SESQUIOXIDE see BMG000
BORON TRIBROMIDE see BMG400
BORON TRIFLUORIDE see BMG700
BORON TRIOXIDE see BMG000

tert-BUTYL PEROXYACETATE see BSC250
o-sec-BUTYLPHENOL see BSE000
4-tert-BUTYLPHENOL see BSE500
p-tert-BUTYLPHENOL (MAK) see BSE500
n-BUTYL PHTHALATE (DOT) see DEH200
BUTYL-2-PROPENOATE see BPW100
p-tert-BUTYLTOLUENE see BSP500
BUTYRIC ACID NITRILE see BSX250
BUTYRONE (DOT) see DWT600
BUTYRONITRILE see BSX250
CADMIUM and COMPOUNDS see CAD000
CALCINED DIATOMITE see SCJ000
CALCINED MAGNESIA see MAH500
CALCIUM CARBIMIDE see CAQ250
CALCIUM CARBONATE see CAO000
CALCIUM CYANAMIDE see CAQ250
CALCIUM CYANIDE see CAQ500
CALCIUM HYDROXIDE see CAT250
CALCIUM OXIDE see CAU500
CALCIUM SILICATE see CAW850
CALCIUM SULFATE see CAX500
CALCIUM(II) SULFATE DIHYDRATE (1-1-2) see CAX750
CALCYANIDE see CAQ500
CALX see CAU500
CAMPHECHLOR see CDV100
CAMPHOR see CBA750
CANE SUGAR see SNH000
CAPROLACTAM see CBF700
omega-CAPROLACTAM (MAK) see CBF700
CAPTAFOL see CBF800
CAPTAN see CBG000
CARBAMALDEHYDE see FMY000
CARBAMATE see FAS000
4-CARBAMIDOPHENYL BIS(o-CARBOXYPHENYLTHIO)ARSENITE see TFD750
CARBAMONITRILE see COH500
CARBARYL see CBM750
CARBAZOTIC ACID see PID000
CARBETHOXY MALATHION see MAK700
CARBOFURAN see CBS275
CARBOLIC ACID see PDN750
CARBOMETHENE see KEU000
CARBON see CBT500
CARBONA see CBY000
CARBON BISULFIDE (DOT) see CBV500
CARBON BLACK see CBT750
CARBON BROMIDE see CBX750
CARBON DIFLUORIDE OXIDE see CCA500
CARBON DIOXIDE see CBU250
CARBON DIOXIDE mixed with NITROUS OXIDE see CBV000
CARBON DISULFIDE see CBV500
CARBON HEXACHLORIDE see HCI000

CARBONIC ACID GAS see CBU250
CARBONIC ANHYDRIDE see CBU250
CARBONIC OXIDE see CBW750
CARBON MONOXIDE see CBW750
CARBON NITRIDE see COO000
CARBON OXYCHLORIDE see PGX000
CARBON TET see CBY000
CARBON TETRABROMIDE see CBX750
CARBON TETRACHLORIDE see CBY000
CARBONYL CHLORIDE see PGX000
CARBONYL FLUORIDE see CCA500
CARBORUNDUM see SCQ000
CARBOXYETHANE see PMU750
CATECHOL see CCP850
CAUSTIC POTASH see PLJ500
CAUSTIC SODA see SHS000
CELLOSOLVE (DOT) see EES350
CELLOSOLVE ACETATE (DOT) see EES400
CELLULOSE, POWDERED see CCU150
CEMENT, PORTLAND see PKS750
CESIUM HYDRATE see CDD750
CESIUM HYDROXIDE see CDD750
CHA see CPF500
CHALCEDONY see SCI500, SCJ500
CHANNEL BLACK see CBT750
CHARCOAL BLACK see CBT500
CHEMICAL MACE see CEA750
CHERTS see SCJ500
CHINA CLAY see KBB600
CHINESE WHITE see ZKA000
CHLORACETONE see CDN200
CHLORALLYLENE see AGB250
CHLORDANE see CDR750
CHLORDECONE see KEA000
CHLORETHYLENE see VNP000
CHLORIDE of PHOSPHORUS see PHT2756
CHLORIDE of SULFUR (DOT) see SON510
CHLORINATED BIPHENYL see PJL750
CHLORINATED CAMPHENE see CDV100
CHLORINATED DIPHENYL OXIDE see CDV175
CHLORINATED HYDROCHLORIC ETHER see DFF809
CHLORINE see CDV750
CHLORINE CYANIDE see COO750
CHLORINE DIOXIDE see CDW450
CHLORINE FLUORIDE see CDX750
CHLORINE FLUORIDE OXIDE see PCF750
CHLORINE OXIDE see CDW450
CHLORINE OXYFLUORIDE see PCF750
CHLORINE PEROXIDE see CDW450
CHLORINE TRIFLUORIDE see CDX750
CHLORMETHINE see BIE250

CYCLOPENTADIENYLMANGANESE TRICARBONYL see CPV000
CYCLOPENTANE see CPV750
CYCLOTRIMETHYLENETRINITRAMINE see CPR800
CYNOGAN see BMM650
2,4-D see DAA800
DAB see DOT300
DALAPON (USDA) see DGI400
DASANIT see FAQ800
DBCP see DDL800
DBP see DEH200
DCA see DFE200
DCB see DEP600, DEV000
DCEE see DFJ050
DCM see MJP450
DCMU see DXQ500
DDC see DQY950
DDM see MJQ000
DDT see DAD200
DDVP see DGP900
DEA see DHF000
DEAE see DHO500
DECABORANE see DAE400
DECACHLOROTETRAHYDRO-4,7-METHANOINDENEONE see KEA000
DEHP see DVL700
DEK see DJN750
DEMETON METHYL see MIW100
DEMETON-O + DEMETON-S see DAO600
DEMOX see DAO600
DETA see DJG600
DFA see DVX800
DGE see DKM200
DIACETONE ALCOHOL see DBF750
DIACETYL PEROXIDE (MAK) see ACV500
DIAMIDE see HGS000
2,4-DIAMINOANISOLE see DBO000
4,4′-DIAMINO-3,3′-DICHLORODIPHENYLMETHANE see MJM200
4,4′-DIAMINO-3,3′-DIMETHYLDIPHENYL see TGJ750
4,4-DIAMINODIPHENYL ETHER see OPM000
2,4′-DIAMINODIPHENYLMETHANE see MJP750
1,2-DIAMINOETHANE see EEA500
2,4-DIAMINOTOLUENE see TGL750
o-DIANISIDINE see DCJ200
o-DIANISIDINE DIHYDROCHLORIDE see DOA800
DIANISIDINE DIISOCYANATE see DCJ400
DIAPADRIN see DGQ875
DIATO BLUE BASE B see DCJ200
DIATOMACEOUS EARTH see DCJ800
DIATOMITE see DCJ800
DIAZINON see DCM750
DIAZOIMIDE see HHG500
DIAZOMETHANE see DCP800

DIBENZO-1,4-THIAZINE see PDP250
DIBENZOYL PEROXIDE (MAK) see BDS000
DIBORANE see DDI450
DIBORON HEXAHYDRIDE see DDI450
1,2-DIBROMO-3-CHLOROPROPANE see DDL800
1,2-DIBROMO-2,2-DICHLOROETHYL DIMETHYL PHOSPHATE see NAG400
DIBROMODIFLUOROMETHANE see DKG850
1,2-DIBROMOETHANE (MAK) see EIY500
2-N-DIBUTYLAMINOETHANOL see DDU600
N,N-DI-n-BUTYLAMINOETHANOL (DOT) see DDU600
β-N-DIBUTYLAMINOETHYL ALCOHOL see DDU600
DIBUTYLATED HYDROXYTOLUENE see BFW750
2,6-DI-tert-BUTYL-p-CRESOL (OSHA, ACGIH) see BFW750
DIBUTYLNITROSOAMINE see BRY500
DI-tert-BUTYL PEROXIDE (MAK) see BSC750
DIBUTYL PHENYL PHOSPHATE see DEG600
DIBUTYL PHOSPHATE see DEG700
DIBUTYL PHTHALATE see DEH200
DICHLORANTIN see DFE200
DICHLOROACETYLENE see DEN600
o-DICHLOROBENZENE see DEP600
p-DICHLOROBENZENE see DEP800
1,2-DICHLOROBENZENE (MAK) see DEP600
1,4-DICHLOROBENZENE (MAK) see DEP800
3′,3′-DICHLOROBENZIDINE see DEQ600
3,3′-DICHLOROBENZIDINE DIHYDROCHLORIDE see DEQ800
3,3′-DICHLORO-(1,1′-BIPHENYL)-4,4′-DIAMINE DIHYDROCHLORIDE see DEQ800
1,4-DICHLORO-2-BUTENE see DEV000
1,4-DICHLOROBUTENE-2 (MAK) see DEV000
2,2-DICHLORO-1,1-DIFLUOROETHYL METHYL ETHER see DFA400
DICHLORODIFLUOROMETHANE see DFA600
1,3-DICHLORO-5,5-DIMETHYL HYDANTOIN see DFE200
3,5-DICHLORO-2,6-DIMETHYL-4-PYRIDINOL see CMX850
DICHLORO DIPHENYL OXIDE see DFE800
DICHLORODIPHENYLTRICHLOROETHANE (DOT) see DAD200
1,1-DICHLOROETHANE see DFF809
1,2-DICHLOROETHANE see EIY600
1,1-DICHLOROETHYLENE see VPK000
1,2-DICHLOROETHYLENE see DFI100, DFI200
cis-DICHLOROETHYLENE see DFI200
trans-1,2-DICHLOROETHYLENE (MAK) see ACK000
DICHLOROETHYL ETHER see DFJ050
β,β′-DICHLOROETHYL ETHER (MAK) see DFJ050
2,2′-DICHLOROETHYL SULPHIDE (MAK) see BIH250
DICHLOROETHYNE see DEN600
DICHLOROFLUOROMETHANE see DFL000
DICHLOROMETHANE (MAK, DOT) see MJP450
sym-DICHLOROMETHYL ETHER see BIK000
DICHLOROMONOFLUOROMETHANE (OSHA, DOT) see DFL000
DICHLORONITROETHANE see DFU000
1,1-DICHLORO-1-NITROETHANE see DFU000

3-(3,4-DICHLOROPHENOL)-1,1-DIMETHYLUREA see DXQ500
2,4-DICHLOROPHENOXYACETIC ACID (DOT) see DAA800
DICHLOROPHENYL ETHER see DFE800
1,2-DICHLOROPROPANE see PNJ400
cis-1,3-DICHLOROPROPENE see DGH200
trans-1,3-DICHLOROPROPENE see DGH000
α-DICHLOROPROPIONIC ACID see DGI400
2,2-DICHLOROPROPIONIC ACID see DGI400
cis-1,3-DICHLOROPROPYLENE see DGH200
trans-1,3-DICHLOROPROPYLENE see DGH000
DICHLOROTETRAFLUOROETHANE see DGL600
DICHLOROTETRAFLUOROETHANE (OSHA, ACGIH) see FOO509
1,2-DICHLORO-1,1,2,2-TETRAFLUOROETHANE (MAK) see FOO509
α,α-DICHLOROTOLUENE see BAY300
2,2-DICHLOROVINYL DIMETHYL PHOSPHATE see DGP900
DICHLORVOS see DGP900
DICOBALT OCTACARBONYL see CNB500
DICROTOPHOS see DGQ875
1,3-DICYANOBENZENE see PHX550
1,4-DICYANOBUTANE see AER250
DICYANOGEN see COO000
DICYANOMETHANE see MAO250
DICYCLOPENTADIENE see DGW000
DICYCLOPENTADIENYL IRON (OSHA, ACGIH) see FBC000
DIELDRIN see DHB400
DIETHANOLAMINE see DHF000
DIETHYL see BOR500
DIETHYLAMINE see DHJ200
2-DIETHYLAMINOETHANOL see DHO500
β-DIETHYLAMINOETHYL ALCOHOL see DHO500
DIETHYLCARBAMOYL CHLORIDE see DIW400
DIETHYLCARBAMYL CHLORIDE see DIW400
DIETHYLCARBINOL (DOT) see IHP010
DIETHYLENE DIOXIDE see DVQ000
DIETHYLENE IMIDOXIDE see MRP750
DIETHYLENE OXIDE see TCR750
DIETHYLENETRIAMINE see DJG600
DIETHYL ESTER SULFURIC ACID see DKB110
N,N-DIETHYLETHANAMINE see TJO000
DIETHYL ETHER (DOT) see EJU000
O,O-DIETHYL-S-(2-ETHTHIOETHYL) PHOSPHORODITHIOATE see DXH325
O,O-DIETHYL-S-ETHYLTHIOMETHYL DITHIOPHOSPHONATE see PGS000
DIETHYL 4-(2-ISOPROPYL-6-METHYLPYRIMIDINYL)PHOSPHOROTHIONATE see DCM750
DIETHYL KETONE see DJN750
DIETHYL MERCAPTOSUCCINATE-O,O-DIMETHYL THIOPHOSPHATE see MAK700
O,O-DIETHYL-O-p-(METHYLSULFINYL)PHENYL THIOPHOSPHATE see FAQ800
O,O-DIETHYL-O-4-NITROPHENYLPHOSPHOROTHIOATE see PAK000
DIETHYLNITROSAMINE see NJW500
DIETHYL PHTHALATE see DJX000
DIETHYL SULFATE see DKB110
DIFLUORODIBROMOMETHANE see DKG850

1,1-DIFLUOROETHYLENE (DOT, MAK) see VPP000
1,2-DIFLUORO-1,1,2,2-TETRACHLOROETHANE see TBP050
DIFOLATAN see CBF800
DIFONATE see FMU045
DIGLYCIDYL ETHER see DKM200
9,10-DIHYDRO-8a,10-DIAZONIAPHENANTHRENE DIBROMIDE see DWX800
2,5-DIHYDROFURAN-2,5-DIONE see MAM000
DIHYDROGEN DIOXIDE see HIB000
p-DIHYDROXYBENZENE see HIH000
1,3-DIHYDROXYBENZENE see REA000
2,4-DIHYDROXY-2-METHYLPENTANE see HFP875
DIISOBUTYL KETONE see DNI800
1,6-DIISOCYANATOHEXANE see DNJ800
1,5-DIISOCYANATONAPHTHALENE see NAM500
2,4-DIISOCYANATOTOLUENE see TGM750
2,6-DIISOCYANATOTOLUENE see TGM800
DIISOPROPYLAMINE see DNM200
DIISOPROPYL OXIDE see IOZ750
DIMAZINE see DSF400
3,3′-DIMETHOXYBENZIDINE DIHYDROCHLORIDE see DOA800
3,3′-DIMETHOXYBENZIDINE-4,4′-DIISOCYANATE see DCJ400
3,3′-DIMETHOXY-4,4′-BIPHENYLENE DIISOCYANATE see DCJ400
DIMETHOXY-DDT see MEI450
DIMETHOXYMETHANE see MGA850
3-(DIMETHOXYPHOSPHINYLOXY)N-METHYL-cis-CROTONAMIDE see MRH209
N,N-DIMETHYLACETAMIDE see DOO800
DIMETHYLACETONE see DJN750
DIMETHYLAMINE see DOQ800
DIMETHYLAMINOAZOBENZENE see DOT300
4-DIMETHYLAMINOAZOBENZENE see DOT300
(DIMETHYLAMINO)BENZENE see DQF800
N,N-DIMETHYLANILINE see DQF800
DIMETHYLBENZENE see XGS000
m-DIMETHYLBENZENE see XHA000
o-DIMETHYLBENZENE see XHJ000
p-DIMETHYLBENZENE see XHS000
N,N-DIMETHYLBENZENEAMINE see DQF800
DIMETHYL-1,2-BENZENEDICARBOXYLATE see DTR200
α,α-DIMETHYLBENZYL HYDROPEROXIDE (MAK) see IOB000
DIMETHYLCARBAMODITHIOIC ACID, IRON COMPLEX see FAS000
DIMETHYL CARBAMOYL CHLORIDE see DQY950
N,N-DIMETHYLCARBAMOYL CHLORIDE (DOT) see DQY950
DIMETHYLCARBINOL see INJ000
DIMETHYL-1-CARBOMETHOXY-1-PROPEN-2-YL PHOSPHATE see MQR750
DIMETHYLCHLOROETHER see CIO250
3,3′-DIMETHYLDIPHENYLMETHANE-4,4′-DIISOCYANATE see MJN750
1,1-DIMETHYLETHYLAMINE see BPY250
1,1-DIMETHYLETHYL HYDROPEROXIDE see BRM250
DIMETHYL FORMAL see MGA850
DIMETHYLFORMAMIDE see DSB000
N,N-DIMETHYLFORMAMIDE (DOT) see DSB000

2,6-DIMETHYL-4-HEPTANONE see DNI800
1,1-DIMETHYLHYDRAZINE see DSF400
1,2-DIMETHYLHYDRAZINE see DSF600
sym-DIMETHYLHYDRAZINE see DSF600
DIMETHYLHYDRAZINE UNSYMMETRICAL (DOT) see DSF400
DIMETHYL KETONE see ABC750
N,N-DIMETHYLMETHANAMINE see TLD500
DIMETHYLMETHANE see PMJ750
O,O-DIMETHYL-p-4-(METHYLMERCAPTO)-3-METHYLPHENYL THIOPHOSPHATE see FAQ999
DIMETHYLNITROMETHANE see NIY000
O,O-DIMETHYL-O-(p-NITROPHENYL) PHOSPHOROTHIOATE see MNH000
DIMETHYLNITROSAMINE see NKA600
DIMETHYLPHENYLAMINE see XMA000
DIMETHYLPHTHALATE see DTR200
2,2-DIMETHYLPROPANE (DOT) see NCH000
DIMETHYL SULFATE see DUD100
DIMETHYL TRICHLOROPHENYL THIOPHOSPHATE see RMA500
DIMETHYL VICLOGEN CHLORIDE see PAJ000
DIMETHYL VIOLOGEN see PAI990
DIMETHYL YELLOW see DOT300
DINITOLMIDE see DUP300
DINITROBENZENE see DUQ180
m-DINITROBENZENE see DUQ200
o-DINITROBENZENE see DUQ400
p-DINITROBENZENE see DUQ600
1,2-DINITROBENZENE see DUQ400
1,3-DINITROBENZENE see DUQ200
DINITROBENZOL SOLID (DOT) see DUQ180
2,4-DINITRO-o-CRESOL see DUT115
DINITROGEN MONOXIDE see NGU000
1,5-DINITRONAPHTHALENE see DUX700
DINITROPHENYLMETHANE see DVG600
1,3-DINITROPYRENE see DVD400
1,6-DINITROPYRENE see DVD600
1,8-DINITROPYRENE see DVD800
3,5-DINITRO-o-TOLUAMIDE see DUP300
DINITROTOLUENE see DVG600
2,3-DINITROTOLUENE see DVG800
2,4-DINITROTOLUENE see DVH000
2,5-DINITROTOLUENE see DVH200
2,6-DINITROTOLUENE see DVH400
3,4-DINITROTOLUENE see DVH600
DINOC see DUT115
DIOCTYL PHTHALATE see DVL700
DI-sec-OCTYL PHTHALATE see DVL700
DIOFORM see DFI100
DIOXANE see DVQ000
1,4-DIOXANE (MAK) see DVQ000
2,3-p-DIOXANE-S,S-BIS(O,O-DIETHYLPHOSPHOROITHIOATE) see DVQ709
DIOXATHION see DVQ709

DIPA see DNM200
DIPHENYL see BGE000
DIPHENYLAMINE see DVX800
N,N-DIPHENYLANILINE see TMQ500
DIPHENYLBENZENE see TBC600
m-DIPHENYLBENZENE see TBC620
p-DIPHENYLBENZENE see TBC750
1,2-DIPHENYLBENZENE see TBC640
DIPHENYL ETHER see PFA850
DIPHENYL METHANE DIISOCYANATE see MJP400
DIPHOSPHORIC ACID THETRAETHYL ESTER see TAL570
DIPHOSPHORUS PENTOXIDE see PHS250
DIPROPYLENE GLYCOL METHYL ETHER see DWT200
DIPROPYLENE GLYCOL MONOMETHYL ETHER see DWT200
DIPROPYL KETONE see DWT600
DIPROPYL METHANE see HBC500
DI-n-PROPYLNITROSAMINE see NKB700
DIQUAT see DWX800
DISODIUM DIFLUORIDE see SHF500
DISTILLED MUSTARD see BIH250
DISULFIRAM see DXH250
DISULFOTON see DXH325
DISULFURAM see DXH250
DISYSTOX see DXH325
DITHANE A-4 see DUQ600
DITHIOCARBONIC ANHYDRIDE see CBV500
DITHIODEMETON see DXH325
DIURON see DXQ500
DIVINYL see BOP500
DIVINYLBENZENE see DXQ745
DMA see DOO800, DOQ800
DMCC see DQY950
DMF see DSB000
DMH see DSF600
DMNA see NKA600
DMS see DUD100
DMSP see FAQ800
DMTP see FAQ999
DNT see DVH000
2,3-DNT see DVG800
2,5-DNT see DVH200
2,6-DNT see DVH400
3,4-DNT see DVH600
DNTP see PAK000
DODECYL MERCAPTAN see LBX000
DOLOCHLOR see CKN500
DOP see DVL700
D.O.T. see DUP300
DOWANOL DPM see DWT200
DOWPON see DGI400
DOWTHERM see PFA860

FLUORINE MONOXIDE see ORA000
FLUORINE OXIDE see ORA000
FLUOROCARBON-12 see DFA600
FLUOROCARBON-22 see CFX500
FLUOROCARBON 113 see FOO000
FLUOROCARBON 114 see FOO509
FLUOROCARBON-115 see CJI500
FLUORODICHLOROMETHANE see DFL000
FLUOROETHENE see VPA000
FLUOROETHYLENE see VPA000
FLUOROPHOSGENE see CCA500
FLUOROTANE see HAG500
FLUOROTRICHLOROMETHANE (OSHA) see TIP500
FLUOROXENE see TKB250
FONOFOS see FMU045
FORMAL see MGA850
FORMALDEHYDE see FMV000
FORMALDEHYDE CYANOHYDRIN see HIM500
FORMALIN (DOT) see FMV000
FORMAMIDE see FMY000
FORMIC ACID see FNA000
FORMIC ACID, ETHYL ESTER see EKL000
FORMIC ALDEHYDE see FMV000
N-FORMYLDIMETHYLAMINE see DSB000
FORMYL TRICHLORIDE see CHJ500
FREON 11 see TIP500
FREON 21 see DFL000
FREON 22 see CFX500
FREON 30 see MJP450
FREON 31 see CHI900
FREON 112 see TBP050
FREON 113 see FOO000
FREON 114 see FOO509
FREON 115 see CJI500
FREON 13B1 see TJY100
FREON 12-B2 see DKG850
FREON F-12 see DFA600
FREON 113TR-T see FOO000
FUMED SILICA see SCH000
FUNDASOL see BAV575
FURADAN see CBS275
2-FURALDEHYDE see FPQ875
2-FURANCARBINOL see FPU000
2-FURANCARBOXALDEHYDE see FPQ875
FURFURAL see FPQ875
FURFURYL ALCOHOL see FPU000
FURNACE BLACK see CBT750
2-FURYLCARBINOL see FPU000
α-FURYLCARBINOL see FPU000
FUSED QUARTZ see SCK600
FUSED SILICA (ACGIH) see SCK600

GASOLINE see GBY000
GERMANE (DOT) see GEI100
GERMANIUM HYDRIDE see GEI100
GERMANIUM TETRAHYDRIDE see GEI100
GLACIAL ACETIC ACID see AAT250
GLASS FIBERS see FBQ000
GLUCINUM see BFO750
GLUTARALDEHYDE see GFQ000
GLYCERIN see GGA000
GLYCERINE see GGA000
GLYCEROL see GGA000
GLYCEROL TRICHLOROHYDRIN see TJB600
GLYCEROL TRINITRATE see NGY000
GLYCIDOL see GGW500
GLYCIDYL ALCOHOL see GGW500
GLYCINOL see EEC600
GLYCOL ALCOHOL see EJC500
GLYCOL BROMIDE see EIY500
GLYCOL DICHLORIDE see EIY600
GLYCOL DINITRATE see EJG000
GLYCOL ETHYLENE ETHER see DVQ000
GLYCOL MONOCHLOROHYDRIN see EIU800
GLYCOL MONOETHYL ETHER ACETATE see EES400
GLYCOLONITRILE see HIM500
GRAIN ALCOHOL see EFU000
GRAPHITE (MAK) see CBT500
GUTHION (DOT) see ASH500
GYPSUM see CAX500, CAX750
HAFNIUM see HAC000
HALANE see DFE200
HALOCARBON 112a see TBP000
HALOCARBON 1132A see VPP000
HALON 1011 see CES650
HALOTHANE see HAG500
HALOWAX see TBR000, TIT500
HCB see HCC500
HCBD see HCD250
HCCPD see HCE500
HCN see HHS000
HELOTHION see SOU625
HEPTACHLOR see HAR000
HEPTANE see HBC500
1-HEPTANETHIOL see HBD500
2-HEPTANONE see MGN500
3-HEPTANONE see EHA600
4-HEPTANONE see DWT600
HEPTYL HYDRIDE see HBC500
HEPTYL MERCAPTAN see HBD500
HERBIZOLE see AMT100
HEXACARBONYLCHROMIUM see HCB000
HEXACHLORBUTADIENE see HCD250

HEXACHLOROBENZENE see HCC500
HEXACHLORO-1,3-BUTADIENE (MAK) see HCD250
HEXACHLOROCYCLOPENTADIENE see HCE500
HEXACHLOROEPOXYOCTAHYDRO-ENDO,EXO-DIMETHANONAPHTHALENE see DHB400
HEXACHLOROETHANE see HCI000
HEXACHLORONAPHTHALENE see HCK500
HEXACHLOROPHENYL ETHER see CDV175
HEXADRIN see EAT500
HEXAFLUOROACETONE see HCZ000
HEXAHYDROBENZENAMINE see CPF500
HEXAHYDROBENZENE see CPB000
HEXAHYDROCRESOL see MIQ745
HEXAHYDROMETHYLPHENOL see MIQ745
HEXAHYDRO-N-NITROSOPYRIDINE see NLJ500
HEXAHYDROTOLUENE see MIQ740
HEXAMETHYLENE see CPB000
HEXAMETHYLENEDIISOCYANATE (DOT) see DNJ800
1,6-HEXAMETHYLENE DIISOCYANATE (MAK) see DNJ800
HEXAMETHYL PHOSPHORAMIDE see HEK000
HEXAMETHYLPHOSPHORIC ACID TRIAMIDE (MAK) see HEK000
HEXANAPHTHENE see CPB000
n-HEXANE see HEN000
HEXANES (FCC) see HEN000
1-HEXANETHIOL see HES000
2-HEXANONE see HEV000
HEXANONE ISOXIME see CBF700
HEXOGEN (explosive) see CPR800
HEXOLITE see CPR800
HEXONE see HFG500
sec-HEXYL ACETATE see HFJ000
HEXYLENE GLYCOL see HFP875
HEXYL MERCAPTAN see HES000
HMDI see DNJ800
HMPA see HEK000
HMPT see HEK000
HS see HGW500
HYDRATED LIME see CAT250
HYDRAZINE see HGS000
HYDRAZINE-BENZENE see PFI000
HYDRAZINE HYDRATE see HGU500
HYDRAZINE HYDROCHLORIDE see HGV000
HYDRAZINE HYDROGEN SULFATE see HGW500
HYDRAZINE MONOHYDRATE see HGU500
HYDRAZINE SULFATE (1–1) see HGW500
HYDRAZINOBENEZENE see PFI000
HYDRAZOIC ACID see HHG500
HYDRAZOMETHANE see DSF600, MKN000
HYDRAZONIUM SULFATE see HGW500
HYDROBROMIC ACID see HHJ000
HYDROCHLORIC ACID see HHL000
HYDROCHLORIC ETHER see EHH000

HYDROCYANIC ACID see HHS000
HYDROCYANIC ACID, POTASSIUM SALT see PLC500
HYDROCYANIC ACID, SODIUM SALT see SGA500
HYDROFLUORIC ACID see HHU500
HYDROGEN ANTIMONIDE see SLQ000
HYDROGEN ARSENIDE see ARK250
HYDROGENATED TERPHENYLS see HHW800
HYDROGEN AZIDE see HHG500
HYDROGEN BROMIDE (OSHA ACGIH, MAK, DOT) see HHJ000
HYDROGEN CHLORIDE see HHX000
HYDROGEN CYANAMIDE see COH500
HYDROGEN CYANIDE (OSHA, ACGIH) see HHS000
HYDROGEN DIOXIDE see HIB000
HYDROGEN FLUORIDE (OSHA, ACGIH, MAK, DOT) see HHU500
HYDROGEN NITRATE see NED500
HYDROGEN PEROXIDE see HIB000
HYDROGEN PHOSPHIDE see PGY000
HYDROGEN SELENIDE see HIC000
HYDROGEN SULFIDE see HIC500
HYDROGEN SULFURIC ACID see HIC500
2-HYDROPEROXY-2-METHYLPROPANE see BRM250
HYDROQUINONE see HIH000
m-HYDROQUINONE see REA000
o-HYDROQUINONE see CCP850
HYDROQUINONE MONOMETHYL ETHER see MFC700
HYDROXYACETONITRILE see HIM500
HYDROXYBENZENE see PDN750
2-HYDROXYBUTANE see BPW750
1-HYDROXY-4-tert-BUTYLBENZENE see BSE500
HYDROXYCYCLOHEXANE see CPB750
3-HYDROXYDIMETHYL CROTONAMIDE DIMETHYL PHOSPHATE see DGQ875
β-HYDROXYETHYL ISOPROPYL ETHER see INA500
4-HYDROXY-4-METHYLPENTANONE-2 see DBF750
1-HYDROXYMETHYLPROPANE see IIL000
2-HYDROXY-2-METHYLPROPIONITRILE see MLC750
1-HYDROXYPROPANE see PND000
3-HYDROXY-1-PROPANESULPHONIC ACID SULFONE see PML400
2-HYDROXYPROPANOIC ACID, BUTYL ESTER see BRR600
3-HYDROXYPROPENE see AFV500
2-HYDROXYPROPYL ACRYLATE see HNT600
m-HYDROXYTOLUENE see CNW750
o-HYDROXYTOLUENE see CNX000
p-HYDROXYTOLUENE see CNX250
2-HYDROXYTRIETHYLAMINE see DHO500
2-HYDROXY-1,3,5-TRINITROBENZENE see PID000
HYVAR see BMM650
IGE see IPD000
ILLOXOL see DHB400
2-IMIDAZOLIDINETHIONE see IAQ000
4,4′-(IMIDOCARBONYL)BIS(N,N-DIMETHYLAMINE) MONOHYDROCHLORIDE see IBA000
4,4′-(IMIDOCARBONYL)BIS(N,N-DIMETHYLANILINE) see IBB000

MAAC see HFJ000
MACE (lacrimator) see CEA750
MAGNESIA see MAH500
MAGNESIA WHITE see CAX750
MAGNESITE see MAC650
MAGNESIUM CARBONATE see MAC650
MAGNESIUM OXIDE see MAH500
MALATHION see MAK700
MALEIC ACID ANHYDRIDE (MAK) see MAM000
MALEIC ANHYDRIDE see MAM000
MALONIC DINITRILE see MAO250
MALONONITRILE see MAO250
MANGANESE see MAP750
MANGANESE CYCLOPENTADIENYL TRICARBONYL see CPV000
MANGANESE OXIDE see MAU800
MANGANESE TETROXIDE see MAU800
MANGANESE TRICARBONYL METHYLCYCLOPENTADIENYL see MAV750
MAOH see MKW600
MAPP see MFX600
MARBLE see CAO000
MBC see BAV575
MBK see HEV000
MBOCA see MJM200
MBOT see MJO250
MCB see CEJ125
MCT see CPV000
MDA see MJQ000
MDI see MJP400
MEK see MKA400
MEKP see MKA500
MEK PEROXIDE see MKA500
MENDRIN see EAT500
MERCAPTOACETIC ACID see TFJ100
1-MERCAPTODODECANE see LBX000
MERCAPTOPHOS see DAO600, FAQ999
3-MERCAPTOPROPANOL see PML500
MERCURY see MCW250
MERCURY, METALLIC (DOT) see MCW250
MERCURY ZINC CHROMATE COMPLEX see ZJA000
MEREX see KEA000
MERPAN see CBG000
MESITYL OXIDE see MDJ750
METALLIC ARSENIC see ARA750
METASILICIC ACID see SCL000
METHACRYLIC ACID see MDN250
METHACRYLIC ACID, METHYL ESTER (MAK) see MLH750
METHANAL see FMV000
METHANAMIDE see FMY000
METHANAMINE (9CI) see MGC250
METHANE TETRACHLORIDE see CBY000
METHANE TETRAMETHYLOL see PBB750

METHANETHIOL see MLE650
METHANE TRICHLORIDE see CHJ500
METHANOIC ACID see FNA000
METHANOL see MGB150
METHENYL TRIBROMIDE see BNL000
METHOMYL see MDU600
o-METHOXYANILINE see AOV900
4-METHOXYBENZENAMINE see AOW000
METHOXYCARBONYLETHYLENE see MGA500
METHOXYCHLOR see MEI450
METHOXY-DDT see MEI450
2-METHOXYETHANOL (ACGIH) see EJH500
METHOXY ETHER of PROPYLENE GLYCOL see PNL250
2-METHOXYETHYL ACETATE (ACGIH) see EJJ500
METHOXYFLURANE see DFA400
4-METHOXYPHENOL see MFC700
o-METHOXYPHENYLAMINE see AOV900
4-METHOXY-m-PHENYLENEDIAMINE see DBO000
1-METHOXY-2-PROPANOL see PNL250
METHYL ACETATE see MFW100
METHYL ACETONE (DOT) see MKA400
METHYL ACETYLENE see MFX590
METHYL ACETYLENE-PROPADIENE MIXTURE see MFX600
METHYLACETYLENE-PROPADIENE, STABILIZED (DOT) see MFX600
β-METHYL ACROLEIN see COB250
METHYL ACRYLATE see MGA500
METHYLACRYLONITRILE see MGA750
METHYLAL see MGA850
METHYL ALCOHOL see MGB150
METHYL ALDEHYDE see FMV000
METHYLAMINE see MGC250
METHYL AMYL ACETATE (DOT) see HFJ000
METHYL AMYL ALCOHOL see MKW600
METHYL n-AMYL KETONE see MGN500
METHYLANILINE see MGN750
4-METHYLANILINE see TGR000
m-METHYLANILINE see TGQ500
N-METHYL ANILINE (MAK) see MGN750
2-METHYLAZIRIDINE see PNL400
METHYLBENZENE see TGK750
N-METHYL-BIS(2-CHLOROETHYL)AMINE (MAK) see BIE250
METHYL BROMIDE see MHR200
2-METHYLBUTANE see EIK000
3-METHYL BUTANOL see IHP000
3-METHYL BUTAN 2-ONE (DOT) see MLA750
3-METHYLBUTYL ACETATE see IHO850
1-METHYL-4-tert-BUTYLBENZENE see BSP500
METHYL n-BUTYL KETONE (ACGIH) see HEV000
METHYLCARBAMATE-1-NAPHTHALENOL see CBM750
METHYL CELLOSOLVE ACETATE (OSHA, DOT) see EJJ500
METHYL CELLOSOLVE (OSHA, DOT) see EJH500

METHYL CHLORIDE see MIF765
2-METHYLCHLOROBENZENE see CLK100
METHYL CHLOROFORM see MIH275
METHYLCHLOROMETHYL ETHER (DOT) see CIO250
METHYLCHLOROPINDOL see CMX850
METHYL CYANIDE see ABE500
METHYL 2-CYANOACRYLATE see MIQ075
METHYLCYCLOHEXANE see MIQ740
METHYLCYCLOHEXANOL see MIQ745
2-METHYLCYCLOHEXANONE see MIR500
1-METHYLCYCLOHEXAN-2-ONE see MIR500
METHYLCYCLOPENTADIENYL MANGANESE TRICARBONYL (OSHA) see MAV750
METHYL DEMETON see MIW100
METHYLDINITROBENZENE see DVG600
1-METHYL-2,4-DINITROBENZENE see DVH000
2-METHYL-1,3-DINITROBENZENE see DVH400
2-METHYL-1,4-DINITROBENZENE see DVH200
4-METHYL-1,2-DINITROBENZENE see DVH600
1-METHYL-2,3-DINITRO-BENZENE (9CI) see DVG800
METHYLENEBIS(ANILINE) see MJQ000
2,4′-METHYLENEBIS(ANILINE) see MJP750
4,4′-METHYLENE BIS(2-CHLOROANILINE) see MJM200
4,4′-METHYLENEBIS(o-CHLOROANILINE) see MJM200
METHYLENE BIS(4-CYCLOHEXYLISOCYANATE see MJM600
4,4′-METHYLENE BIS(N,N′-DIMETHYLANILINE) see MJN000
4,4′-METHYLENEBIS(N,N-DIMETHYL)BENZENAMINE see MJN000
METHYLENEBIS(4-ISOCYANATOBENZENE) see MJP400
5,5′-METHYLENEBIS(2-ISOCYANATO)TOLUENE see MJN750
4,4′-METHYLENEBIS(2-METHYLANILINE) see MJO250
4,4′-METHYLENEBIS(2-METHYLBENZENAMINE) see MJO250
METHYLENE BISPHENYL ISOCYANATE see MJP400
METHYLENE CHLORIDE see MJP450
METHYLENE CHLOROBROMIDE see CES650
METHYLENE CYANIDE see MAO250
2,4′-METHYLENEDIANILINE see MJP750
4,4′-METHYLENEDIANILINE see MJQ000
METHYLENE DIMETHYL ETHER see MGA850
4,4′-METHYLENE DI-o-TOLUIDINE see MJO250
METHYL ESTER of o-SILICIC ACID see MPI750
METHYL ETHANOATE see MFW100
1-METHYLETHYLAMINE see INK000
2-(1-METHYLETHYL)BENZENAMINE see INX000
METHYL ETHYLENE OXIDE see PNL600
METHYLETHYLENIMINE see PNL400
1-(METHYLETHYL)-ETHYL 3-METHYL-4-(METHYLTHIO)PHENYL PHOSPHORAMIDATE see FAK000
METHYL ETHYL KETONE see MKA400
METHYL ETHYL KETONE PEROXIDE see MKA500
METHYLETHYLMETHANE see BOR500
N,N-METHYLETHYLNITROSAMINE see MKB000
N-(1-METHYLETHYL)-2-PROPANAMINE see DNM200

METHYL FORMATE see MKG750
METHYL FORMATE (DOT) see MKG750
3-METHYL-5-HEPTANONE see EGI750
2-METHYL-5-HEXANONE see MKW450
METHYL HYDRAZINE see MKN000
1-METHYL HYDRAZINE see MKN000
METHYLHYDRAZINE HYDROCHLORIDE see MKN250
METHYL IODIDE see MKW200
METHYL ISOAMYL KETONE see MKW450
METHYL ISOBUTENYL KETONE see MDJ750
METHYL ISOBUTYL CARBINOL see MKW600
METHYLISOBUTYLCARBINOL ACETATE see HFJ000
METHYL ISOBUTYL KETONE (ACGIH, DOT) see HFG500
METHYL ISOCYANATE see MKX250
METHYL ISOPROPYL KETONE see MLA750
METHYL KETONE see ABC750
2-METHYLLACTONITRILE see MLC750
METHYL MERCAPTAN see MLE650
METHYLMERCURY see MLF550
METHYLMERCURY(II) CATION see MLF550
METHYLMERCURY ION see MLF550
METHYL METHACRYLATE see MLH750
N-METHYLMETHANAMINE see DOQ800
METHYL METHANOATE see MKG750
METHYL-α-METHYLACRYLATE see MLH750
METHYL-N-((METHYLCARBAMOYL)OXY)THIOACETIMIDATE see MDU600
METHYL-2-METHYL-2-PROPENOATE see MLH750
6-METHYL-3-NITROANILINE see NMP500
2-METHYLNITROBENZENE see NMO525
4-METHYLNITROBENZENE see NMO550
m-METHYLNITROBENZENE see NMO500
N-METHYL-N-NITROSOANILINE see MMU250
N-METHYL-N-NITROSOBENZENAMINE see MMU250
METHYL ORTHOSILICATE see MPI750
METHYL OXIRANE see PNL600
METHYL PARATHION see MNH000
2-METHYLPENTANE see IKS600
2-METHYL-2,4-PENTANEDIOL see HFP875
4-METHYL-2-PENTANOL (MAK) see MKW600
2-METHYL-2-PENTEN-4-ONE see MDJ750
4-METHYL-2-PENTYL ACETATE see HFJ000
METHYL PENTYL KETONE see MGN500
2-METHYL-m-PHENYLENE ISOCYANATE see TGM800
METHYL PHOSPHATE see TMD250
METHYL PHTHALATE see DTR200
2-METHYL PROPANOL see IIL000
2-METHYLPROPENENITRILE see MGA750
METHYL PROPENOATE see MGA500
2-METHYLPROPENOIC ACID see MDN250
2-METHYLPROPIONITRILE see IJX000
2-METHYLPROPYL ACETATE see IIJ000

1-METHYLPROPYLAMINE see BPY000
β-METHYLPROPYL ETHANOATE see IIJ000
METHYL PROPYL KETONE (ACGIH, DOT) see PBN250
3-(N-METHYLPYRROLIDINO)PYRIDINE see NDN000
N-METHYLPYRROLIDONE see MPF200
1-METHYL-2-PYRROLIDONE see MPF200
METHYL SILICATE see MPI750
METHYL STYRENE see VQK650
α-METHYL STYRENE see MPK250
METHYL STYRENE (mixed isomers) see MPK500
METHYL SULFATE (DOT) see DUD100
METHYL SYSTOX see MIW100
METHYL THIRAM see TFS350
METHYL TOLUENE see XGS000
p-METHYLTOLUENE see XHS000
2-METHYL-p-TOLUIDINE see XMS000
METHYLTRICHLOROMETHANE see MIH275
METHYL TUADS see TFS350
METHYL VIOLOGEN (2+) see PAI990
METILTRIAZOTION see ASH500
METRIBUZIN see MQR275
MEVINPHOS see MQR750
MIAK see MKW450
MIBC see MKW600
MICA see MQS250
MICA SILICATE see MQS250
MICHLER'S BASE see MJN000
MICHLER'S KETONE see MQS500
MINERAL NAPHTHA see BBL250
MINERAL OIL see MQV750
MINERAL OIL, WHITE (FCC) see MQV750
MINERAL SPIRITS see PCT250
MIPK see MLA750
MIRBANE OIL see NEX000
MME see MFC700, MLH750
MMH see MKN000
MMT see MAV750
MNA see MMU250
MNT see NMO500
MOCA see MJM200
MOLYBDATE see MRC250
MOLYBDENUM see MRC250
MONOBROMOETHANE see EGV400
MONOCHLOROACETYL CHLORIDE see CEC250
MONOCHLOROBENZENE see CEJ125
MONOCHLORODIFLUOROMETHANE see CFX500
MONOCHLORODIMETHYL ETHER (MAK) see CIO250
MONOCHLORO DIPHENYL OXIDE see MRG000
MONOCHLOROETHYLENE (DOT) see VNP000
MONOCHLOROMETHANE see MIF765
MONOCHLOROMONOFLUOROMETHANE see CHI900

MONOCHLOROPENTAFLUOROETHANE (DOT) see CJI500
MONOCHLOROPHENYLETHER see MRG000
MONOCROTOPHOS see MRH209
MONOETHYLAMINE (DOT) see EFU400
MONOISOBUTYLAMINE see IIM000
MONOMETHYL ANILINE (OSHA) see MGN750
MONOSILANE see SDH575
MORPHOLINE see MRP750
MOTOR FUEL (DOT) see GBY000
MPK see PBN250
MPP see FAQ999
MURIATIC ACID (DOT) see HHL000
MUSTARD GAS see BIH250
MUTHMANN'S LIQUID see ACK250
MUTOXIN see DAD200
MXDA see XHS800
NALED see NAG400
5-NAN see NEJ500
NAPHTHA see NAI500, PCS250, ROU000
NAPHTHALENE see NAJ500
1,5-NAPHTHALENE DIISOCYANATE see NAM500
NAPHTHALENE OIL see CMY825
NAPHTHALINE see NAJ500
NAPHTHA SAFETY SOLVENT see SLU500
1-NAPHTHYLAMINE see NBE000
α-NAPHTHYLAMINE see NBE000
β-NAPHTHYLAMINE see NBE500
2-NAPHTHYLPHENYLAMINE see PFT500
α-NAPHTHYLTHIOUREA (DOT) see AQN635
NAXOL see CPB750
NDEA see NJW500
NDELA see NKM000
NDMA see NKA600
NEA see NKD000
NEMA see MKB000
NEMACUR see FAK000
NEMATOCIDE see DDL800
NEON see NCG500
NEOPENTANE see NCH000
NEOPRENE see NCI500
NIAX CATALYST ESN see NCS000
NIBREN WAX see TIT500
NICKEL CARBONYL see NCZ000
NICKEL and COMPOUNDS see NCW500
NICKEL TETRACARBONYL see NCZ000
NICOTINE see NDN000
NIRAN see CDR750
NITRAPYRIN (ACGIH) see CLP750
NITRIC ACID see NED500
NITRIC OXIDE see NEG100
5-NITROACENAPHTHENE see NEJ500

5-NITROACENAPHTHYLENE see NEJ500
2-NITRO-4-AMINOPHENOL see NEM480
p-NITROANILINE see NEO500
4-NITROANILINE (MAK) see NEO500
3-NITRO-3-AZAPENTANE-1,5-DIISOCYANATE see NHI500
NITROBENZENE see NEX000
2-NITRO-1,4-BENZENEDIAMINE see ALL750
4-NITROBIPHENYL see NFQ000
NITROCARBOL see NHM500
p-NITROCHLOROBENZENE see NFS525
NITROETHANE see NFY500
NITROGEN DIOXIDE see NGR500
NITROGEN FLUORIDE see NGW000
NITROGEN MONOXIDE see NEG100
NITROGEN MUSTARD see BIE250
NITROGEN OXIDE see NGU000
NITROGEN PEROXIDE, LIQUID (DOT) see NGR500
NITROGEN TRIFLUORIDE see NGW000
NITROGLYCERIN see NGY000
NITROGLYCERIN mixed with ETHYLENE GLYCOL, DINITRATE (1–1) see NGY500
NITROIMINODIETHYLENEDIISOCYANIC ACID see NHI500
NITROMETHANE see NHM500
1-NITRONAPHTHALENE see NHQ000
2-NITRONAPHTHALENE see NHQ500
α-NITRONAPHTHALENE see NHQ000
β-NITRONAPHTHALENE see NHQ500
1-NITROPROPANE see NIX500
2-NITROPROPANE see NIY000
3-NITROPYRENE see NJA000
N-NITROSODI-n-BUTYLAMINE (MAK) see BRY500
N-NITROSODIETHANOLAMINE (MAK) see NKM000
N-NITROSODIETHYLAMINE see NJW500
N-NITROSODIISOPROPYLAMINE see NKA000
N-NITROSODIMETHYLAMINE see NKA600
N-NITROSODI-i-PROPYLAMINE (MAK) see NKA000
N-NITROSODI-N-PROPYLAMINE see NKB700
N-NITROSO-N-ETHYL ANILINE see NKD000
N-NITROSOETHYLPHENYLAMINE (MAK) see NKD000
NITROSOIMINO DIETHANOL see NKM000
N-NITROSOMETHYLETHYLAMINE (MAK) see MKB000
N-NITROSOMETHYLPHENYLAMINE (MAK) see MMU250
4-NITROSOMORPHOLINE see NKZ000
N-NITROSOMORPHOLINE (MAK) see NKZ000
N-NITROSOPIPERIDINE see NLJ500
N-NITROSOPYRROLIDINE see NLP500
m-NITROTOLUENE see NMO500
o-NITROTOLUENE see NMO525
p-NITROTOLUENE see NMO550
mixo-NITROTOLUENE see NMO600
5-NITRO-o-TOLUIDINE see NMP500
3-NITROTOLUOL see NMO500

PARRAFIN OIL see MQV750
PCB (DOT, USDA) see PJL750
PCB's see PJM500, PJN000
PCL see HCE500
PCM see PCF300
PCP see PAX250
PDCB see DEP800
PENTA see PAX250
PENTABORANE see PAT750
PENTACARBONYLIRON see IHG500
PENTACHLORO DIPHENYL OXIDE see PAW250
PENTACHLOROETHANE see PAW500
PENTACHLORONAPHTHALENE see PAW750
PENTACHLOROPHENOL see PAX250
PENTACHLOROPHENYL CHLORIDE see HCC500
PENTAERYTHRITOL see PBB750
PENTALIN see PAW500
PENTAMETHYLENE see CPV750
PENTANAL see VAG000
PENTANE see PBK250
tert-PENTANE (DOT) see NCH000
1,5-PENTANEDIONE see GFQ000
1-PENTANETHIOL see PBM000
3-PENTANOL see IHP010
2-PENTANONE see PBN250
3-PENTANONE see DJN750
PENTOLE see CPU500
2-PENTYL ACETATE see AOD735
n-PENTYL ACETATE see AOD725
PENTYL MERCAPTAN see PBM000
PERACETIC ACID (MAK) see PCL500
PERCHLORETHYLENE see PCF275
PERCHLOROBUTADIENE see HCD250
PERCHLOROETHYLENE see PCF275
PERCHLOROMETHYL MERCAPTAN see PCF300
PERCHLORYL FLUORIDE see PCF750
PERFLUOROAMMONIUM OCTANOATE see ANP625
PERLITE see PCJ400
PERMANENT WHITE see BAP000
PERMANGANATE of POTASH (DOT) see PLP000
PEROXYACETIC ACID see PCL500
PETROL (DOT) see GBY000
PETROLATUM, LIQUID see MQV750
PETROLEUM DISTILLATE see PCS250
PETROLEUM DISTILLATES (NAPHTHA) see NAI500
PETROLEUM GAS, LIQUEFIED see LGM000
PETROLEUM PITCH see ARO500
PETROLEUM SPIRIT (DOT) see NAI500
PETROLEUM SPIRITS see PCT250
PGDN see PNL000
PGE see PFH000

PHENANTHRENE see PCW250
PHENANTRIN see PCW250
PHENOL see PDN750
PHENOL GLYCIDYL ETHER (MAK) see PFH000
PHENOTHIAZINE see PDP250
PHENOXYBENZENE see PFA850
PHENYLAMINE see AOQ000
N-PHENYLANILINE see DVX800
PHENYLBENZENE see BGE000
PHENYLCHLOROMETHYLKETONE see CEA750
4-PHENYLDIPHENYL see TBC750
m-PHENYLENEBIS(METHYLAMINE) see XHS800
p-PHENYLENEDIAMINE see PEY500
m-PHENYLENE ISOCYANATE see BBP000
PHENYL-2,3-EPOXYPROPYL ETHER see PFH000
PHENYLETHANE see EGP500
PHENYL ETHER see PFA850
PHENYL ETHER-BIPHENYL MIXTURE see PFA860
PHENYL ETHER HEXACHLORO see CDV175
PHENYL ETHER MONO-CHLORO see MRG000
PHENYL ETHER PENTACHLORO see PAW250
PHENYL ETHER TETRACHLORO see TBP250
PHENYL GLYCYDYL ETHER see PFH000
PHENYLHYDRAZINE see PFI000
PHENYLHYDRAZINE HYDROCHLORIDE see PFI250
PHENYLHYDRAZINIUM CHLORIDE see PFI250
PHENYL MERCAPTAN see PFL850
N-PHENYLMETHYLAMINE see MGN750
PHENYLMETHYLNITROSAMINE see MMU250
N-PHENYL-β-NAPHTHYLAMINE see PFT500
p-PHENYL-NITROBENZENE see NFQ000
PHENYLPHOSPHINE see PFV250
2-PHENYLPROPENE see MPK250
β-PHENYLPROPYLENE see MPK250
PHORATE see PGS000
PHOSDRIN (OSHA) see MQR750
PHOSGENE see PGX000
PHOSPHINE see PGY000
PHOSPHORIC ACID see PHB250
PHOSPHORIC ACID, DIBUTYL PHENYL ESTER see DEG600
PHOSPHORIC ANHYDRIDE see PHS250
PHOSPHORIC CHLORIDE see PHR500
PHOSPHORIC SULFIDE see PHS000
PHOSPHORUS CHLORIDE see PHT275
PHOSPHORUS OXYCHLORIDE see PHQ800
PHOSPHORUS OXYTRICHLORIDE see PHQ800
PHOSPHORUS PENTACHLORIDE see PHR500
PHOSPHORUS PENTASULFIDE see PHS000
PHOSPHORUS PENTOXIDE see PHS250
PHOSPHORUS PERCHLORIDE see PHR500
PHOSPHORUS TRICHLORIDE see PHT275

PHOSPHORUS TRIHYDRIDE see PGY000
PHOSPHORUS (red) see PHO500
PHOSPHORUS (yellow) see PHO750
PHOSPHORYL CHLORIDE see PHQ800
PHTHALIC ACID ANHYDRIDE see PHW750
PHTHALIC ANHYDRIDE see PHW750
m-PHTHALODINITRILE see PHX550
PHTHALOL see DJX000
PICLORAM see PIB900
PICRIC ACID see PID000
PICRYLMETHYLNITRAMINE see TEG250
PINDONE see PIH175
PIPERAZINE DIHYDROCHLORIDE see PIK000
PITCH see CMZ100
PITCH, COAL TAR see CMZ100
PIVAL see PIH175
2-PIVALYL-1,3-INDANDIONE see PIH175
PLASTER of PARIS see CAX500
PLATINUM see PJD500
PNA see NEO500
PNB see NFQ000
PNCB see NFS525
PNT see NMO550
POLYCHLORINATED BIPHENYL (AROCLOR 1242) see PJM500
POLYCHLORINATED BIPHENYL (AROCLOR 1254) see PJN000
POLYCHLORINATED BIPHENYLS see PJL750
POLYCHLORINATED CAMPHENES see CDV100
POLY(CHLOROETHYLENE) see PKQ059
POLYVINYL CHLORIDE see PKQ059
PORTLAND CEMENT see PKS750
POTASSIUM CYANIDE see PLC500
POTASSIUM HYDRATE (DOT) see PLJ500
POTASSIUM HYDROXIDE see PLJ500
POTASSIUM PERMANGANATE see PLP000
PPD see PEY500
PRECIPITATED SILICA see SCL000
PRIMATOL see ARQ725
PROPANE see PMJ750
1,2-PROPANEDIOL-1-ACRYLATE see HNT600
PROPANE SULTONE see PML400
1,3-PROPANE SULTONE (MAK) see PML400
PROPANETHIOL see PML500
1,2,3-PROPANETRIOL see GGA000
2-PROPANOL see INJ000
n-PROPANOL see PND000
PROPANONE see ABC750
2-PROPANONE see ABC750
PROPARGYL ALCOHOL see PMN450
2-PROPENAL see ADR000
PROPENAMIDE see ADS250
2-PROPENAMIDE see ADS250

2-PROPENENITRILE see ADX500
2-PROPENOIC ACID see ADS750
2-PROPENOIC ACID, ETHYL ESTER (MAK) see EFT000
PROPENOIC ACID METHYL ESTER see MGA500
PROPENOL see AFV500
PROPINE see MFX590
β-PROPIOLACTONE see PMT100
PROPIONE see DJN750
PROPIONIC ACID see PMU750
β-PROPIONOLACTONE see PMT100
PROPIONONITRILE see PMV750
PROPOXUR see PMY300
n-PROPYL ACETATE see PNC250
n-PROPYL ALCOHOL see PND000
sec-PROPYL ALCOHOL (DOT) see INJ000
2-PROPYLAMINE see INK000
PROPYLCARBINOL see BPW500
PROPYL CYANIDE see BSX250
PROPYLENE DICHLORIDE see PNJ400
PROPYLENE GLYCOL DINITRATE see PNL000
PROPYLENE GLYCOL-1,2-DINITRATE see PNL000
PROPYLENE GLYCOL METHYL ETHER see PNL250
PROPYLENE GLYCOL MONOACRYLATE see HNT600
PROPYLENE GLYCOL MONOMETHYL ETHER see PNL250
PROPYLENE IMINE see PNL400
PROPYLENE OXIDE see PNL600
PROPYL HYDRIDE see PMJ750
PROPYL MERCAPTAN see PML500
n-PROPYL NITRATE see PNQ500
PROPYNE see MFX590
PROPYNE mixed with PROPADIENE see MFX600
2-PROPYN-1-OL see PMN450
PRUSSIC ACID (DOT) see HHS000
PRUSSITE see COO000
PSEUDOCUMIDINE see TLG250
PURE QUARTZ see SCI500
PYRANTON see DBF750
PYRENE see PON250
PYRETHRINS see POO250
PYRETHRUM (ACGIH) see POO250
PYRIDINE see POP250
β-PYRINE see PON250
PYRINEX see CMA100
PYROCATECHOL see CCP850
PYROMUCIC ALDEHYDE see FPQ875
PYROPENTYLENE see CPU500
QUARTZ see SCJ500
QUARTZ GLASS see SCK600
QUICKLIME (DOT) see CAU500
QUICK SILVER see MCW250
QUINONE see QQS200

TETRACHLOROETHYLENE (DOT) see PCF275
TETRACHLORONAPHTHALENE see TBR000
TETRADIOXIN see TAI000
TETRAETHYL DITHIOPYROPHOSPHATE see SOD100
TETRAETHYL LEAD see TCF000
TETRAETHYL ORTHOSILICATE (DOT) see EPF550
TETRAETHYLPLUMBANE see TCF000
TETRAETHYL PYROPHOSPHATE see TAL570
TETRAETHYL SILICATE (DOT) see EPF550
TETRAFLUORODICHLOROETHANE see DGL600
TETRAFLUOROSULFURANE see SOR000
1,2,3,4-TETRAHYDROBENZENE see CPC579
TETRAHYDRO-p-DIOXIN see DVQ000
TETRAHYDROFURAN see TCR750
TETRAHYDRO-p-ISOXAZINE see MRP750
TETRAHYDRO-N-NITROSOPYRROLE see NLP500
TETRAMETHYLDIAMINOBENZOPHENONE see MQS500
p,p-TETRAMETHYLDIAMINODIPHENYLMETHANE see MJN000
TETRAMETHYLENE CYANIDE see AER250
TETRAMETHYL LEAD see TDR500
TETRAMETHYLPLUMBANE see TDR500
TETRAMETHYLSILICATE see MPI750
TETRAMETHYLSUCCINONITRILE see TDW250
TETRANITROMETHANE see TDY250
TETRASODIUM PYROPHOSPHATE see TEE500
TETRYL see TEG250
THALLIUM see TEI000
THF see TCR750
4,4′-THIOBIS(6-tert-BUTYL-m-CRESOL) see TFC600
4,4′-THIOBIS(2-tert-BUTYL-5-METHYLPHENOL) see TFC600
THIOCARBAMIZINE see TFD750
4,4′-THIODIANILINE see TFI000
THIODIPHENYLAMINE see PDP250
THIOGLYCOLIC ACID see TFJ100
THIONYL CHLORIDE see TFL000
THIOPHENOL (DOT) see PFL850
THIOPHOSPHORIC ANHYDRIDE see PHS000
THIOSULFAN see EAQ750
THIRAM see TFS350
TIN and COMPOUNDS see TGB250
TINNING GLUX (DOT) see ZFA000
TITANIUM DIOXIDE see TGG760
TITANIUM OXIDE see TGG760
TMA see TLD500
TMAN see TKV000
TML see TDR500
TMP see TMD250
TNT see TMN490
TOCP see TMO600
o-TOLIDINE see TGJ750
TOLUENE see TGK750

TOLUENE-2,4-DIAMINE see TGL750
TOLUENE-2,4-DIISOCYANATE see TGM750
2,4-TOLUENEDIISOCYANATE see TGM750
TOLUENE-2,6-DIISOCYANATE see TGM800
2,6-TOLUENE DIISOCYANATE see TGM800
TOLUENE HEXAHYDRIDE see MIQ740
m-TOLUIDINE see TGQ500
o-TOLUIDINE see TGQ750
p-TOLUIDINE see TGR000
p-TOLUOL see CNX250
TOLUOL (DOT) see TGK750
TOLYL CHLORIDE see BEE375
o-TOLYL CHLORIDE see CLK100
o-TOLYL PHOSPHATE see TMO600
TORDON see PIB900
TOXAPHENE see CDV100
TOXICHLOR see CDR750
TOXILIC ANHYDRIDE see MAM000
TREMOLITE ASBESTOS see ARM250
TRIATOMIC OXYGEN see ORW000
TRIAZOIC ACID see HHG500
TRIAZOLAMINE see AMT100
TRIBROMOMETHANE see BNL000
TRIBUTYL PHOSPHATE see TIA250
TRICHLOROACETIC ACID see TII250
1,2,4-TRICHLOROBENZENE see TIK250
2,3,4-TRICHLOROBUTENE-1 see TIL360
1,1,1-TRICHLOROETHANE see MIH275
1,1,2-TRICHLOROETHANE see TIN000
TRICHLOROETHYLENE see TIO750
TRICHLOROFLUOROMETHANE see TIP500
TRICHLOROHYDRIN see TJB600
TRICHLOROMETHANE see CHJ500
TRICHLOROMETHANE SULFENYL CHLORIDE see PCF300
TRICHLOROMETHYLBENZENE see BFL250
N-TRICHLOROMETHYLTHIO-3A,4,7,7A-TETRAHYDROPHTHALIMIDE see CBG000
TRICHLORONAPHTHALENE see TIT500
TRICHLORONITROMETHANE see CKN500
2,4,5-TRICHLOROPHENOXYACETIC ACID see TAA100
1,2,3-TRICHLOROPROPANE see TJB600
1,1,2-TRICHLORO-1,2,2-TRIFLUOROETHANE (OSHA, ACGIH, MAK) see FOO000
TRIDYMITE see SCI500, SCK000
TRIETHYLAMINE see TJO000
1,1,1-TRIFLUORO-2-BROMO-2-CHLOROETHANE see HAG500
TRIFLUOROBROMOMETHANE see TJY100
(2,2,2-TRIFLUOROETHOXY)ETHENE see TKB250
2,2,2-TRIFLUOROETHYL VINYL ETHER see TKB250
TRIIODOMETHANE see IEP000
TRIMANGANESE TETROXIDE see MAU800
TRIMELLIC ACID ANHYDRIDE see TKV000
TRIMELLITIC ANHYDRIDE see TKV000

# Appendix II
# CAS Number Cross-reference

50-00-0 see FMV000
50-29-3 see DAD200
50-32-8 see BCS750
50-78-2 see ADA725
51-75-2 see BIE250
51-79-6 see UVA000
53-96-3 see FDR000
54-11-5 see NDN000
55-18-5 see NJW500
55-38-9 see FAQ999
55-63-0 see NGY000
56-23-5 see CBY000
56-38-2 see PAK000
56-81-5 see GGA000
57-12-5 see COI500
57-14-7 see DSF400
57-24-9 see SMN500
57-50-1 see SNH000
57-57-8 see PMT100
57-74-9 see CDR750
58-89-9 see BBQ500
59-88-1 see PFI250
59-89-2 see NKZ000
60-11-7 see DOT300
60-29-7 see EJU000
60-34-4 see MKN000
60-35-5 see AAI000
60-57-1 see DHB400
61-82-5 see AMT100
62-53-3 see AOQ000
62-73-7 see DGP900
62-74-8 see SHG500
62-75-9 see NKA600
63-25-2 see CBM750
64-17-5 see EFU000
64-18-6 see FNA000
64-19-7 see AAT250
64-67-5 see DKB110
67-48-1 see CMF750
67-56-1 see MGB150
67-63-0 see INJ000
67-64-1 see ABC750
67-66-3 see CHJ500
67-72-1 see HCI000
68-11-1 see TFJ100
68-12-2 see DSB000
71-23-8 see PND000
71-36-3 see BPW500
71-43-2 see BBL250
71-55-6 see MIH275
72-20-8 see EAT500
72-43-5 see MEI450
74-83-9 see MHR200
74-86-2 see ACI750
74-87-3 see MIF765
74-88-4 see MKW200
74-89-5 see MGC250
74-90-8 see HHS000
74-93-1 see MLE650
74-96-4 see EGV400
74-97-5 see CES650
74-98-6 see PMJ750
74-99-7 see MFX590
75-00-3 see EHH000
75-01-4 see VNP000
75-02-5 see VPA000
75-04-7 see EFU400
75-05-8 see ABE500
75-07-0 see AAG250
75-08-1 see EMB100
75-09-2 see MJP450
75-12-7 see FMY000
75-15-0 see CBV500
75-21-8 see EJN500
75-25-2 see BNL000
75-31-0 see INK000
75-34-3 see DFF809
75-35-4 see VPK000
75-38-7 see VPP000
75-43-4 see DFL000
75-44-5 see PGX000
75-45-6 see CFX500
75-47-8 see IEP000
75-50-3 see TLD500
75-52-5 see NHM500
75-55-8 see PNL400
75-56-9 see PNL600
75-61-6 see DKG850
75-63-8 see TJY100
75-64-9 see BPY250

# Appendix III
# DOT Number Cross-reference

0072 see CPR800
0118 see CPR800
0143 see NGY000
0154 see PID000
0208 see TEG250
0209 see TMN490
1001 see ACI750
1005 see AMY500
1008 see BMG700
1009 see TJY100
1011 see BOR500
1013 see CBU250
1015 see CBV000
1016 see CBW750
1017 see CDV750
1018 see CFX500
1020 see CJI500
1026 see COO000
1028 see DFA600
1029 see DFL000
1032 see DOQ800
1036 see EFU400
1037 see EHH000
1040 see EJN500
1045 see FEZ000
1048 see HHJ000
1050 see HHL000
1051 see HHS000
1052 see HHU500
1053 see HIC500
1060 see MFX600
1061 see MGC250
1062 see MHR200
1063 see MIF765
1064 see MLE650
1065 see NCG500
1067 see NGR500
1070 see NGU000
1075 see LGM000
1075 see PMJ750
1076 see PGX000
1079 see SOH500
1080 see SOI000
1083 see TLD500
1085 see VMP000
1086 see VNP000
1089 see AAG250
1090 see ABC750
1092 see ADR000
1093 see ADX500
1098 see AFV500
1100 see AGB250
1104 see AOD725
1104 see AOD735
1105 see IHP000
1110 see MGN500
1111 see PBM000
1114 see BBL250
1115 see PCT250
1120 see BPW500
1120 see BPW750
1120 see BPX000
1123 see BPU750
1123 see BPV000
1123 see BPV100
1123 see IIJ000
1125 see BPX750
1125 see BPY000
1125 see BPY250
1131 see CBV500
1134 see CEJ125
1135 see EIU800
1136 see CMY825
1143 see COB250
1145 see CPB000
1146 see CPV750
1148 see DBF750
1154 see DHJ200
1155 see EJU000
1156 see DJN750
1157 see DNI800
1158 see DNM200
1159 see IOZ750
1160 see DOQ800
1163 see DSF400
1165 see DVQ000
1170 see EFU000
1171 see EES350
1172 see EES400
1173 see EFR000
1175 see EGP500
1184 see EIY600
1185 see EJR000
1188 see EJH500
1189 see EJJ500
1190 see EKL000
1193 see MKA400
1198 see FMV000
1199 see FPQ875
1203 see GBY000
1206 see HBC500
1208 see HEN000
1208 see IKS600
1212 see IIL000
1213 see IIJ000
1214 see IIM000
1219 see INJ000
1220 see INE100
1221 see INK000
1223 see KEK000
1229 see MDJ750
1230 see MGB150
1231 see MFW100
1232 see MKA400
1233 see HFJ000
1234 see MGA850
1235 see MGC250
1239 see CIO250
1243 see MKG750
1244 see MKN000
1245 see HFG500
1247 see MLH750
1249 see PBN250
1255 see NAI500
1256 see NAI500
1257 see GBY000
1259 see NCZ000
1261 see NHM500
1262 see OCU000
1262 see TLY500
1265 see EIK000
1265 see NCH000
1265 see PBK250
1268 see PCS250
1271 see NAI500